MORE ADVANCE PRAISE FOR *BIG CHICKEN*:

"Maryn McKenna's enthralling book is ostensibly about chicken but is really about us: the foolish choices we have made and the happier, healthier future that awaits if we liberate this most American of foods from the drug fix we have imposed on it. Her deep, careful reporting respects every nuance but builds to a clarion call that is as persuasive as it is profound."

—*Dan Fagin, author of the Pulitzer Prize–winning* Toms River

"Maryn McKenna has told an important and frightening story— and told it compellingly. As she makes clear, getting antibiotics out of routine food production will make our food tastier and safer and help fight against antibiotic resistance around the world."

—*Thomas R. Frieden, M.D., former director of the Centers for Disease Control and Prevention*

"Maryn McKenna is the leading journalist worldwide on antibiotic over-use and resistance. In *Big Chicken* she tells a crucial part of that story: the vast misuse and overuse of antibiotics in industrial farming. This clear, urgent explanation of how we got here and what's at risk should be required reading for anyone who wants to see change happen."

—*Lance B. Price, Ph.D., founder and director of the Antibiotic Resistance Action Center*

"Always curious, never pedantic, Maryn McKenna shows empathy for man and sympathy for fowl, while giving voice to scientists and farmers who have concluded that antibiotic-drugged chickens imperil the American diet."

—*John T. Edge, author of* The Potlikker Papers

"This important book is a must-read for anyone wanting to understand why our approach to producing food is unsustainable—and the changes we must make if we don't want to return to a pre-antibiotic era."

—*Richard E. Besser, M.D., president and CEO of the Robert Wood Johnson Foundation*

BIG
CHICKEN

The Incredible Story of How Antibiotics
Created Modern Agriculture
and Changed the Way the World Eats

MARYN McKENNA

NATIONAL
GEOGRAPHIC
Washington, D.C.

Published by National Geographic Partners, LLC
1145 17th Street NW Washington, DC 20036

Library of Congress Cataloging-in-Publication Data
Names: McKenna, Maryn.
Title: Big chicken : the incredible story of how antibiotics created modern
    agriculture and changed the way the world eats / Maryn McKenna.
Description: Washington, D.C. : National Geographic, [2017] | Includes
    bibliographical references and index.
Identifiers: LCCN 2017011586 | ISBN 9781426217661 (hardcover : alk. paper)
Subjects: LCSH: Antibiotics in animal nutrition. | Drug resistance in
    microorganisms. | Chickens--Microbiology. | Poultry--Marketing. |
    Chickens--Marketing.
Classification: LCC SF98.A5 M35 2017 | DDC 636.5/0895329--dc23
LC record available at https://lccn.loc.gov/2017011586

Since 1888, the National Geographic Society has funded more than 12,000 research, exploration, and preservation projects around the world. National Geographic Partners distributes a portion of the funds it receives from your purchase to National Geographic Society to support programs including the conservation of animals and their habitats.

National Geographic Partners
1145 17th Street NW
Washington, DC 20036-4688 USA

Become a member of National Geographic and activate your benefits today at natgeo .com/jointoday.

For information about special discounts for bulk purchases, please contact National Geographic Books Special Sales: specialsales@natgeo.com

For rights or permissions inquiries, please contact National Geographic Books Subsidiary Rights: bookrights@natgeo.com

Interior design: Katie Olsen

Printed in the United States of America

17/QGT-LSCML/1

For Bob Lauder

On this earth there are plagues and there are victims, and it's up to us, as much as possible, not to join forces with the plagues.
—Albert Camus, *The Plague,* 1947

I think any industry producing meat for almost the price of bread has got a big future.
—Henry Saglio, to the U.S. House of Representatives, 1957

CONTENTS

PROLOGUE

EVERY YEAR I SPEND SOME TIME in a tiny apartment in Paris, seven stories above the mayor's offices of the 11th arrondissement. The Place de la Bastille—the spot where the French Revolution sparked political change that transformed the world—is a 10-minute walk down a narrow street that threads between student nightclubs and Chinese fabric wholesalers. Twice a week, hundreds of Parisians crowd down it, heading to the marché de la Bastille, stretched out along the center island of the Boulevard Richard Lenoir.

Blocks before you reach the market, you can hear it: a low hum of argument and chatter, punctuated by dollies thumping over the curbstones and vendors shouting deals. But even before you hear it, you can smell it: the funk of bruised cabbage leaves underfoot, the sharp sweetness of fruit sliced open for samples, the iodine tang of seaweed propping up rafts of scallops in broad rose-colored shells. Threaded through them is one aroma that I wait for. Burnished and herbal, salty and slightly burned, it has so much heft that it feels physical, like an arm slid around your shoulders to urge you to move a little faster. It leads to a tented booth in the middle of the market and a line of customers that wraps around the tent poles and trails down the market alley, tangling with the crowd in front of the flower seller.

In the middle of the booth is a closet-size metal cabinet, propped up on iron wheels and bricks. Inside the cabinet, flattened chickens are speared on rotisserie bars that have been turning since before dawn. Every few minutes, one of the workers detaches a bar, slides off its dripping bronze contents, slips the chickens into flat foil-lined bags, and hands them to the customers who have persisted to the head of the line. I can barely wait to get my chicken home.

The skin of a *poulet crapaudine*—named because its spatchcocked outline resembles a *crapaud,* a toad—shatters like mica; the flesh underneath, basted for hours by the birds dripping onto it from above, is pillowy but springy, imbued to the bone with pepper and thyme. The first time I ate it, I was stunned into happy silence, too intoxicated by the experience to process why it felt so new. The second time, I was delighted again—and then, afterward, sulky and sad.

I had eaten chicken all my life: in my grandmother's kitchen in Brooklyn, in my parents' house in Houston, in a college dining hall, friends' apartments, restaurants and fast-food places, trendy bars in cities and old-school joints on back roads in the South. I thought I roasted a chicken pretty well myself. But none of them were ever like this, mineral and lush and direct. I thought of the chickens I'd grown up eating. They tasted like whatever the cook added to them: canned soup in my grandmother's fricassee, her party dish; soy sauce and sesame in the stir-fries my college housemate brought from her aunt's restaurant; lemon juice when my mother worried about my father's blood pressure and banned salt from the house. This French chicken tasted like muscle and blood and exercise and the outdoors. It tasted like something that it was too easy to pretend it was not: like an animal, like a living thing.

We have made it easy not to think about what chickens were before we find them on our plates or pluck them from supermarket cold cases. I live, most of the time, less than an hour's drive from Gainesville, Georgia, the self-described Poultry Capital of the World, where the modern chicken industry was born. Georgia raises 1.4 billion broilers a year, making it the single biggest contributor to the almost 9 billion birds raised each year in the United States; if it were an independent nation, it would rank in chicken production somewhere near China and Brazil. Yet you could drive around for hours without ever knowing you were in the heart of chicken country, unless you happened to get behind a truck heaped with crates of birds on their way from the remote solid-walled barns they are raised in to the gated slaughter plants where they are turned into meat.

That first French market chicken opened my eyes to how invisible chickens had been for me, and after that, my job began to show me what that invisibility had masked. My house is less than two miles from the front gate of the Centers for Disease Control and Prevention (CDC), the federal agency that sends disease detectives racing to outbreaks all over the world. For more than a decade, one of my obsessions as a journalist has been following them on their investigations—and in long late-night conversations in the United States and Asia and Africa, with physicians and veterinarians and epidemiologists, I learned that the chickens that had surprised me and the epidemics that fascinated me were more closely linked than I had ever realized.

I discovered that the reason American chicken tastes so different from those I ate everywhere else was that in the United States, we breed for everything but flavor: for abundance, for consistency, for speed. Many things made that transformation possible. But as I came to understand, the single biggest influence was that, consistently over decades, we have been feeding chickens, and almost every other meat animal, routine doses of antibiotics on almost every day of their lives. Antibiotics do not create blandness, but they created the conditions that allowed chicken to be bland, allowing us to turn a skittish, active backyard bird into a fast-growing, slow-moving, docile block of protein, as muscle-bound and top-heavy as a bodybuilder in a kids' cartoon.

At this moment, most meat animals, across most of the planet, are raised with the assistance of doses of antibiotics on most days of their lives: 63,151 tons of antibiotics per year, about 126 million pounds. Farmers began using the drugs because antibiotics allowed animals to convert feed to tasty muscle more efficiently; when that result made it irresistible to pack more livestock into barns, antibiotics protected animals against the likelihood of disease. Those discoveries, which began with chickens, created "what we choose to call industrialized agriculture," a poultry historian living in Georgia proudly wrote in 1971. Chicken prices fell so low that it became the meat that Americans eat more than any other—and the meat most likely to transmit foodborne illness, and also antibiotic resistance, the greatest slow-brewing health crisis of our time.

As I began to piece this together—first with perplexity and then with disbelief—I learned that a few far-seeing scientists had warned from the start of the unintended consequences of farm antibiotics. But I also found that in just the past few years, those warnings finally have been heard. Under pressure from chefs and consumers and out of a growing sense of its own neglected responsibility, industrial poultry production has begun relinquishing antibiotic use and reshaping how it farms.

This is a story woven from two parallel narratives: how we entered into routine antibiotic use and then questioned it, and how we created industrial chicken and then reconsidered it—and what those histories reveal about what we elevate, and sacrifice, when we decide how to raise our food. I spoke, and often traveled great distances to meet, with farmers in a dozen states and several foreign countries, and with chemists, lawyers, historians, microbiologists, bureaucrats, disease detectives, politicians, chefs, and stylish French poultry sellers.

There was a time when any chicken was as safe and as full of integrity and flavor as my sidewalk Paris *poulet*. With enough attention to the pressure of markets, the protein needs of the world, the risks of diseases, the welfare of animals, and our authentic desire to eat something delicious, it can be again.

PART 1

How Chicken Became Essential

CHAPTER 1

ILLNESS, AND
A BAD YEAR

RICK SCHILLER HAD NEVER FELT SO SICK.

Schiller was 51, and a big man, 6'1" and 230 pounds. He had a black belt in tae kwan do; he worked out; he had never been hospitalized in his life. But on the last morning in September in 2013, he was lying on a gurney in an emergency room in his hometown south of San Jose, California, broiling with fever, writhing in pain, and staring in disbelief at his right leg. It was swollen to three times its normal size, purple and hot, so puffed and hard with inflammation it felt like it would pop.

It was the leg that had forced Schiller to the emergency room. Pain like fire had woken him at 3 a.m., and when he pulled back the covers to see what was wrong, he yelled. His fiancée, Loan Tran, shrieked. They rushed out of the house, him hopping in his underwear and bracing his hands against the walls, her trying to fold down the seat in his sports car to make room for the rigid, log-like limb. At the hospital, a crowd of health care workers levered him out of the car, threw him onto a gurney, skidded him into a room, and plugged in fluids and morphine. It was before dawn on a Monday morning, normally a quiet time in an ER, and a physician arrived quickly, carrying a sterilized syringe in a tray.

The resident told Schiller that the swelling was so severe they were afraid his skin would split. "I have to tap your leg to relieve the pressure," she said. Schiller nodded, gritting his teeth. She punctured the taut surface with the tip of the needle and slid it in, expecting a gush of blood or pus that would push the plunger back into her hand. Nothing came out. She frowned, asked a nurse for a syringe with a wider needle, and probed again, looking for a pool of blood trapped in a blood vessel, or a pocket of infection making his leg balloon. Still nothing. She asked for one more syringe— Schiller remembers the needle being the width of a pencil lead— swabbed and pierced his skin a third time, and then tugged the plunger back gently. He heard her gasp. He looked down. The barrel was filled with something red and heavy. He thought it looked like meat.

Hours later—packed in ice for the fever, muzzy headed from sedatives, with the leg as hard as ever—Schiller tried to piece together what had happened. He thought it had started 10 days earlier with a late-night fast-food snack: a sandwich, tacos, and a milkshake. It had tasted funny, and he hadn't finished it, which was rare for him. He'd started throwing up after midnight and had been sick ever since, with vomiting and explosive diarrhea, so nauseated he could barely keep down water.

Between that night and this morning, he had gone to a local emergency room and had also sought help from his primary care doctor. His doctor took a stool sample, in case Schiller had picked up a gut bug, but said he ought to be better in a few days. He didn't get better. He lay flat on the couch, staggering unsteadily between living room and bathroom and barely eating. Just a day earlier, though, he had turned a corner: His appetite perked up. He asked his fiancée to make him some soup and ate a few spoonfuls and a couple of crackers. Then he had conked out, exhausted again, until the throbbing in his leg woke him.

Tapping his leg with the needles hadn't revealed what was wrong, and neither did a quickly ordered ultrasound, or sending him to radiology for an emergency MRI. There was no abscess that could be emptied, no blood clot to be dissolved, nothing to explain the roaring fever and swelling. Now the staff was waiting for the drugs they had given him to start working and for the test results to arrive, so that someone could make decisions about what they should try next.

Schiller shook with pain and exhaustion, huddled in a nest of blankets in an exam room while the ER woke up around him. He had kept his phone when they took his clothes away, and he slid it out and thumbed open the recording app. His voice was rough from vomiting and fear, but he tried to keep it steady. "Q," he said, using his pet name for his fiancée, "this is my last will and testament. I think I'm going to die."

—⚬⚬⚬—

WHEN SCHILLER WOKE UP LATER that day in a hospital bed, his leg was still painful and hard. It was obvious that somewhere in his body, he was harboring an infection that had spilled over into his bloodstream. His immune system had recognized the invader and reacted to it; that response triggered the fever and the inflammation that disrupted his circulation and puffed up his leg. When they admitted him, the medical staff had plugged broad-spectrum antibiotics into his IV, drugs that can counter many types of disease-causing bacteria. Now it was a waiting game to see whether those drugs had any effect, and whether the hospital's lab could culture any bacteria from his blood and determine a better treatment.

After another day, the swelling abated. Schiller was up, bracing himself against the bed and trying to put weight on the bad leg, when his cell phone rang. It was his primary care doctor, calling with the results of the test he had ordered when Schiller had arrived with vomiting and diarrhea, before the emergency dash to the ER.

His doctor said: "Do you realize you've got salmonella poisoning?"

Schiller answered: "Do you realize I'm in the hospital and I almost died?"

His doctor hung up and rang the hospital physician overseeing Schiller's care. They would not have to wait any longer for test results now. *Salmonella* is a common cause of foodborne illness that every year sickens 1 million people in the United States and almost 100 million around the world. Most people recover after a week of misery; but each year in the United States, some unlucky thousands end up in the hospital, and almost 400 people die. Knowing for sure what was making him sick allowed the medical staff to tailor his treatment. A few days later, sore and shaky and depleted, but with the fever gone and the leg almost usable, Schiller went home.

All along, Schiller had blamed his illness on the fast-food meal he consumed the day his symptoms started. A few weeks later, a call from an investigator from the state health department shifted his understanding of how he got sick. Her name was Ada Yue, and she wanted to know more about his infection. He told her about the fast food, how he ate it one evening and started throwing up the same night, but over the phone, he could hear her shaking her head. "The timing doesn't work out," she told him. "It takes longer than that."

Yue explained that salmonella takes days to develop in the body's system once the person has swallowed whatever food is contaminated with it; it wasn't possible to develop symptoms as serious as his in just a few hours. So she wanted to ask him a few questions about where he had shopped and eaten in the weeks before he got sick. It turned out to be a lot of questions. When he asked why so many, she said other people, in other towns in California, had gotten sick at about the same time, and it was possible that the same food had caused the illnesses in all of them. The Centers for Disease Control and Prevention, the U.S. federal agency that monitors the occurrence of illness around the country, was working with the state health department to try to narrow down the possibilities. They were zeroing in on a few foods that might be the culprits. She wanted to know if Schiller could recall any details about grocery shopping just before he fell ill. She especially wanted to know whether he had bought any chicken.

—◊—

UNTIL THAT PHONE CALL, Schiller had no way of knowing that he was not alone in his illness. He was part of a foodborne epidemic, one of the largest and longest on record. Before it was over, it would stretch across 29 states and Puerto Rico and sicken

634 known victims, along with possibly thousands more whose illnesses were never diagnosed.

The first sign of something going wrong arrived a few months before he got sick, in June 2013. A computer program operated by the CDC pinged an alert: Something was happening in salmonella in the western states. There were an unusually high number of cases of a particular strain, *Salmonella* Heidelberg, and a particular type within that strain that researchers knew as 258.

The CDC program that detected the anomaly, PulseNet, could not do much more than sound an alarm about a possible outbreak because it had no details on the cases. PulseNet does not interview patients or doctors. Instead it sifts through images of patterns made by the DNA of foodborne organisms, taken from patients when they are diagnosed. PulseNet is named for the lab technique that produces the patterns: pulsed-field gel electrophoresis (PFGE), which breaks apart an organism's DNA and uses electrical current to tug the genetic material through a sheet of gel. The pattern PFGE produces looks like an inventory bar code, and like a bar code, it can accommodate many subtle differences, making it a good vehicle for differentiating the many strains and subtypes of foodborne organisms. Epidemiologists refer to the bar codes as fingerprints—and just like a perpetrator's fingerprint left at a crime scene, PFGE patterns can help scientists recognize when an organism has caused an outbreak.

Once, it was easy to know when a food was responsible for illnesses, because cases naturally clustered near each other. If a hundred people who drank from the same well or ate at the same church supper got sick, someone in that community would notice and tell someone in authority. But across the latter half of the 20th century, food production became more complicated: first through better shipping, then through corporate consolidations, and finally through the kind of economic maneuvers that made it reasonable to raise and slaughter an animal on one coast and eat it on the other side of the country, or grow and harvest fruit in one hemisphere and ship it halfway around the world to be sold. If a food was contaminated at the place where it was killed or packed or processed, and then was distributed over hundreds or thousands of miles, the cases of illnesses it caused would appear random. PulseNet's ability to compare DNA fingerprints drew links between them, even if they were far apart in distance or time.

By the time Schiller lurched into the San Jose emergency room, the CDC was pursuing a trail of clues. Its epidemiologists knew that 278 people had fallen ill since March. They were as young as infants and as old as 93, and they lived in 17 states—all over the country, as far south as Florida and as far east as Connecticut. No one had died, but almost half of the victims had been hospitalized, an unusually high percentage for salmonella. In bacterial samples taken from victims and grown in labs to be analyzed, the same fingerprints kept recurring. More than 100 of the sick had filled out lengthy questionnaires like the one Schiller had answered to narrow down possible culprits. The food that kept surfacing was chicken.

The Food and Drug Administration (FDA) had delved into its own records, analyses of foodborne-illness bacteria retrieved from meat purchased in supermarkets around the country, and had found the same genetic fingerprint in *Salmonella* from chicken. And the U.S. Department of Agriculture (USDA) was zeroing in on a slaughter plant that might be responsible—a plant that belonged to a company that packaged the brand of chicken that the sick people had eaten and that the FDA's database had recorded.

There was an extra aspect to the outbreak that made the investigation feel urgent. The *Salmonella* bacteria responsible were not only causing more serious illness than was usual. They were also displaying antibiotic resistance to a wide range of common drugs: ampicillin, chloramphenicol, gentamicin, kanamycin, streptomycin, sulfa drugs, and tetracycline. The epidemic that had swept Schiller into its grip was a demonstration of how bacteria resistant to antibiotics, which the United Nations calls "the greatest and most urgent global risk," are spreading by means of food.

—⬯—

FOR MOST PEOPLE, antibiotic resistance is a hidden epidemic, unless they have the misfortune to contract an infection themselves or have a family member or friend unlucky enough to become infected. Drug-resistant infections have no celebrity spokespeople, negligible political support, and few patients' organizations advocating for them. If we think of resistant infections, we imagine them as something rare, occurring to people unlike us, whoever we are: people who are in nursing homes at the end of their lives, or dealing with the drain of chronic illness, or in intensive-care units after terrible trauma. But resistant infections are a vast and common problem that occur in every part of daily life: to children in day care, athletes playing sports, teens going for piercings, people getting healthy in the gym. And though common, resistant bacteria are a grave threat and getting worse. They are responsible for at least 700,000 deaths around the world each year: 23,000 in the United States, 25,000 in Europe, more than 63,000 babies in India. Beyond those deaths, bacteria that are resistant to antibiotics cause millions of illnesses—two million annually just in the United States—and cost billions in health care spending, lost wages, and lost national productivity. It is predicted that by 2050, antibiotic resistance will cost the world $100 trillion and will cause a staggering 10 million deaths per year.

Disease organisms have been developing defenses against the antibiotics meant to kill them for as long as antibiotics have existed. Penicillin arrived in the 1940s, and resistance to it swept the world in the 1950s. Tetracycline arrived in 1948, and resistance was nibbling at its effectiveness before the 1950s ended. Erythromycin was discovered in 1952, and erythromycin resistance arrived in 1955. Methicillin, a lab-synthesized relative of penicillin, was developed in 1960 specifically to counter penicillin resistance, yet within a year, staph bacteria developed defenses against it as well, earning the bug the name MRSA, methicillin-resistant *Staphylococcus aureus*. After MRSA, there were the ESBLs, extended-spectrum beta-lactamases, which defeated not only penicillin and its relatives but also a large family of antibiotics called cephalosporins. And after cephalosporins were undermined, new antibiotics were achieved and lost in turn.

Each time pharmaceutical chemistry produced a new class of antibiotics, with a new molecular shape and a new mode of action, bacteria adapted. In fact, as the decades passed, they seemed to adapt faster than before. Their persistence threatened to inaugurate a post-antibiotic era, in which surgery could be too dangerous to attempt and ordinary health problems—scrapes, tooth extractions, broken limbs—could pose a deadly risk.

For a long time, it was assumed that the extraordinary unspooling of antibiotic resistance around the world was due only to misuse of the drugs in medicine: to parents begging for the drugs even though their children had viral illnesses that antibiotics could not help; physicians prescribing antibiotics without check-ing to see whether the drug they chose was a good match; people stopping their prescriptions halfway through the prescribed course because they felt better, or saving some pills for friends without health insurance, or buying antibiotics over the counter, in the many countries where they are available that way, and dosing themselves.

But from the earliest days of the antibiotic era, the drugs have had another, parallel use: in animals that are grown to become food. Eighty percent of the antibiotics sold in the United States and more than half of those sold around the world are used in animals, not in humans. Animals destined to be meat routinely receive antibiotics in their feed and water, and most of those drugs are not given to treat diseases, which is how we use them in people. Instead, antibiotics are given to make food animals put on weight more quickly than they would otherwise, or to protect food ani-mals from illnesses that the crowded conditions of livestock pro-duction make them vulnerable to. And nearly two-thirds of the antibiotics that are used for those purposes are compounds that are also used against human illness—which means that when resistance against the farm use of those drugs arises, it undermines the drugs' usefulness in human medicine as well.

Resistance is a defensive adaptation, an evolutionary strategy that allows bacteria to protect themselves against antibiotics' power to kill them. It is created by subtle genetic changes that allow organisms to counter antibiotics' attacks on them, altering their cell walls to keep drug molecules from attaching or pene-trating, or forming tiny pumps that eject the drugs after they have entered the cell. What slows the emergence of resistance is using an antibiotic conservatively: at the right dose, for the right length of time, for an organism that will be vulnerable to the drug, and not for any other reason. Most antibiotic use in agriculture violates those rules. Resistant bacteria are the result.

Experimenters began trying the then new miracle drugs in animals almost as soon as antibiotics were achieved in the lab in the 1940s—and concern about that use dates just as far back. From the start, in protests that were downplayed for decades, a few perceptive researchers warned that resistant bacteria would arise in livestock, find a way off farms, and move silently through the wider world. The shortest route off farms is in the meat that animals eventually become: In the year Schiller got sick, 26 percent of the *Salmonella* found on supermarket chicken by government testing was resistant to at least three separate families of antibiotics. But resistant bacteria also leave farms in manure, in storm runoff, in groundwater, in dust, and via the skin and clothes and microbial hitchhikers of people who work on farms and live there. When those organisms escape, they disperse in a manner that is impossible to track, and they cause illness and alarm far from the farms where they originated.

—m—

WHILE THE CDC'S SCIENTISTS were tracing the salmonella outbreak that had Schiller in its grip, a set of researchers on the other side of the world were pursuing another resistant pathogen. Scientists in China had been running a project in which they checked hogs raised on intensive farms—the kind of properties where animals are confined permanently in buildings and antibiotics are used routinely—to see whether they were carrying resistant bacteria. In July 2013, they found a pig outside Shanghai harboring a strain of the bacterium *Escherichia coli* in its manure. That was normal, because *E. coli*'s multitude of strains make their homes in the guts of most animals. But the content of this *E. coli* was unusual and alarming. It was hiding a gene no one had seen before, conferring resistance to a drug called colistin.

If colistin sounds unfamiliar, there is a reason. It is an old drug, discovered in 1949, and for decades medicine disdained it as a clumsy, toxic relic of a cruder era of chemistry. Physicians rarely used it, and no one prescribed it outside of hospitals. But because colistin sat for so long at the back of the shelf, disease organisms never encountered it either, and never developed defenses against it. In the mid-2000s, the advance of resistance undermined a critical, powerful class of drugs called carbapenems, which are used to cure multidrug-resistant organisms—*Klebsiella, Pseudomonas, Acinetobacter*—that cause dire infections in hospitals. Against these newly hardy bacteria, colistin was the only antibiotic that still worked reliably. Suddenly the clunky, unwanted compound from the past was crucial to preserve.

There was one catch, however: While medicine had been disdaining the drug, agriculture had adopted it. Because colistin was an old compound, it was cheap, so it made an inexpensive preventative for the kind of gut and lung infections that could happen to animals in crowded barns. Colistin is not used in animals in the United States, but European and Asian countries used millions of pounds each year. No one thought this was a problem, because medicine had not wanted the drug and because resistance seemed unlikely to develop, requiring a genetically tricky maneuver that no one had ever observed.

But what the Chinese researchers found in 2013 upended the serene assumption that colistin was secure. The new gene they found in the pig was on a plasmid, self-contained loops of DNA inside a cell that spread not just by inheritance when cells divide, but by jumping from one bacterium to another. This meant that colistin resistance could be disseminating through the bacterial world unnoticed—and in fact, it was. Within three years, epidemiologists in Asia, and in Africa, Europe, and South America, would identify the resistance-conferring gene in animals, the environment, and people in more than 30 countries.

That included the United States. The resistance-conferring gene, dubbed MCR, appeared first in a woman in Pennsylvania who was carrying it unknowingly, and then in men in New York and New Jersey who also did not know they were carriers, and then in a Connecticut toddler, and more. None of those people were ill with colistin-resistant infections; that was true for most of the people carrying the rogue gene. Rather, they were an epidemic waiting to ignite, held in abeyance because the use of colistin in medicine was still rare. The spread of colistin resistance across the world was a time bomb with a fuse of uncertain length, and it had been created and disseminated by antibiotic use on farms.

—⚏—

ONE OTHER THING WAS HAPPENING in autumn 2013 while the CDC was grappling with drug-resistant salmonella and the Chinese microbiologists were pursuing colistin resistance. The U.S. government, for the first time in its history, was moving to put federal controls on agricultural antibiotics.

The United States was late to this. England recognized the danger in the 1960s, and most of Europe followed its lead beginning in the 1980s. Copying those countries, the FDA had made one attempt in 1977, but it was defeated by congressional interference and never tried again. Thirty-six years later, emboldened by the election of President Barack Obama, the agency proposed making one type of agricultural antibiotic use, the weight-conferring doses known as growth promoters, illegal in the United States.

The FDA would have a battle ahead of it. American farm animals were consuming 32.6 million pounds of antibiotics in 2013, four times what human patients were receiving. But the agency also had irrefutable evidence that curbs were needed. Not only was resistance rising; for the first time, no new drugs were entering the market to replace the ones being lost. Pharmaceutical companies were contending that antibiotics were no longer profitable to make, and they had good reasons. By widely accepted industry math, it takes 10 to 15 years, and about a billion dollars, to get a new drug to market—but resistance was using up antibiotics so quickly that companies could not recoup their investments or collect any profit before a drug lost effectiveness. And if a new drug was so effective that medicine elected not to deploy it but to hold it on the shelf against a future emergency, then the companies made back nothing at all.

The FDA locked in its new policy in December 2013. It gave agriculture three years to adapt to relinquishing growth promoters and bringing other antibiotic use under the control of veterinarians. Its reforms became final on January 1, 2017, but how effective they might be will not be proven for years.

THE EVENTS OF AUTUMN 2013—the massive salmonella outbreak, the recognition of colistin resistance, and the U.S. government's belated attempt to exert some control over farm antibiotic use—mark a turn in a story that has been unfolding for almost 70 years. Antibiotics were first added to animal feeds in the late 1940s, in the surge of soaring confidence in science that followed the end of World War II. They remained a crucial component of meat production for decades even though more and more voices warned against the practice: first lone scientists who were mocked for raising the alarm, then small report-writing committees, then large medical societies, and finally governments, attempting to defy one of the largest, most globalized industries in the world.

Antibiotics have been so difficult to root out of modern meat because, in a crucial way, they created it. The drugs made it irresistible to load more animals into barns and protected animals and their growers from the consequences of that crowding. The escalating spiral of production drove down prices, making meat a cheap commodity, but it also drove down profits, undermining independent farmers and fostering the growth of global corporations.

That history, beginning with the positive effects of antibiotics on agriculture and continuing with their much more abundant negative ones, is most evident in the parallel story of poultry. Chickens were the first animals to receive what came to be called growth promoters, and the first in which scientists demonstrated that daily antibiotic doses could protect birds from diseases of close confinement. They are the animals most transformed by the post–World War II mission to feed the world at any cost. Today, a meat chicken's slaughter weight is twice what it was 70 years ago and is achieved in half the time. Across those decades, chickens went from a scarce and expensive Sunday treat to the meat that Americans eat more often than any other and that is growing fastest in consumption around the world. Until recently, the transformation of chicken was a point of pride. "Here is good news for both farmers and meat eaters. Antibiotics provide more meat with less feed," *Fortune* magazine reported in 1952. The USDA boasted in 1975: "Broiler production is industrialized in much the same way as the production of cars."

Yet after the events of 2013, chicken turned against its own history. Some of the largest production companies in the industry renounced antibiotic use. Some of the largest food service retailers in the United States committed to carrying only birds raised without routine drugs. Medical centers, college campuses, school systems, and restaurant chains joined the refusal, pushed by advocates and by parents who had awakened to the danger to their kids. At a point where the cattle and hog industries were digging in their heels to resist FDA policies, poultry rushed to the front of the line and called it a parade.

The intertwined histories of the advent of agricultural antibiotics, and the rise and transformation of chicken, are mostly a story of hubris: of the romance of innovation, the seduction of profit, and the failure to anticipate unintended consequences. But they also tell a story of how an industry can assess its own dark past and adjust its path, and they offer hope that food production in the rest of the world might be kept from sickening millions more victims like Rick Schiller, and from making the same mistakes that the United States and Europe made.

To understand both stories, it is necessary to go back to their beginnings: to the earliest days of the antibiotic era, and a moment when it was urgent to find a new way to feed the world.

CHAPTER 2

BETTER LIVING THROUGH CHEMISTRY

IN ALL THE ARGUMENTS that erupted afterward, everyone agreed on this: The chickens gained weight.

It was Christmas Day, 1948. The streets of Pearl River, New York, a small town 20 miles outside Manhattan on the New Jersey border, were very quiet, and the halls of Lederle Laboratories even more so. There was a skeleton crew at the 500-acre campus: the minimum necessary number of laboratory staff nipping in and out to monitor equipment and make sure that experimental animals were fed. Thomas Jukes did not plan to spend much time there himself. He had told his lab assistant to take the holiday off, arguing that the task that had to be performed on this day would take only a few minutes. All he needed was to slip into the animal colony, corner the 133 juvenile chickens that made up his experiment, and weigh them. He expected that it would not take long.

He probably did not expect to change the world.

Jukes was British, slender and dark haired, with alert eyes behind oversized glasses: an energetic, self-made man who had left home at age 17, immigrated to Canada, and worked on a farm and in factories in Detroit to amass enough money for college. He had earned a degree at Ontario Agriculture College—sleeping, at one point, on a cot in the poultry building—and, after that, a Ph.D. in biochemistry at the University of Toronto's medical school, studying the immune systems of chickens and ducks. In 1933, he migrated across the continent for a postdoctoral appointment at the University of California, Berkeley. But the collapse of government budgets in the depths of the Great Depression drained the grant he hoped would support his research; it was canceled after one year. He scrounged for a replacement job at the university's College of Agriculture and for money from the U.S. Department of Agriculture to study poultry nutrition. Despite the career detour and the thin funding, he accomplished notable work, identifying which vitamins had to be added to chicken feed to allow birds to thrive on a manufactured diet.

In the 1930s, that was a crucial question. Until World War I, almost every farmer had kept a few hens to produce eggs and had eaten the hens when their egg-laying abilities were exhausted. Now, though, chicken meat and eggs were becoming a crop, a farm's reason for existence rather than a farm by-product. Instead of a few at a time, chickens were being raised by the thousands, and instead of wandering around a barnyard, they were sequestered indoors, unable to reach the grains and grubs they would have scratched up from the ground. To survive, the birds needed synthetic nutrition, and the poultry industry growing up around them needed a parallel industry of experts to supply it.

Jukes became one of those experts, and Lederle, the company that recruited him away from California with the promise of a lab and staff, was becoming a leader in the business of bringing men with his expertise to the aid of poultry science. Lederle was not an agricultural company, strictly speaking. It was a pharmaceutical company, one of the first manufacturers of antibiotics.

Jukes joined Lederle in 1942. Fourteen years earlier, a Scottish researcher named Alexander Fleming had discovered that specks of mold growing on a dish of staph bacteria secreted a chemical that killed the organisms around them. Two years earlier, Fleming's discovery had been turned into a drug by researchers Howard Florey and Ernst Chain. Their experiments on mice showed that the compound produced by the mold could kill bacteria that were infecting animals without harming the animals themselves, something that had never been seen before. One year previously, the new drug, dubbed penicillin in honor of the blue-green mold it came from, *Penicillium notatum*, had almost saved the life of a 43-year-old British police constable, Albert Alexander, who had scratched his face on a rosebush while gardening. Alexander was riddled with staph and strep, oozing pus from abscesses on his face and scalp; already, doctors had been forced to remove one of his eyes. In February 1941, he began receiving injections of the scarce new drug. Within a week, he was almost completely recovered, but then the tiny supply of penicillin available in England ran out. The infection recurred, and Alexander died—demonstrating both the deadly persistence of infections, and also that at last there might be something to hold them at bay.

Three months after Alexander's death, Florey and Chain smuggled the mold that produced penicillin out of England, sneaking it out between Nazi attacks. They hoped that American industry, not yet drawn into World War II, might possess the money and capacity to manufacture enough of the drug to make a difference. In 1942, the same year Jukes moved to Pearl River, penicillin brought a New Haven nurse, Anne Sheafe Miller, back from the edge of death, where she had hovered for a month after contracting an infection following a miscarriage. Then it saved the lives of more than one hundred victims of the Cocoanut Grove nightclub disaster in Boston, one of the worst fires in U.S. history, by preventing infections from taking hold in severe burns. That was enough proof of the power of penicillin for the U.S. government to invest in its production and send millions of doses out onto the battlefields of World War II, saving untold thousands with a speed that seemed miraculous.

Penicillin's success ignited a hunger for additional antibiotics, the name scientists gave to compounds manufactured by some organisms in order to kill other organisms. And it ignited another hunger as well: for the profit these new drugs could bring. Penicillin was never proprietary; Fleming and his collaborators shared the formula and method for making it with several firms at the same time so the maximum possible amount of drug could be made for the war effort. Fortunes waited for whoever found the next miracle drugs and patented them.

In 1943, Selman Waksman and his student Albert Schatz isolated streptomycin, the first antibiotic able to cure tuberculosis, from an organism found in "heavily manured" soil in New Jersey. In 1947, Paul Burkholder crystallized chloramphenicol, the first antibiotic that could vanquish typhoid, from a bacterium living in a Venezuelan compost heap. Other researchers and the companies they worked for were desperate to find their own sources of new drugs: Pfizer Inc. and Eli Lilly and Company sent sterile sample tubes around the world, begging missionaries and military personnel to spoon up and send home any mold or dirt that looked promising. Lederle's chief pathologist, Benjamin Duggar, had once been employed at the University of Missouri, and he asked a former colleague to send him random scoops of dirt from the campus. One tube, dug from a field where the agricultural school grew varieties of forage grass, contained a bacterium that exuded a golden-yellow chemical. In tests, the compound killed a wide array of disease bacteria—more than penicillin could manage, and different ones from those that streptomycin could kill. American Cyanamid Company, Lederle's parent corporation, jubilantly filed for a patent in February 1948. In a nod to the compound's color—and maybe to the income he hoped would flow from it—Duggar dubbed the fungus *Streptomyces aureofaciens,* "gold-making." He called the compound "Aureomycin." Later, it would be known as chlortetracycline, the first of the entire family of tetracycline drugs.

Jukes was not part of Lederle's antibiotic effort; he had been hired to work on nutrition. He and his colleagues figured out how to synthesize folic acid, a vitamin that prevents devastating birth defects, and while they were tinkering with it, they developed methotrexate, one of the first cancer chemotherapy drugs. But Jukes was still interested in what chickens needed to eat in order to thrive in confinement—and by an accident of history, that question was even more important than it had been 15 years earlier. World War II had spurred such demand for protein that production of chicken almost tripled, rising to more than one billion pounds of chicken meat per year. But when the war ended, the poultry market that it had guaranteed collapsed. Producers, forced to cope with more chickens than they had demand for, struggled to cut their costs. En masse, they switched their birds' diet from vitamin-rich fishmeal—ground-up anchovies netted off the southern California coast—to much cheaper soybeans. Chickens did not do well on soybeans, though. They grew slowly, and the eggs that hens laid were thin-shelled and did not hatch. Even when vitamins were added into their feed, as Jukes had learned to do in his first job, the birds did not thrive. People talked about needing to add a nutritious boost, an "animal protein factor."

Then Merck & Company, a Lederle rival, announced its researchers had found it. Merck was making streptomycin, Waksman's drug, brewing it out of *Streptomyces griseus,* the bacterium he had harvested from the manured patch of soil near the Rutgers University campus. Merck researchers said that a by-product of the brewing process made chickens do better, even when they were fed the low amounts of protein now present in conventional feed. Earlier in the century, researchers had identified and learned how to synthesize vitamins: B_2, B_3, B_5, B_6. Merck's scientists identified their new compound as the last in that series: vitamin B_{12}.

Jukes wondered if Lederle's own bacterium—*Streptomyces aureofaciens,* the source of Aureomycin and a distant relative of the species the Merck scientists used—could perform the same trick. That was what brought him to his office on Christmas morning, on a mild, dry day with just a dusting of snow. A few weeks earlier, he had set up an experiment to test whether Lederle possessed an animal protein factor of its own. Today he would find out.

He had chosen a small group of six-month-old hens and roosters from the birds the company raised to use in research. He fed them a specially mixed diet, low in nutrients so that the chicks they produced would be feeble; that way, it would be easier to distinguish the effect of any additives. When the hens laid eggs, he hatched the chicks in an incubator, divided them into groups of 12, and sequestered one dozen to keep as a control group. Those birds got the same deficient diet as their parents. The other groups got different doses of supplements, precisely measured: six different amounts of liver extract, a natural but expensive source of B_{12}; six different amounts of synthetic vitamin B_{12}; liver extract with additional sources of nutrition such as alfalfa or fish extracts or distillers' grains, left over from producing alcohol; or tiny portions of the mash, or growth medium, that Aureomycin had been brewed in.

Christmas Day was the 25th day since the chicks had hatched, the point where Jukes had determined he would weigh them and assess the experiment. Almost all of the deprived control chicks had died, but he expected that. Almost all of the rest, though, had lived, proving that the supplements he was giving them had contributed to their diets something the chicks needed to thrive. One by one, he weighed them. The three survivors of the control group were small and sickly: They weighed just 110 grams, not quite 4 ounces. The birds that got the crystalline vitamin looked healthy, pink skin showing between the shafts of the red feathers sprouting on their heads and wings; they weighed from 179 to 203 grams. The ones that received the liver extract weighed, at the highest dose, 216 grams.

Then he moved to the pens containing the chicks that had been given the antibiotic discards. He had dosed them at four different levels, so much mash per kilo of feed. He weighed the four groups, averaged the weights within each of them, recorded the numbers, and looked. And then looked again. The birds that had gotten the highest dose of Aureomycin were the heaviest in the room. They weighed 277 grams, almost 10 ounces: two and a half times the control chicks' weight, a third more than the chicks receiving Merck's B$_{12}$, and a quarter more than the chicks receiving the liver extract, an expensive ingredient that no farmer could afford.

The chicks had gotten to that weight with the help of 60 grams of the mash containing a trace amount of Aureomycin. Sixty grams is two ounces, and two ounces is nothing: a handful of pennies, two slices of bread, an egg. Yet that tiny weight would exert enough force to alter the entire structure of agriculture—and affect land use, labor relations, international trade, animal welfare, and the diet of most of the world.

—✶—

IT ISN'T CLEAR HOW QUICKLY Jukes and his research partner, E. L. Robert Stokstad, understood what caused the "growth-promoting" effect Jukes found. When he wrote a description of the experiment a few months later, he guessed "the 'animal protein factor' consisted of vitamin B$_{12}$ plus some factor as yet unidentified." Within a year, though, they were sure: What caused the chicks to put on weight was not the vitamin, but tiny doses of Lederle's antibiotic, persisting in the discards of production.

To understand what they had done, it helps to know a little about how antibiotics were made—and are made, even today. It's a lot like brewing beer. You start with an organism that makes the compound you want, add it to a solution of water and sugars (extracted from grains, if you are making beer), and let the mixture ferment. The organisms consume nutrients in the growth medium and excrete the by-products of their digestion: alcohol and carbon dioxide in the case of beer yeast and raw antibiotics in the case of *Streptomyces*. If you are making beer, you drain off the liquid by-product and flavor and bottle it. If you are aiming for a drug, there is an extra step: first pumping out the liquid and then chemically extracting the antibiotic compound. When all of that is done, what remains afterward is the liquid the antibiotic came out of, plus a sticky mash of leftover sugars and the exhausted remains of the microorganisms you started with.

Through the ages, brewers have dried their fermentation leftovers and sold them for livestock feed. Jukes and his colleagues saw an economic opportunity in their leftovers too: money for nothing, the basis of a whole new industry, extracted from what Lederle was throwing away.

In Jukes's first experiment, he fed the chicks tiny portions of the leftover mash, dried and ground. But he apparently suspected there might be Aureomycin left over in the fermentation liquid too. Lederle used the solvent acetone to wash impurities out of the fermented liquid. He captured that solvent solution before it was dumped and dried it in a giant furnace, called a "tank house," where the company burned the corpses of horses that had been infected with diseases in order to produce antibody serums from their blood. (He joked years later that he promised to assume responsibility if the highly flammable acetone blew up. It wasn't much of a joke: He actually had burned down an outbuilding at the University of California, Davis while heat-treating experimental diets for chicks.) Once he rescued trace amounts of Aureomycin from the solvent, and dried and powdered it, he used it as a feed additive too—and achieved even better results than his first attempt, boosting his experimental chicks' weight another 25 percent to 368 grams, or 13 ounces. With that confirmation of the "growth-promoting" effect he had perceived in the first round of his experiment, Jukes passed samples to scientists he knew at state agricultural colleges throughout the United States and asked them to conduct experiments of their own. His colleagues were astounded: They reported back that small doses of Aureomycin not only cured a bloody diarrhea that would have killed young pigs but also tripled their rate of growth and boosted the weight of turkey chicks, called poults.

Word got around. So many researchers asked for Aureomycin residue that Jukes ran through the Pearl River plant's fermentation by-products faster than they were being made. He resorted to plundering the company dump for any discarded fermentation vessels, including reused glass Coke bottles, that might contain a few precious grams of drug. Scientists were not the only ones who wanted it; farmers began clamoring for it too. Lederle began drying the fermentation residue and selling it, and when that proved too slow to keep up with demand, the company began pumping the raw brine left over from fermentation into railroad tank cars and sold it by the carload. Demand was so intense that a senator from Nebraska sent an official complaint that the farmers in next-door Iowa were getting more of the product than his own constituents. The vice president, Alben Barkley, requested and received a load to feed to livestock on his family's Kentucky farm. In Austin, Minnesota—home of the giant hog company Hormel Foods, maker of SPAM—a pharmacist somehow diverted a shipment of residue, which he packaged and resold. He made enough money on the deal to retire, quickly, to Florida.

As a pharmaceutical manufacturer, Lederle had to report any new drug and any new use of a drug to the FDA. It had done that when Aureomycin was first achieved, registering it appropriately as a human medication. But when it came to Aureomycin in animal feed, the company was cagey. Phrasing its statements carefully, Lederle said the fermentation products it was selling by the ton were intended as vitamin supplements. That might have been true; it was possible that fermentation for Aureomycin was producing B_{12}, though Lederle never tested its new product to find out. But it was also disingenuous. The wording of a patent application the company filed in September 1949 for adding Aureomycin to animal feed—as a drug, not a vaguely described "source of vitamin B_{12}"—proves they understood what was going on. Yet Jukes and Stokstad did not publicly acknowledge how their discovery worked until April 1950, at the annual meeting of the American Chemical Society. A *New York Times* reporter happened to be covering the conference. The following morning, his story trumpeted the news on the front page: "'Wonder Drug' Aureomycin Found to Spur Growth 50%":

> The golden-colored chemical aureomycin, life-saving drug of the group known as antibiotics, has been found to be one of the greatest growth-promoting substances so far to be discovered, producing effects beyond those obtainable with any known vitamin . . .
>
> The discovery of the new role for aureomycin, described in the announcement as "spectacular," is believed to "hold enormous long-range significance for the survival of the human race in a world of dwindling resources and expanding populations."

The story contained a clue as to why Aureomycin had become so popular: It was cheap. "Five pounds of an unpurified product, selling at 30 to 40 cents a pound, when added to a whole ton of animal feed . . . 'has increased the rate of growth of hogs by as much as 50 per cent,'" it said. And it ended by declaring, with a boldness that would turn out to be overconfident: "No undesirable side effects have been observed."

—◊—

IN RETROSPECT, Lederle's willingness to dump its new antibiotic into animals without even knowing the dose it was dispensing is startling. But in the context of the time, Jukes's rush to deploy Aureomycin and the company's eagerness to profit from it both make sense. Antibiotics were new, and the whole world was giddily in love with them.

There was a reason they were called miracle drugs. Before penicillin, even minor infections were a death sentence. The devastating illness that ravaged constable Albert Alexander after his encounter with the rosebush before he received penicillin—and again after the drug ran out—was appalling but utterly normal. In the pre-antibiotic era, what should have been minor cuts and scrapes burgeoned into infections so serious they required amputations. Three out of every 10 pneumonia patients died; so did nine women out of every thousand who gave birth. (That was in the cleanest hospitals; often the death toll was higher.) Untreated ear infections ravaged children's hearing, and untreated strep throat led to rheumatic fever, paving the way for heart failure later in life. With no way of controlling it, bacterial meningitis killed children with convulsions or left them neurologically damaged. One out of every six soldiers wounded on the battlefield died there, and huge numbers—one out of three, in some military camps—contracted syphilis and gonorrhea that left them disabled, arthritic, or blind.

The relief of being freed from that burden sparked a joyous overreaction. Penicillin was not only dispensed to patients in hospitals; manufacturers tossed it into ointments, throat lozenges, gum, toothpaste, inhalable powders, even lipstick. Anyone could buy penicillin over the counter in a pharmacy, and huge numbers of people did; it was not restricted to being dispensed by prescription until 1951 (and even then only because overuse was provoking allergies). There was no understanding yet that antibiotics work only against bacterial diseases, since early research reports were full of optimism regarding their effect on viruses. It seemed only smart to take the new drugs for any ailment—foolish, in fact, not to. (Aureomycin's discoverer Duggar recalled years later that his lab assistants—who of all people should have known better— liberated samples of the crude drug "to cure their colds," which an antibiotic cannot do.)

Some researchers did wonder whether, if antibiotics worked such magic for people, they might benefit animals, too. In fact, two years before Jukes's experiment in 1946, a team at the University of Wisconsin had dosed commercially hatched rooster chicks with some of the antibiotics that had been discovered so far: a sulfa drug, newly licensed streptomycin, and a third drug, less powerful, called streptothricin. They were trying to find a way to sterilize the contents of the chickens' guts, which they thought would be useful once they took the birds into the lab to use them for studies. So they were surprised to find that both the sulfa drug and the streptomycin increased the birds' weight; when the chickens were slaughtered at 28 days old, they weighed from 240 to 300 grams. Perplexingly—perhaps because Wisconsin was already a center for vitamin research—the group put the research aside, never following up on what they found. But after Lederle announced its Aureomycin results in 1950, researchers piled on; antibiotics in animal diets became an enormous research focus in drug companies and at almost every university with an agricultural college. When Jukes counted the published research in 1955 in order to write a summary of what had been discovered so far, he found that almost 400 scientific papers about feeding antibiotics to animals had been published in just five years. The market for antibiotic-laced feed had boomed too. U.S. farmers already were giving livestock 490,000 pounds of antibiotics per year.

Almost no one seems to have thought this was a bad idea.

That is odd. From the earliest moments of the antibiotic era, there had been rumblings of concern about how long the effectiveness of the wonder drugs would last. In December 1940, before any human had received penicillin, two of Fleming's collaborators wrote to a medical journal that they had observed the common gut and lab bacterium *Escherichia coli* developing defenses against the new drug. In 1945, a few months before receiving the Nobel Prize in Medicine for discovering penicillin, Fleming warned an audience in New York about the consequences of deploying the drug carelessly. The *New York Times* quoted him:

> The greatest possibility of evil in self-medication is the use of too-small doses, so that, instead of clearing up the infection, the microbes are educated to resist penicillin and a host of penicillin-fast organisms is bred out which can be passed on to other individuals and perhaps from there to others until they reach someone who gets a septicemia or a pneumonia which penicillin cannot save.

In such a case the thoughtless person playing with penicillin treatment is morally responsible for the death of the man who finally succumbs to infection with the penicillin-resistant organism. I hope this evil can be averted.

Fleming was prescient, and unheeded. By 1947, a hospital in London was experiencing an outbreak of staph infections that did not respond to penicillin. By 1953, the same resistant bug sparked an epidemic in Australia, and in 1955, it crossed to the United States, infecting more than 5,000 mothers who had given birth in hospitals near Seattle and their newborns too. Those illnesses marked the start of the lethal game of leapfrog that organisms and antibiotics have been engaged in ever since. Researchers direct a drug against them; they evolve a defense against it; other researchers produce a new drug; bacteria evolve a defense again.

Fleming had warned specifically against underdosing. In medicine, then and now, antibiotics are prescribed in amounts that account for variations in the defenses of the bacteria, minor spelling errors in their genetic code that occur randomly as they reproduce. Some of those changes lessen an organism's chance of survival, but others improve its ability to protect itself from another bacterium or from a drug, and doses need to be large enough and long enough to make sure that even better-defended bacteria are killed. Fleming's concern was that less-than-lethal doses of antibiotic would create a Darwinian battleground: killing the weaker bacteria but allowing the stronger ones, whose changes conferred some resistance to the drug's attack, to survive—and to multiply into the living space that the death of the weak ones had cleared.

The antibiotics being administered to animals for growth promotion could not strictly be called underdosing—but only because the animals getting them were healthy. (There was no illness present that required a dose of any size.) But the amounts being given were minuscule, the equivalent of 10 grams of drug per ton of feed. Jukes was not concerned. Others at Lederle, however, were. Three decades later, Jukes revealed that staff veterinarians worried that the company's new product would encourage antibiotic resistance. They "strongly opposed" selling Aureomycin as a growth promoter, Jukes wrote, but were overruled by Wilbur Malcolm, the general manager of Lederle's corporate parent, American Cyanamid. "Competition was right on our heels," Jukes said.

In hindsight, that decision is extraordinary. Evidence was already accumulating that administering Aureomycin, even in the tiny doses of growth promotion, caused bacteria in the guts of the animals that received the drug to become drug resistant. Jukes, in fact, assumed this was part of the process. If bacteria in his chickens' intestines had not become resistant, he argued, they would all have died under the assault of the antibiotics—and then the birds would have died too because they needed the help of those bacteria to extract nourishment from their food. Instead, the gut bacteria thrived, and so did the chickens. Explaining it years later, Jukes called it "illogical." "We were not prepared for . . . the fact that the changed and resistant flora were in some way beneficial," he wrote. It did not worry him that continual dosing throughout an animal's entire life would sustain resistance in those bacteria. He assumed the process of growth promotion contained a built-in safety valve: If resistant bacteria in an animal's system burgeoned beyond some undefined point, the antibiotic doses would stop working, animals would stop gaining weight, and farmers would abandon the drugs. But as far as he could see, the opposite was happening. Animals not only gained weight after they were fed growth promoters; their weight gain continued even when the drugs were discontinued.

Whatever resistant bacteria were developing within the animals, Jukes observed, they posed no risk to livestock. He did not ask whether they posed a risk to humans.

Viewed from the perspective of the ways that government agencies operate now, with adversarial administrative processes that seem to grind on for years, what happened next is stunning. When the first antibiotics, including Aureomycin, arrived between 1945 and 1948, the FDA had viewed them as benefiting the public and worked with pharma companies to quickly get them licensed. The agency took the same attitude regarding antibiotics in animal feed: It accepted the companies' assertions that growth promoters were safe. In 1951, with no advance public notice and without holding a hearing, the FDA approved Aureomycin and five other antibiotics for use as growth promoters in animal feed. In the text of the order, the administrator of the Federal Security Administration, which oversaw the FDA at the time, said it "was drawn in collaboration with interested members of the affected industries and . . . it would be against public interest to delay."

It would not take long—only about a decade—to determine whether the public interest really had been served.

—⟶m⟵—

LOOKING BACK AT THE RESEARCH that was published as agriculture embraced growth promoters makes it clear that no one quite understood how the drugs worked. Some researchers hypothesized that the antibiotic doses encouraged animals to retain water, or affected the rate at which fat was stored, or cured subclinical infections, ones that were not causing visible symptoms but were a drain on animals' metabolisms. Jukes himself believed the drugs were affecting the bacteria that reside permanently in animals' intestines: their intestinal flora, or what we would now call their gut microbiome. The many functions those bacteria perform would not be understood for decades; in the 1950s, scientists did not possess the necessary molecular tools even to identify most of them. Jukes based his intuition on several observations. The growth-promoting effect did not work at all in "germ-free" chickens, ones that did not possess gut flora because they were hatched into a controlled sterile environment and fed a sterilized diet. Antibiotics did not work as well in normal chickens raised in very hygienic conditions, such as a recently cleaned barn, as they did in barns where litter or manure had been allowed to accumulate. And they did not work in runts. Growth promoters helped animals that had been deprived of nutrition to put on weight, but they could not turn a genetically undersized animal into a normal one.

As more scientists studied the problem, they recognized that using growth promoters did not change the overall count of bacteria in the intestines; that is, the drugs were not killing intestinal flora, and they were not encouraging greater amounts of bacteria to grow. But to the degree that researchers could extract and study the gut contents of slaughtered animals—a challenging task, because those bacteria do not all thrive in lab conditions—they perceived that the drugs did seem to change the balance of bacteria present, encouraging some species to reproduce and discouraging the growth of others. The antibiotics also seemed to change the physiology of the gut, thinning the lining through which nutrients are absorbed. Researchers suggested, but could not prove, that this helped animals extract more nutrients from their feed. They thought the normally thick gut walls found in chickens that did not receive growth promoters might let less nutrition through.

But none of these studies of the gut or its contents could pinpoint any downside to growth promoters, and thus researchers began to wonder whether humans could benefit from them too. Between 1950 and 1955, experimenters fed routine doses of antibiotics to small groups of premature babies, trying to get them up to healthier normal weights as fast as possible. Other researchers concocted other human trials of growth promoters, ones that are unnerving by modern ethical standards. Antibiotics were given, for a few weeks and up to several years, to people who had no capacity to consent, including developmentally disabled children held in a eugenics institution in Florida and undernourished poor children in Guatemala and Kenya. In the largest such test, 220 recruits at the Great Lakes Naval Training Center in Illinois—who, by virtue of their service oaths, could not really refuse—took daily doses of antibiotics for almost two months. Fortunately for the subjects, none of the trials reported adverse effects, and in all of them, the growth promoters worked. All of the recipients, adults and children, put on muscle, and the children grew taller as well.

Those results made researchers even more confident that antibiotics in animals had no negative effects, and that led in turn to the drugs' most out-there use: keeping food from going bad. In several countries, led by the United States, experimenters added antibiotics to the cold-water tanks in which fish were held on fishing boats and to the ice on which fish were stored in processing plants. They washed spinach in a streptomycin solution after harvesting. They painted the drugs on the outside of cuts of meat and mixed them into ground beef. Researchers infused antibiotics into cow carcasses after the animals were slaughtered and injected the abdomens and blood vessels of cattle before they were killed. (They concluded that the volume of antibiotics needed to perfuse an entire cow was too expensive, and that it was too hard to hold the animal still long enough for the drugs to work.) They also investigated putting single large doses of antibiotics into chickens' drinking water just before slaughtering them, as well as increasing the ratio of growth promoters in feed from Jukes's original 10 grams per ton to 1,000 grams or more. That line of research had to be abandoned: There was so much antibiotic in the feed that it moved out of the gut into the birds' muscles, leaving drug residues that exceeded federal food safety standards.

The experiments led to a kind of category creep for Aureomycin. The drug that first had been used only to make animals grow faster began to be used to protect them from diseases as well. This required a larger dose. Lederle salesmen began to tell farmers that they should give chickens not 10 grams of Aureomycin per ton of feed but up to 200 grams, a 20-fold increase. The FDA blessed the practice in April 1953, extending approval of Aureomycin from just growth promotion to prevention as well—once more without any advance notice and without a public hearing.

This was huge—and not just for Lederle, which instantly could count on bigger sales. It gave farmers permission to use much more Aureomycin than they had been, along with all the other antibiotics that had been approved for growth promotion two years earlier; the FDA endorsed all of them for preventive use too. But it also insulated unscrupulous or inexperienced or careless producers from the consequences of farming badly. They could squeeze animals in more tightly, clean their barns less frequently, scrimp on nutrition, turn a blind eye to pests—and know they were protected against the diseases that would otherwise have resulted, because the antibiotic doses protected their livestock from the start. The decision opened the door to industrial-scale production and the animal welfare abuses it would one day be accused of. And though it would take years before anyone put the pieces together, it would increase antibiotic resistance as well.

To the end of his long life—he died in 1999, at 93, after a third career in which he returned to the University of California, Berkeley as a molecular biologist—Jukes championed his invention and refused to acknowledge any downside. That may have been possessiveness, or arrogance, or just cussedness: He seems to have taken pleasure in defying received wisdom. Sometimes he was right. Jukes challenged chemist and double Nobel Prize winner Linus Pauling for championing megadoses of vitamin C to prevent colds and ameliorate cancers, a piece of medical advice that was hugely popular in the 1970s and now has been conclusively disproved. But he also derided regulation of dangerous food additives and called organic food a "myth." He opposed federal action that prevented beef cattle from getting the estrogen compound DES, even though it was known to cause cancer in the daughters of women who were given it in pregnancy. He directed special rage at the ban on the pesticide DDT and the hugely influential 1962 book, *Silent Spring,* which provoked it. He railed that the federal government had caved to "that segment of society represented by the antivivisectionists, anti-fluoridationists and organic farmers." In a parody published in the journal *Chemical Week,* he mocked its author, Rachel Carson, as a writer of "science fiction horror stories."

It is possible that Jukes was a strict empiricist, uninterested in any precautionary principle that might deprive society of the benefits of science. But also, he clearly loathed red tape. Writing about growth promoters almost 40 years after he recognized them, he said: "If such a discovery were made . . . in 1985, there would be round after round of committee meetings, and plans would be made to cope with various FDA roadblocks. Long-term and short-term toxicity tests would be started. Metabolites and residues would be isolated and identified. Above all, the product would be tested for carcinogenicity. Finally, the FDA would refuse permission to market it."

Writing for both scientific journals and newspapers—his zestful polemics appeared, among other places, in *Science, Nature,* the *Journal of the American Medical Association,* and the *New York Times*—Jukes dismissed any concerns about livestock antibiotics and the concentrated confinement farms they would create. "The use of antibiotics for farm animals does not present a hazard to public health," he declared in the *New England Journal of Medicine* in 1970. Appearing at the New York Academy of Sciences in 1971, he asked: "Do we have so many cattle, pigs, and chickens that we ignore the need for feeding them by the most economical means? I do not think so." He maintained in the *New York Times* in 1972 that "antibiotics saved $414 million . . . for producers of meat." In 1992, when he was 86—in what may have been his last writing on the issue—he said, "The urban public has been encouraged by the animal-rights movement to believe that animals 'feel better' when they have more space for roaming. But how do we know this is true? Humans voluntarily jam themselves together to watch sports or in social gatherings. The larger and denser the crowd, the more successful the event."

Jukes's unwavering belief in farm antibiotics as a public good attracted many adherents. But slowly, over decades, a more complicated story would unfold.

CHAPTER 3

MEAT FOR THE PRICE OF BREAD

THOMAS JUKES'S RESEARCH—his insight into the vitamins poultry needed to live in confinement and the antibiotics that could make confinement profitable and safe—would turn chickens, the random inhabitants of every American barnyard, into chicken, the market juggernaut. Within a few decades, Americans would eat more chicken than any other meat. The backyard bird would become the most-studied, most-crossbred, most industrialized meat animal on the planet. But for all that to happen, there would need to be a chicken industry—and in the 1940s, that was just beginning.

Before chemists learned how to synthesize vitamins, there was no sustained trade in meat chickens. Birds for eating were byproducts of egg production: "spent" hens—those too old to lay eggs reliably and tough from chasing chicks around a barnyard—or unneeded roosters, which were fed for a few months until they gained a few pounds of flesh and then sold as tender young "spring" chickens. (Hens and roosters hatch in a roughly 50:50 ratio, but a farmer who wants fertile eggs but minimal fighting in a flock will discard most of the males.) Spring chickens were a luxury dish, capitalized and highlighted when they were listed on menus, and the demand for them hinted a bigger market might be waiting. But farmers lacked a way to keep young chickens alive through the winters; indoor chickens contracted colorfully named diseases like "leg weakness," "slipped-tendon," and "curl-toe paralysis." Synthetic vitamin supplements solved those problems, kept birds alive in confinement, launched the industry that gave Jukes his start, and created an entire new field of agriculture: raising "broiler" chickens as a product separate from egg production. In 1909, in the entire United States, 154 million chickens were sold for meat, alive or slaughtered; by 1949—after vitamins but before growth promoters—the total was 588 million.

The birthplace of broilers was the Delmarva Peninsula, which describes almost all of Delaware, the eastern part of Maryland, and a snippet of Virginia, south of New Jersey and east of the Chesapeake Bay. According to industry lore, the founder was an egg producer, Cecile (Mrs. Wilmer) Steele. In 1923, the mail-order hatchery from which she bought laying-hen chicks accidentally sent her 10 times what she had ordered: 500 instead of 50. With no other option, she decided to try raising and selling them for meat. She received 60 cents per pound for the birds—five times what she would have gotten for them if they had lived out their lives and been sold as spent hens—and ordered 1,000 as a second round. Each time she raised and sold a flock, she expanded her business. Soon her neighbors copied her. Chickens looked like a reliable alternative to the main local crop, strawberries, which were fragile and vulnerable to insects and storms. Within a decade, there were at least 500 Delmarva broiler farmers like Mrs. Steele, and they dominated the East Coast market: Washington, Philadelphia, and especially New York City.

New York was poultry's niche market, for a complex set of reasons. The city was the home of nearly half the Jewish immigrants to the United States, almost two million, making it effectively the largest Jewish city in the world. Observing the Jewish Sabbath properly included serving something celebratory, but the dishes that Christians would have chosen for an equivalent Sunday dinner were not an option. Pork was forbidden, of course; beef was permissible, but whether the animal had been slaughtered in a kosher manner was difficult to verify and subject to fraud. Chickens could be transported into the city alive and killed in front of a customer, which guaranteed dietary purity. And chicken was luxurious and special. That was the meaning behind the promise made in the 1928 election, "A chicken in every pot"—which, though it is always attributed to Herbert Hoover, was actually made on his behalf in campaign ads by an association called the Republican Business Men. Broiler chickens fit the Jewish community's need for a trustworthy holiday meal so well that catering to the city became a huge stimulus to Delmarva production. Delaware farmers not only converted en masse to poultry raising, but also banded together to build the first large-scale plants for killing and packaging birds.

By 1942, Delmarva was producing almost 90 million broilers a year. At that same moment, though, the United States was a year into World War II and deploying millions of troops who needed to be fed. The War Food Administration, the agency in charge of rationing food for civilians and provisioning the armed forces, looked at Delmarva and saw not just a source of protein but a unique opportunity to control production. The peninsula is small and surrounded by water, and land access is limited to a few roads—all of which made it possible for the government to monitor what was going in and out and to block black-market deal making. The agency co-opted the peninsula's entire chicken production, cutting off its existing customers and funneling all of its birds into the military supply chain.

For Delmarva, it was a blow that would take decades to recover from. But it was an opportunity that other farmers—and one visionary chicken feed dealer—had been waiting for.

—⁓—

THE FORCIBLE REMOVAL of Delmarva from the American poultry market was the first good thing that happened to the northeastern counties of Georgia in as long as anyone there could remember. For almost a century, dating back to the Civil War, the hill country tailing off from the Appalachians had been unable to catch a break.

First, the small subsistence farms that dotted the hillsides had been ravaged by retreating Confederate troops and Union Army squads pursuing them. To build the properties back up, self-sufficient farmers who had always resisted debt resorted to growing cotton on credit, an exploitative system dependent on a crop that did badly in the thin, stony mountain soil. A tornado ripped through the region in 1903, killing about 100 people. The cotton market plunged in 1920, slashing farmers' meager earnings in half. A boll weevil infestation arrived in 1921. The stock market crash in 1929 forced cotton prices down to almost one-eighth of what they had been 10 years before. In 1933, New Deal austerity measures compelled landowners to plow under one-third of their acreage, cutting their own income and devastating the tenant farmers who relied on the crops they grew to pay back loans to their landlords. In 1936, a devastating double tornado, much larger than the one in 1903, roared across the hills, killing hundreds of people and flattening most of Gainesville, Georgia, the area's railroad stop and main market town.

One of the local businesses caught in that decelerating economic spiral was a seed and feed store belonging to Jesse Dickson (or Dixon) Jewell. Jewell was 34 the year the tornado came, but its destruction was one more disaster in a life that had already seen too many. His father, who started the family store, died when Jewell was seven; his stepfather, who took over its management, died when Jewell was 28, and his exhausted mother handed the business to him to run. The repeated crashes of the local economy had cut sales to almost nothing, and to keep his wife and daughters fed, Jewell cast about for alternatives. There was already some chicken raising going on in the Georgia hills, but it was small-scale and seasonal to fill out farms' income. In the year before the tornado, the 30 northeastern counties had produced only 500,000 birds.

Jewell thought that could be expanded, and he had an idea how to do it. From the time of the Civil War (which in Georgia was still called the War of Northern Aggression, and not as a joke), small-scale farmers had worked land they did not own in an arrangement called a crop lien—or, more often, sharecropping. They rented small plots of land from large landowners and borrowed tools, seed, and fertilizer, or the money to pay for them. When the crop was ready, they sold it, paid the landowner the rents they owed, and the inevitably high interest, and lived on what little remained. The system was always resented and frequently abusive; it was notorious for never allowing farmers to climb out of debt. But it was familiar: It had been the dominant form of organization for farm labor in the South for a long time.

Jewell adapted it to poultry. He persuaded local farmers to switch to growing chickens: He would bring the birds, they would raise them on credit, and they would earn their pay—minus expenses—when the birds were sold. He drove his family's feed trucks to a hatchery to buy chicks and then used the trucks to deliver the birds to the farms. He supplied feed on the same kind of lien and offered cash loans so farmers could convert their barns to chicken houses. Once their broilers were market size, he picked up the birds in his family's trucks and drove them to market, anywhere from Atlanta, 60 miles away, to Miami, 700 miles.

Jewell's new idea saved the family business, but Jewell had bigger plans. Reports of how Delmarva farmers organized their production cooperatively—before the government intruded—had been published in poultry magazines, and he modeled his next steps on what they had done. First, he wangled feed on consignment from a mill across the nearby Tennessee border, getting credit from that business in the same way he extended it to his growers. This gave him a larger stock to dispense than his family business could supply—he bought five tons at a time—and that in turn allowed him to recruit more growers and buy more chicks. In 1940, he founded his own hatchery, and in 1941, he established his own processing plant to slaughter and package the chickens his client farmers raised. Georgia broiler sales rose to 3.5 million birds in 1940 and 10 million in 1942. North Georgia's farmers— even ones who distrusted Jewell's scheme because they held hard old grudges against the sharecropping system—began to turn a profit for the first time in their lives.

As devastating as it was to the Delaware farmers, the government's appropriation of Delmarva's chickens was an extraordinary gift to Georgia. Outside of Delmarva, poultry production was not under government control and chicken meat was not rationed. On the contrary, the government encouraged families to eat chicken and eggs in order to spare beef and pork for the troops. It also bought chickens for military use, at a guaranteed price that was above what the civilian market paid. Arkansas was also becoming a center of chicken raising, coming along to challenge Georgia in filling the hole Delaware left in the market. But Georgia kept pulling ahead: 17 million broilers in 1943, 24 million in 1944, and almost 30 million in 1945. Most of that was due to Jewell. Throughout the war, he kept reclaiming pieces of the chicken business that had belonged to subcontractors, from growing grains that could be milled into feed, to rendering slaughter discards to sell to other industries, to writing marketing brochures. He rewrote his agreements with his growers, shortening the length of contracts so they would be forced to renegotiate more often and working out a new formula in which they were paid not per chicken but by how much weight the chickens had gained relative to how much feed they had eaten.

The war was a gift to Georgia's new chicken industry, but the end of hostilities almost doomed it. The government pulled out of purchasing, abandoning its contracts, releasing the entire capacity of Delmarva back onto the market, and also setting free the curbs it had placed on sales of pork and beef. The new infrastructure of the poultry industry wobbled, unable to sustain its expenses under the onslaught of sliding prices. Georgia's roaring chicken trade faltered. For the first time since the Gainesville tornado, the area produced fewer birds than it had the year before.

The crisis and price collapse came at just the right time to persuade the industry to embrace Jukes's discovery that minuscule antibiotic doses would put weight on the birds. And the final piece of Jewell's reorganization of the poultry business made it certain that farmers would use growth promoters. In 1954, he cut out the last middleman, building his own mill to grind and blend the feed that his contract farmers would now be compelled to use for the birds he brought them. Jewell had created the model of the modern poultry company: a vertically integrated corporation that contributed the hens, the chicks, the feed, the supplements, the transportation, the processing, and the distribution and sale. All that remained to farmers were the property the birds were raised on, the labor they invested to accomplish that, the debt they undertook to upgrade it, and the risk that the birds would not do well.

The first years of the new arrangement were profitable. People joked that north Georgia held more Cadillacs than Texas. By 1950, Georgia was producing almost 63 million broilers, and Gainesville dubbed itself the Poultry Capital of the World. By 1954, the production of the entire industry across the United States would top one billion broilers for the first time. There seemed to be no limit to what organized, streamlined, technological poultry raising could achieve. But another crash was coming.

—◊—

IN SPRING 1957, a Who's Who of the poultry industry gathered in a wood-paneled, gilt-finished hearing room on Capitol Hill. The crowd included the chief executives of the largest hatching companies in the country, the largest feed companies, and the largest meat processors. There were government economists and statisticians, and bankers and insurance managers. On the dais, next to the chairman's gavel, sat stacks of letters and telegrams, and a murmuring crowd filled the chairs set aside for the public. They were all there for the same reason: to figure out how poultry, so gorgeously promising after Jukes's experiment nine years ago, had gone so badly wrong.

Just one year earlier, farmers had produced record-high numbers of chickens, and eggs and turkeys too. Yet poultry prices were crashing. At the start of the 1950s, farmers had received 48 cents for every dozen eggs; they got 31 cents now. The price for turkey had been 37 cents per pound; now it was 26. The price for meat chickens had dropped lowest of all, from 29 cents to 20. A government economist explained the forces behind the price drop. Since Jukes's discovery of growth promoters, chicken production in the United States had tripled, from 371 million birds in 1948 to 1.34 billion in the year before the hearing. But chicken consumption was barely budging: It had been stuck at less than 25 pounds per person per year—sometimes much less—for the past 20 years. In contrast, Americans had just eaten more red meat in one year than at any other time since the government began counting: 167 pounds per person. Spread out across the year, that was the equivalent of almost half a pound of meat every day. But people were eating that same amount of chicken just once a week, the equivalent of one person's serving from the Sunday roast bird.

Production was rising, but consumption was flat. It was no wonder prices were skidding. "This industry is sick," declared Congressman Perkins Bass from New Hampshire. "Scores of once healthy poultry businesses in New Hampshire have . . . been destroyed or imperiled." Peter Chichester, vice president of the Maryland feed manufacturer Dietrich & Gambrill, called the situation "disastrous."

The mission of the subcommittee that had summoned the crowd of witnesses was not to draft legislation but to research problems so that other congressmen could propose legislative fixes. Over nine long days, the hearings exposed that broiler growing had become a gold rush. Easy credit—and in some cases, insurance so generous that it covered any losses, even from low prices—had persuaded existing farmers to expand and enticed thousands of new producers into raising chickens. Thanks to antibiotics and better breeding, those new and expanded farms were able to raise more birds at less cost per bird.

"It now takes half the amount of feed to grow a much bigger and better bird," Representative Timothy Sheehan of Illinois explained. "You are going to have a never-ending cycle of constantly improving the feed and improving the bird—more meat from less birds."

Witnesses said the only thing keeping the industry from collapsing was Jewell's new business model, which dozens of others had copied. The contracts offered by the "integrators," as they were being called—the men who had organized poultry into corporations that controlled chickens from before hatching to after slaughter—were insulating individual farmers from the consequences of being overextended. Jewell, who was among the witnesses, testified that the interplay of the parts of his company protected him from being harmed by low prices. His growers might decide to raise too many chickens, which endangered their individual earnings—but the mill that sold them the chicken feed made more money because it was feeding more birds, and the slaughter plant did better because it was processing more. Therefore, local economies benefited. In north Georgia, once desperately poor, loans and deposits held in the local banks had soared, and feed mills, rendering plants, equipment dealers, and farm supply stores were jostling to move into the area. He told the committee: "We think the industry can work out its own problems if given time."

The economists and bankers in the room disagreed. They said that riding out economic storms would be possible only for very large companies, because they could balance losses in one part of the firm against profits in another part, while maintaining enough capital to hold them all up. (Jewell was one of those. He said he had 400 farmers under contract.) The experts forecast a ruthless future in which the stress of low prices would force independent farmers to become contractors and compel small integrators to sell themselves to larger ones. "Only the very best managers appear to be able to continue a profitable operation—and in the case of the broiler growers we believe less than 5 percent are now continuing independent operations," Mark Witmer of the Eastern States Farmers' Exchange warned. Chichester, from the feed mill, predicted: "The large ones are willing to continue to lose in the hope the others will drop out."

Representative Horace Seely-Brown, Jr., of Connecticut objected that poultry was "going to end up with one man or ten people sitting in control of the entire industry." Jewell responded, with a coolness that even today leaps off the bound pages of the transcript, "That is the way the auto industry is today, isn't it? I don't know what the outcome of it is going to be, but I guess that some atomic bomb will come along and knock them into pieces again."

That was a bold joke to make, deep in the paranoia of the Cold War. (Five months later, Russia would launch Sputnik I.) But U.S.

car manufacturers had already collapsed into the Big Three, and poultry would follow the same path. In 1950, there were 1.6 million poultry farms in America, most of them still independent; 50 years down the road, 98 percent of them would vanish. Today there are about 25,000 U.S. farms raising poultry, almost all operating under contracts with the integrators that survived consolidation: Tyson Foods, Sanderson Farms, Pilgrim's, and others —altogether, just 35 firms. (Jewell's own firm, J. D. Jewell, did not survive. It was sold in the 1970s to a company that itself was sold in 2012.)

In 1957, the people fretting about the future of the poultry industry could see those consolidations coming, but no one could agree how to stave it off. Mildred Neff Stetzel, who had been running a broiler hatchery since 1918—first near Ross, Iowa, and then in Paris, Arkansas—urged the congressional committee to protect smaller businesses. She described listening to a rival hatchery's manager boasting at a sales meeting that they could add another million chicks in the coming year, figuring the number of birds they sold would offset any price collapse to which they might be contributing. She asked Congress to preserve a space for small proprietors by limiting how many birds the industry could hatch in any year.

The breeder company representatives who appeared after her—all men, all from companies many times bigger than hers— rejected government controls. They argued that any breeder who refused to supply a hatchery out of public interest would be supplanted by another more willing to sell, and said the same thing would happen to any hatchery that refused to send chicks out to farmers to be raised. It was important, they said, to let the market shake out the poor performers and allow the canny businessmen to thrive. Robert C. Cobb, Jr., vice president of Cobb's Pedigreed Chicks of Massachusetts (which would grow into one of the top breeder companies worldwide), summed up the feeling: "A man has a right to go bust with his eyes open."

But none of them thought they would. Like Jukes almost a decade before, the leaders of the big new poultry companies argued that they were serving society, as well as their own profits: furthering science, feeding the country, making protein cheap. Another of the breeders, Henry Saglio of Connecticut, whose company Arbor Acres would battle Cobb for dominance for decades, expressed the annoyance they all seemed to feel over Congress's scrutiny. He meant it sincerely, though now we can hear the unintended irony: "I think any industry producing meat for almost the price of bread has got a big future."

Before it could achieve the future its leaders envisioned, the poultry industry had to solve the market imbalance it was mired in. Put simply, there were two options: produce less chicken or sell more. Yet the industry's leaders had just told Congress that they would not volunteer to grow fewer birds and would resist any federal attempts to force them. Their strategy for hauling the poultry industry out of the crash everyone could see coming was really just a belief: that people would keep buying more chicken.

A few of them did sense the industry was making a mistake. D. H. McVey—a vice president of General Mills, Inc., one of the largest feed suppliers—was alarmed by the complacency. "Far too often we as producers, which includes the hatcheryman, the farmer, as well as the feed manufacturer, look upon ourselves as controlling the rate of production of broilers, turkeys, or eggs," he warned. "This may be true from the physical production standpoint, but in the larger sense, the group that controls the rate of consumption of our products is the American homemaker. She is the one who decides which food we shall enjoy and how often."

There were plenty of housewives in America; the year of the hearings, 1957, was the peak birth year of the baby boom. But they were not purchasing enough poultry to keep the industry afloat, and no one could figure out how to convince them to buy more. Four years after the hearing, *Harper's Magazine* described a "chicken explosion" of overproduction. "The industry has increased . . . output more than fifty-fold since 1929," the magazine said, describing poultry falling into a "quagmire of 'overproduction' and 'excess capacity.'" The author, an economist, perceived the same mismatch of supply and demand that Congress had been worried about: too many chickens in barns, not enough chickens on plates. He wrote: "There is some limit to the amount of chicken any one person will eat in a year."

And then someone figured out how to lift that limit. An earnest scientist who worked nowhere near the poultry-growing centers of Georgia, Arkansas, and Delmarva got enough distance on the problem to understand that people *could* be made to eat more chicken more often. He wanted to improve the financial prospects for chicken farmers. He ended up changing what millions of people eat.

—◆—

HERE WAS THE DIFFICULTY: Getting people to eat more chicken really meant getting people to cook more chicken. The men who headed the poultry companies had not thought that through, because buying and cooking were considered women's work.

In 1960, four-fifths of the broilers shipped to retail were sold whole; only a few supermarkets and bespoke butchers sold chickens precut into parts. Buying a whole chicken meant that the woman purchasing it—and at that time, it almost always was a woman—would either have to break it down at home, a messy job requiring good knives, or cook it intact. Cooking it whole meant roasting. Dealing with the portions required pan frying or braising, or broiling if the birds were young. All of those methods were time-consuming, and none of them fit the scarce hours and changing tasks of women who were breaking free of 1950s conformity and moving into the workforce.

Plus, chicken had a size problem. Even in the 1960s, a bird was too big for one couple to eat at a single meal but too small for a family's weeknight supper. And, most difficult to overcome, chicken had a predictability problem. If a woman planning meals for her family wanted to feed them beef, she could grill a steak, braise a brisket, or fry a hamburger; if she wanted to give them pork, she could roast a loin or griddle pork chops or bake a ham. Chicken, whole or in parts, roast or sautéed, was really still just chicken. It needed a tinkerer, someone who could recast the same protein into different flavors and easier-to-cook shapes, the way that fish sticks did for fish when they debuted in 1953.

Chicken found its disrupter in an inventive professor from the upper edge of New York State, the son of a family who owned a fruit farm, named Robert Baker. Baker planned to go into the family business; he went to Cornell University, the biggest local school, to earn a degree in fruit agriculture. Once he arrived there, though, his aspiration became returning as a professor. That required going somewhere else to work first, so he took a job as a cooperative extension agent, one of the federally funded liaisons between land-grant universities and their local communities. He went to Pennsylvania for a master's degree and to Indiana for a Ph.D., and he returned gratefully to northern New York in 1949 to join the Cornell faculty.

Considering the prominence of Delmarva and the emergence of Georgia, upstate New York might not seem an obvious place to do poultry research. But Cornell was one of the original land-grant colleges, founded in 1865, and the home of the first poultry science department in a U.S. university. It had an explicit mission to support agriculture in its area, and that included its poultry farmers, who were suffering twice over from the lingering postwar dip in sales and the geographical slide of the industry toward the South. In 1959, Cornell asked Baker to develop products that would help increase local farmers' sales of chicken. The university built him a food science laboratory in the basement of a low modern building at the center of the agricultural college campus and gave him research assistants to help his work move along.

Baker's first experiments focused on marketing: packaging broilers with containers of barbecue sauce to suggest how supermarket shoppers should cook them and showcasing small eggs, which languished in stores, in a "Kid's Pack" carton marked "especially for children." But having been born not just on a farm but immediately before the Great Depression, Baker was innately thrifty, and he itched to find better uses for little-valued spent hens and the disdained parts of broilers: back, necks, even the skin. He designed the first machine for separating meat left on the carcass from the underlying bones and experimented with grinding scraps into binders for sausage and hash. He produced chicken hot dogs made from hen meat, testing varied casing colors and spicing formulas on high school students and different packages and promotional schemes in supermarkets. (As part of the marketing trials, he named some boxes "chicken franks" and others "bird dogs" and measured the sales. Male shoppers preferred "bird dogs," but "chicken franks" narrowly won.) He developed a chicken bologna that he called "chickalona," chicken breakfast sausage, chicken burger patties, chicken spaghetti sauce, and a bake-and-serve chicken meatloaf frozen into an aluminum pan. Baker worked as well on egg products and later with fish. But the product that would transform the poultry industry emerged from his lab early in his research program, in 1963. Baker called them chicken sticks.

Later, the world would know them as chicken nuggets.

Chicken sticks were not exactly like fish sticks, because the fish variety were made out of whole muscle, sawed plank by plank from solid-frozen blocks. But doing that with a chicken would have yielded just a few slices from each breast; it would have wasted the rest of the carcass, and Baker abhorred waste. He turned instead to the blending he had been experimenting with, aiming to make a product composed from the breast but reshaped. That presented several food engineering challenges: first, keeping the meat in one piece without putting a sausage-like casing around it, and then devising a coating that would stand up to the shrinkage of freezing and the steam that would be released when the products cooked. Baker and his graduate student, Joseph Marshall, solved the first problem by grinding raw breasts with salt and vinegar, drawing out a sticky matrix of protein, and adding a binder of powdered milk and grain; the second, by shaping the sticks, freezing them, coating them with batter, and freezing them a second time. They designed an attractive box with a cellophane window and a dummy label that said nothing about Cornell, and placed them in five local supermarkets for a marketing test. In the first week, they sold 200 boxes.

Baker never patented his invention, and neither did Cornell. On the contrary, the university distributed his recipes, techniques, packaging designs, and marketing strategies for free in monthly bulletins that it published for decades. They were mailed to other universities and to about 500 food companies; no one who is now at Cornell can guess how widely they were distributed. "He literally gave ideas away," his former student, Robert Gravani—now a Cornell professor himself—told me in 2013. "And other people patented them." Baker went on to chair the department of poultry science and to consult for food companies. Many of his graduates went to work for them.

There is no absolute proof that Baker's invention inspired the McNugget, the category-defining product that McDonald's Corporation introduced in 1980, 17 years after his concept was published. (The official history, told in a company biography that McDonald's cooperated with, is that the McNugget arose from an exchange of ideas between founder Ray Kroc, chairman Fred Turner, and René Arend, the executive chef.) But the timing is telling. Three years before the McNugget debuted—and exploded every sales record at the Knoxville, Tennessee, McDonald's locations where it was secretly launched—the Senate Select Committee on Nutrition, headed by Senator George McGovern, triggered an earthquake in American diets by publishing the first ever Dietary Goals for the United States. The goals, the forerunners of today's contentious Dietary Guidelines, exposed scientists' worries about rising rates of heart disease and recommended for the first time that Americans eat less saturated fat, which was immediately interpreted as a recommendation against red meat. Chicken was the obvious alternative, and the corporation cast a wide net for ideas—which likely was wide enough to scoop up Cornell bulletins resting in their archives.

In the poultry world, it is taken for granted that Baker is the father of the nugget and, even more, of the "further processed" category that his inventions launched. He gave chicken the boost that it needed, transforming it from a single uninspiring food into an enticing, easy-to-eat array. Once people began eating more chicken, poultry's oversupply problem was solved—reversed, in fact: 1977, the year that the Dietary Goals were published, was the first year that Americans ate less beef than they had the year before. Every year after that, meat retreated and poultry advanced. In 1960, around the time Baker began his work, Americans ate 28 pounds of chicken per year; in 2016, we ate more than 92 pounds, the equivalent of 4 ounces every day. The constantly rising demand made growth promoters essential. It also made them normal, so routine that when food companies started introducing antibiotics into our diets, no one thought it was odd.

CHAPTER 4

RESISTANCE BEGINS

SOMETIME IN THE LATE 1950s, a new word began appearing in ads in magazines and newspapers. "Our Poultry Is Acronized!" the Town and Country Market, a local grocery store, announced in the August 1956 edition of the Ukiah, California, *Daily Journal*. A 1957 ad in the Syracuse, New York, *Post-Standard* advised, "There are all sorts of ways to tell about chicken freshness. But there's only one *sure* guide . . . When you spot an Acronized chicken, you know it's the best." A full-page color ad that ran in several women's magazines displayed a glistening, crisp-skinned whole chicken, displayed with elegant geometric candlesticks and an aspirational oversized pepper mill, and promised: "It tastes like fresh chicken bought right at the farm—juicy, wholesome and country-sweet. But the miracle is, today you can get it at your neighborhood food store . . . *Acronized* chicken!"

What on earth was "Acronized"? If a woman reading the ads was perplexed, there were stories in the same newspapers—many of them in the lifestyle sections that were usually called the "women's pages"—adroitly positioned to fill the gap in her knowledge. "Thousands of American homemakers who have never cooked and eaten a fresh-killed chicken will be doing so, thanks to a revolutionary process which helps maintain freshness in such perishables as poultry, fish and meat," the *Odessa American* in Texas told readers in January 1956. "Fresher, tastier poultry on your table, and a virtual end to storage problems. That's what scientists predict for the new 'Acronize' preservation process for poultry," the *Kossuth County Advance* in Algona, Iowa, wrote in May 1957. "What does the term 'Acronize' mean? . . . Acronize is a term that applies to poultry that has a special food grade," Vermont's *Bennington Banner* explained in January 1958. Lower down in each story, Acronize was finally defined: It was a preservation technique that kept meat from spoiling. It was an exclusive process, indicated by a special seal, that only a few slaughterhouses in any area were allowed to perform. It was modern, and scientific, and it was going to change how meat was sold.

In fact, it was antibiotics. Acronizing was an invention of American Cyanamid, the parent company of Lederle. It represented yet another use for chlortetracycline (the drug the company marketed as Aureomycin) beyond growth promoters and beyond the preventive dosing that could protect animals from diseases spreading in barns. Instead of putting antibiotics into the chickens while they were alive, Acronizing applied the drug after birds were slaughtered and eviscerated. Every bird (and, later, fish too) that was advertised as having been Acronized had been soaked in a diluted solution of antibiotics while it was being butchered. The solution contained enough drug to leave a film on the meat. The film lingered while the chickens were packaged for sale, while they sat in stores' refrigerated cases, and all the way into home kitchens, keeping bacteria that would have caused the meat to spoil from growing on its surface. The goal of this process was to extend the time in which raw meat could be offered for sale—not just by a few days, but by a freakishly artificial several weeks and up to a month. (That may be a clue to the invented word with which Lederle named the process: *a + chron,* a combination whose etymology, reaching back to ancient Greek, suggests timelessness, or detached from time.)

The ads touting Acronizing, and especially the stories explaining it, which rolled out across the U.S. media in the late 1950s—in national magazines, and in dozens of newspapers published throughout the country—looked like authentic excitement over yet another gift of science to the postwar world. In reality, they were an aggressive marketing campaign. American Cyanamid had hired a prominent New York advertising firm to place ads across the country, film TV commercials, record radio spots, and get stories into local papers. The firm arranged exclusives for each of the wire services: the Associated Press, Dow Jones, and United Press International all sent out pieces for small papers to use. The campaign staged events where it fed Acronized and roasted chicken to food editors, who praised it to their readers, and the Acronization process received the much trusted Good Housekeeping Seal of Approval, which is still being issued by the same magazine today.

Lederle was able to do all that because, just as with growth promoters and preventive dosing of livestock, the FDA had blessed its activities. At the end of November 1955, without scheduling a hearing or inviting any public scrutiny, the FDA issued yet another

30-word order, published in the *Federal Register*, asserting that "it would be against public interest to delay." The order explicitly permitted that the raw meat going through Acronizing would come out of the process carrying active antibiotic, capable of killing bacteria; that was what was necessary to keep spoilage at bay. Lederle had assured the agency that any antibiotic left on the meat would be destroyed by heat when the meat was cooked, so no one eating it would receive an accidental dose of antibiotics. But there is no indication the company examined what would happen to anyone else exposed to the drug: homemakers handling the raw meat in their kitchens, butchers packaging chickens in stores, or workers who killed the birds, tossed them into the antibiotic-laced baths, and fished them out again.

Lederle had good reason to turn a blind eye to any risks: Acronizing was going to make the company an enormous amount of money. In 1956, one year after the FDA approved the process, *Business Week* predicted: "The annual sales potential for antibiotic food preservatives [is] upwards of $20-million in the U.S., and $200,000-million per year abroad."

—m—

THERE WAS A SUBTEXT to the way that ads stressed the freshness of Acronized chickens. The reality was that poultry often was not wholesome, and buyers frequently were sickened by it. In the 1950s, poultry was responsible for one-third of the foodborne illnesses in the United States, even though people were still eating less chicken than beef. People who slaughtered and handled poultry were in worse shape: Hundreds of workers were being sickened by illnesses such as Newcastle disease, which was fatal to birds and caused eye infections in humans, and psittacosis, which causes fever and a lung inflammation resembling pneumonia. Several dozen poultry workers had died. Meat-cutters' unions trying to move into poultry plants insisted that additional unrecognized deaths were occurring as well.

The low demand and downward price pressure that would bring poultry industry bigwigs to Capitol Hill in 1957 was creating a marketplace that rewarded fraud. There was an incentive to cheat: Poultry had become a huge business. In 1956, gross farm income from poultry was $3.5 billion, three and a half times what it was in the 1930s when Jesse Jewell began demonstrating how consolidation would work. In grisly testimony at a separate series of Senate hearings in 1956, slaughterhouse workers complained they would open trucks coming from farms, expecting to unload live birds for slaughter, and find they were already dead. A farmer whose birds looked sickly might be told by the integrator for whom he was raising them that the meat inspectors at the local plant were too sharp-eyed; if he wanted to sell his birds, he should drive to the next county instead. Meat inspectors themselves said they were pressured by local politicians to be less exacting. Carcasses that they pulled from butchering because the birds looked diseased were taken back, in defiance of their authority, and rewrapped for sale. Representatives for the city of Denver told lawmakers they had purchased three train carloads of poultry to use in public school lunches, found they were spoiled, and refused to accept them—and then were told the birds were going to be resold to Omaha schools.

Those stories made chicken feel like an untrustworthy product, and lacing it with antibiotics gave homemakers confidence that they would not unknowingly feed unhealthy meat to their families. The reassurance fit into a context that anyone who had lived through World War II already understood: Until very recently, the food supply had seemed perilously unstable.

Ten years earlier, just as the war ended, there had been widespread food shortages. Some were deliberate: Germany and Japan had used famine as a weapon, driving people out of lands those regimes had annexed. Others were caused by freak weather events that followed on the war's devastation like a final kick in the ribs. Droughts ruined the wheat harvests in Australia, Argentina, and Africa and the rice harvests in China and India. Japan's rice-growing areas were swamped by typhoons. Even in the United States, some foods were scarce. Supplies of beef, pork, and poultry ran out so quickly that military commanders worried they could not meet soldiers' nutritional needs and newspaper headlines warned of a "meat famine." In February 1946, six months after the war ended, the UN General Assembly urgently warned, in a special assembly devoted to emergency action on famine, "The world is faced with conditions which may cause widespread suffering and death." In some countries, including the war's winners, food rationing would continue into the 1950s.

Making sure that food could be abundant again must have felt like a final rebuke to fascism, and any technology that got more safe food to more people would have felt worth the risk. Looking back, though, it seems that no one thought there was any risk in Acronizing—or in Biostat, an almost identical process concocted by Lederle's rival Pfizer using their drug oxytetracycline. From the mid-1950s through the mid-1960s, hundreds of scientists experimented with coating meats and fish in antibiotic solutions, misting the drugs onto fruits and vegetables and mixing them into milk. Food producers embraced the new science. Researchers promised it would increase fresh meat sales in South America, where refrigerated transport was scarce. It would let Australia sell its beef to other countries, which it had never been able to accomplish because potential buyers were four weeks away by boat, longer than refrigerated meat could last. Canadian fishermen said they were losing almost a quarter of their catch to spoilage before they could get it to market; they hoped Acronizing would preserve it. Norwegian whaling companies boasted they would be able to make whale steaks an everyday meat; out in the North Atlantic, they were shooting whales with antibiotic-packed harpoons.

In the United States, Lederle cleverly managed the mounting demand. Before they could Acronize, poultry processors had to apply to the company for permission and pay a franchise fee. That created an aura of exclusivity at the same time that the public relations campaign was convincing homemakers that buying Acronized chicken was essential. Securing a license to Acronize made chicken processors seem canny and forward-thinking, and they called their local media as soon as the paperwork arrived. The Port Halifax Packing Company in Waterville, Maine, told reporters that Acronizing had increased its poultry sales by 50 percent. Quick Frozen Foods, in Tupelo, Mississippi, said that it was now able to truck refrigerated but not frozen chicken to California—a distance of almost 2,000 miles—without worrying that the meat might spoil. Perry Brothers, a processing plant south of Seattle that Acronized 18,000 chickens a week, told a local newspaper that the process would allow them to send fresh poultry to Alaska and Hawaii by ship instead of air; cheaper transport would let them cut the price per pound by a third, making the meat more affordable. By 1958, more than half of the slaughterhouses in the United States had licenses to Acronize chickens. Fish wholesalers were using the process too.

Maybe, given so much enthusiasm, the companies should have expected that standards would slip. In interviews, American Cyanamid representatives asserted that eaters were at no risk from Acronizing. They said there was so little Aureomycin on treated meat that people would need to eat 450 chickens to consume the equivalent of one prescription dose. That assurance was based on the company's own dosing instructions to slaughterhouses, that the antibiotic solution should be mixed at a proportion of 10 parts of drug per million parts of water. (That was the same ratio, in liquid form, that Jukes originally concocted for growth promoters in feed—which the company equally insisted was safe.) Cyanamid did not seem to notice that its own franchisees were gleefully adding the drug in much greater amounts, telling local newspapers that they used the equivalent of 80 to 100 parts, 10 times more. Things were even less precise for Pfizer's Biostat process. *Business Week* documented that its users did not get any training. Instead, they got a measuring spoon, packaged with the drug—but "no restrictions" on how much drug was to be used.

The FDA's approvals of Lederle's and Pfizer's processes had been based on there being so little drug residue on meat or fish that the heat of cooking would denature it. That could not be guaranteed if the doses used on meat were larger; it was possible that housewives were unwittingly feeding their families tetracycline-laced fish and chicken. And doctors would soon discover that the people responsible for getting those proteins to dinner tables were being exposed to antibiotics in a manner that no one had accounted for.

—᠁—

REIMERT RAVENHOLT, a physician at the Seattle Department of Public Health, was puzzled. It was the winter of 1956, and for weeks now, local doctors had been calling him, describing blue-collar men coming into their offices with hot, red rashes and swollen boils running up their arms. The men were feverish and in so much pain they had to stay home from work, sometimes for weeks.

The puzzle was not what was afflicting them. That was easy to establish: It was *Staphylococcus aureus,* or staph, a common cause of skin infections. Ravenholt happened to have a lot of experience with staph. He was the health department's chief of communicable diseases, the person who recognized and tracked down outbreaks, and for the entire previous year, he had been dealing with a staph epidemic in Seattle's hospitals. The organism had infected 1,300 women immediately after they gave birth, and more than 4,000 newborn babies, killing 24 mothers and children. It was a dreadful episode.

The thing that was keeping Ravenholt up at night now was not the cause of this apparent new outbreak: It was the victims. Medicine already knew that staph could spread rapidly through a hospital, carried unknowingly by health care workers as they went from patient to patient. But outside of hospitals, it was equally taken for granted that staph infections occurred individually and by happenstance. Unless there was an explicit health care connection—a shared nurse or doctor, a crib in a nursery shared by many other newborns—there was no reason to suppose two staph cases were linked. The men coming down with the bug, several a month for five months in a row, were not linked by any hospital or doctor, yet they all had the same pattern of lesions in the same places on their arms and hands.

The outbreak looked like a mystery, one that required a detective. Fortunately, Ravenholt was one. He was a graduate of the Epidemic Intelligence Service, an elite training program for epidemiologists—disease detectives—maintained by the CDC. Ravenholt was one of the first graduates of the two-year program, which was designed to create a rapid-reaction force that could deploy across the country. It had begun in 1951, and Ravenholt entered the next year. When Seattle-area doctors began calling him in 1956, fewer than 100 people in the United States had been schooled, as he had been, in what the CDC called "shoe-leather epidemiology": sleuthing the details of disease outbreaks by leaving the office to meet victims, wherever they happened to be.

Thanks to that training, Ravenholt was equipped to recognize the pattern of an outbreak, even though everything that was known about staph indicated that an outbreak with no hospital connection ought not to exist. The 31-year-old physician called the doctors who had seen the men, pored over the medical records, tracked down the patients, and interviewed them all. It did not take long to discover that they were in fact connected. They had not gone to the same hospital—or any hospital, for that matter— but they did share another institution, one that they visited every day: their workplace. They were slaughter workers at a single poultry processing plant.

Ravenholt called the plant's owners. He half-expected that they would refuse to talk to him and was surprised when they said he could come by. When he got there, they told him why they allowed the visit: They were struggling with poor-quality poultry, sold to them by local farms, that showed the same problems processors would complain about to Congress later that year. They wanted it known that they were doing what they could to get out a clean, quality product, and they felt they were being undermined.

They showed him what they were dealing with. Birds that looked healthy turned out, once killed and defeathered, to be riddled with hidden abscesses, pockets of pus layered in their breast muscles. Ravenholt took some of the pus and cultured bacteria from it. The lesions were caused by staph. He told the owners the bacteria in the abscesses were leaking out when the birds were cut apart, contaminating the ice bath where the just-killed chickens were chilled, and getting into nicks and cuts that the knife-wielding workers naturally accumulated over the course of a workday. Well, that was frustrating, the owners said back to him. They had spent a lot of money and invested a lot of time to add a hygienic new process, called Acronizing, that was supposed to prevent bacterial contamination. They had only installed it in May.

May was when the workers' doctors had started calling.

Ravenholt had never heard of Acronizing before, but he instantly perceived the contradiction. If the point of the antibiotic soak was to kill bacteria that cause spoilage, it also should have killed the staph bacteria that were oozing from the meat and infecting the workers. He asked the plant owners for the names of all the farmers who raised the birds that were killed at the slaughterhouse. There were 21 farms, and he wrote letters to all of them, asking them whether there had been disease outbreaks in any of their poultry. Fifteen wrote back, and all of them assured him that their flocks showed no visible signs of illness. Thirteen of the 15 said they were shocked to hear of the problems, because they were taking special steps to keep their poultry healthy. They were dosing their chickens with Aureomycin to prevent them from developing any disease.

The lab tools that were available in 1956 were much cruder than the ones that exist today; it was more difficult and time-consuming then to distinguish between staph strains or demonstrate that a cluster of cases of illness came from a single source. Ravenholt could not prove in a lab that the antibiotic doses, the chickens' lesions, the antibiotic soaks, and the workers' health problems were linked. But he was confident that what happened had proceeded like this: Drugs in the feed had affected bacteria in the birds, habituating them to antibiotics, and the low dose of the same antibiotics in the chilling bath had eliminated all the bacteria except for the ones that had become resistant. Those had survived to infect the workers who were plunging their hands and arms into the contaminated water.

Ravenholt is in his 90s now and still lives in Seattle. More than 60 years later, his memory of his conclusions then is sharp. "Instead of the old tried-and-true preventatives of contamination, they had switched to these miraculous new drugs that they thought did everything," he told me. "Instead of preventing a problem, it was like putting kerosene on a blaze."

By the time he was done investigating, the problem had spread from one slaughterhouse to several, and fully half the workers in the plants had the same hot, painful abscesses and boils. Even without lab evidence, that was enough to demonstrate that Acronizing was creating a problem. Ravenholt was able to persuade the slaughter plant owners to cease using the antibiotic dunks, and when they stopped, the outbreak did too.

With the outbreak over and other diseases clamoring for his attention, Ravenholt had no reason to keep poking at the issue of the plant workers' illnesses. But the episode nagged at him, and periodically he looped back to the problem of how the men became infected, scrutinizing any blip that suggested farms and slaughterhouses might be conducting illnesses into the city undetected. He conducted a survey of meat cutters in processing plants, asking about lacerations and boils and hospitalizations. The workers he interviewed all told the same story: of skin eruptions that hurt and ached, gave them fevers, kept them away from work, and recurred for years. They believed their problems originated in the meat and fish they were handling. The illnesses had names on the cutting floors, they told him. They were called "pork infection" and "fish poisoning."

Ravenholt thought back to the terrible 1955 hospital outbreak in mothers and babies. He had assumed at the time that the staph ravaging mothers and newborns in Seattle's hospitals had arisen there first and then leaked into the outside world. Now it occurred to him that the bacterial traffic might have gone the other way. Perhaps the virulent staph originated in the meat trade, affected by the antibiotics that the animals consumed while they were living and that they soaked in after they died. Meat cutters were overwhelmingly men, but maybe one of them had brought the bacteria home on his bloody clothing or his soaked boots or in the cuts on his injured hands. Maybe he had passed the bacterium without knowing it to his pregnant wife or girlfriend, and she had carried it innocently into a hospital and sparked an explosion of disease.

It was years later and there was no way to know. And there was not even a ripple of concern yet in the wider world about the possibility of resistant bacteria arising from antibiotic use in food animals. But at the CDC, Ravenholt had learned that diseases could echo in odd ways down the decades of a career; an outbreak that seemed mysterious at the time might eventually be explained by a discovery years later. So he noted his concerns, in case they might be useful in the future. In 1961 he wrote:

> The outbreak of boils among workers in a poultry-processing plant . . . is the only such outbreak in this community in at least the last 15 years . . . That outbreak coincided in time and place with the use of the chlortetracycline process, which was discontinued shortly thereafter . . .
>
> These findings suggest that the use of tetracycline in the processing of poultry somehow caused the outbreak . . . And if so, that possibly hospital outbreaks . . . are in some way, not yet defined, related to the use of tetracycline.

—⚡—

THE OUTBREAK THAT Ravenholt unraveled was a small one, even by the standards of the 1950s, and until he published his description in 1961, it received no attention outside Seattle. But elsewhere in the United States, the problem of disease organisms in food and food workers, and the ways in which antibiotic use might be affecting them, was gathering attention.

The first sign of trouble surfaced in cheese—or rather, in milk that was supposed to become cheese but would not coagulate. The reason was penicillin. Automatic milking machines had recently come on the market, replacing the laborious hand milking that dairy farmers had done for millennia. The suction generated by the machines was tough on cows' udders, bruising them and causing infections. Injecting large doses of penicillin into the teats cured the problem, but the antibiotic lingered in the udder and could contaminate the cow's milk for a while. To prevent any of that penicillin from being consumed accidentally, the FDA required dairymen to throw away any milk that was collected in the first few days after the drugs were injected. (The British government had a similar but looser rule that hinged on how advanced a cow's infection was.) Yet some farmers must have objected to sacrificing that small amount of profit lost in that discarded milk—because starting in the mid-1950s, penicillin allergies in both countries suddenly became much more common.

This was strange timing, because penicillin had just been made prescription only, precisely because enthusiastic buyers of the drug had sensitized themselves into becoming allergic when it was sold over the counter. With the introduction of prescription penicillin, allergies to the drug should have been decreasing. They were not. Doctors reported adults and, even more, children—who drink more milk than adults—breaking out in the kinds of rashes that previously had affected nurses who handled raw penicillin in the early days. In 1956, the FDA tested milk that it bought in supermarkets across the United States and found that more than 11 percent of the samples contained penicillin; some contained so much that the milk could have been administered as a drug. By 1963, the situation was serious enough for the World Health Organization to flag it in a special report.

Other foods were getting close examination as disease vehicles rather than resistance risks. In March 1964, the CDC summoned physicians, epidemiologists, and federal planners to its headquarters in Atlanta to discuss an urgent trend: *Salmonella* infections in the United States had increased 20-fold in 20 years. Eggs seemed to be to blame. In the largest single outbreak, liquid eggs—ones that are broken open, combined, frozen while still raw, and sold to food service companies—sickened more than 800 ill and fragile patients in 22 hospitals. Dr. Alexander Langmuir, the founder of the Epidemic Intelligence Service, who had trained Ravenholt, complained: "It certainly piques our pride that in these days of heart surgery, artificial kidneys and organ transplants, we cannot take dominance over a minuscule little bacillus . . . that gets into our hospitals, causes no end of trouble, and has us stumped."

Foodborne illness outbreaks in institutions—hospitals, prisons, schools—were usually assumed to be the fault of whoever was in the kitchen. The CDC's investigation established that this was wrong. There was no way that identical outbreaks could have happened in so many hospital kitchens at the same time, caused by the same food, and yet be unconnected. Salmonella was not a kitchen problem; it was a food system one. That shift in emphasis enraged the egg industry, in a way that would echo through every foodborne outbreak thereafter, pitting the suffering of the victims against companies' lost sales. After the egg outbreaks were publicized, "There was probably an egg price depression of somewhere near a cent a dozen," Dr. Wade Smith, Jr., a veterinarian with the Tennessee egg producer Blanton-Smith, fumed during the CDC meeting. "A cent a dozen does not sound like very much to those of us who buy a dozen eggs a week. But a cent a dozen for six months is approximately half a bird's production."

The concern for foodborne outbreaks and the new worries over resistant foodborne bacteria forced a reexamination of Acronizing. At the USDA, several scientists who had been monitoring poultry plants—watching how much drug they used in the chilling bath and how long they soaked the birds—went back to their federal laboratory to try to recreate the process. Their results, once they replicated what slaughterhouses were doing, confirmed Ravenholt's suspicions from years before. Acronizing treatment changed the mix of bacteria on the surface of meat, encouraging resistant bacteria to develop and multiply—resistant bacteria that were present only on pieces of meat that had been Acronized.

Everyday shoppers were probably not reading the scientific publications where those results were made public. Nevertheless, in supermarkets and home kitchens, a cultural shift was occurring: Consumers were scrutinizing food additives and losing trust in food production. "We have felt for a long time that something was wrong with the poultry we buy," "A Consumer" wrote to the editor of the Pottstown, Pennsylvania, *Mercury.* "It does not have the good flavor that it had in the past, regardless of how it is prepared. We would like to see a ban on the use of all dyes and preservatives in the food we buy, including the acronizing of chicken." Lois Reed of Twin Falls, Idaho, wrote to the *Montana Standard-Post:* "How about the acronizing of chickens? When you purchase one such chicken you are completely in the dark as to the time it was prepared for market—two days ago—six months ago—who knows? . . . We are doing ourselves and our children a great injustice by being indifferent to these various practices. Our very lives depend upon action now!" "Non-Acronized" began to appear in grocery-store advertisements across the country—including in the Helena, Montana, *Independent Record;* the Bend, Oregon, *Bulletin;* and the Eau Claire, Wisconsin, *Daily Telegram*—as prominently displayed as "Acronized" had been just a few years before. "Even your children can tell the difference," Capuchino Foods promised in the San Mateo, California, *Post.* Colorado and then Massachusetts banned Acronized birds from being sold within their borders.

The weight of negative opinion changed the FDA's mind. In September 1966, the agency canceled the licenses it had granted a decade earlier for Acronizing and the rival process, Biostat. Antibiotics could no longer be added to food as it was packaged. But the agency did nothing about antibiotics fed to animals before they were slaughtered and became food. That was not yet on the public's agenda, and only a few scientists were concerned. One was Marie E. Coates, a scientist at England's National Institute for Research in Dairying, who studied poultry nutrition. In 1962, at a conference on antibiotics and agriculture that was held periodically at the University of Nottingham, she worried aloud:

> Widespread use of antibiotic feed supplements may induce the establishment of strains of organisms resistant to their action. The least harmful result would be the loss of efficiency of antibiotics as growth promoters. A more disastrous consequence might be the development of resistance in pathogens against which antibiotics are at present the only means of defense.

Coates was prescient. Just a few years later, a little more than a hundred miles away, a tragic outbreak would demonstrate that she was right to be afraid.

—⁓—

AT THE TIME, IT SEEMED like a small outbreak. Politicians seeking to dismiss it would later describe it as "unfortunate" and "not of exceptional severity." But it was devastating to Middlesbrough, an ironworking town on the River Tees in North Yorkshire. In October 1967, stomach upset and diarrhea broke out among infants and toddlers there. It seemed normal: Children get "stomach flu"—which isn't really flu, but is caused by viruses and bacteria—millions of times every year, in England and everywhere else. And as was also normal, the most seriously sick children, who had fevers and needed intravenous therapy because vomiting and diarrhea were draining their bodies of fluids, were taken to West Lane, the local hospital.

At the hospital, the sick children were given standard care for stomach flu: fluids, drugs to bring down fever, and tests to see whether the cause of their illness was bacterial or viral. That was important to know, because it would determine the children's treatment; gastroenteritis caused by bacteria is less common than the kind caused by viruses, but it is much more severe, with high fevers that can give babies convulsions. Tests showed the problem stemmed from the common bacterium *E. coli*. The physicians were relieved: that meant they could use an antibiotic to kill the organisms while the babies recovered. They chose an old standby, neomycin, and expected the children to get better in a few days.

The children did not get better. The infections did not respond to the antibiotic. Instead, the kids' fevers shot higher, diarrhea gushed from their bodies, and they dwindled before the doctors' eyes. As more cases came in, sick children were stashed wherever spare beds could be found and transferred from West Lane Hospital to others nearby. The children brought the infection with them, spreading it to additional children who were in other hospitals for other reasons, and sparking an outbreak in a unit that housed developmentally disabled babies.

Doggedly, the physicians tried one drug after another, running through the limited number of antibiotics that were safe to use on young children. It took nine tries before they found one that the bacterium responded to. The *E. coli* causing the stomach flu was resistant to eight different classes of antibiotics. Fifteen babies died.

Where had this bacterium come from, to attack children out of nowhere? And how had it piled resistance on resistance, making it into a fatal superbug? No one who worked in medicine in Yorkshire had seen anything like it before. But in a laboratory at the other end of England, one scientist was sure that he knew the answer.

Ephraim Saul Anderson, known to everyone as "Andy," was the director of one of the British government's national public health laboratories, tucked into a suburb on London's far north side. He was a physician who had powered his way up from immigrant roots in working-class Newcastle—his parents fled anti-Semitism in Estonia just before he was born—to a soaring academic career developing ahead-of-his-time laboratory techniques. He was independent and brusque (later, a colleague would say working with him was like "driving a car without springs across a lava field") and had an epidemiologist's intuition for how diseases could sneak past human defenses.

In the late 1950s Anderson had untangled a perplexing salmonella outbreak of 90 cases that were scattered across the southeastern corner of England like salt spilled on a tablecloth. The sick people were men and women, adults and children, in a half-dozen towns. The disease was severe, and the diarrhea and vomiting from the infection left the victims very ill. One woman died, and another was so overtaxed she suffered a heart attack; one toddler's fever was so high he went into convulsions, and a teenager was left with lifelong arthritis. But despite being so far apart geographically, with no connections between them, all of the victims had the same strain of *Salmonella*.

To track down how this outbreak had happened, Anderson left his lab to sleuth out the connection, starting at markets and pursuing the source all the way back to farms. He discovered that the disease came from veal, which in England was a new source for *Salmonella,* and he also discovered that the cause was a change in how farmers were raising cattle. In earlier years, dairy farmers whose cows had produced unneeded male calves would have sold them directly to a local farmer or butcher. Now, a new profession had arisen: middlemen who drove a circuit of farms to buy calves and deposit them at a temporary depot until the depot had enough animals to make an auction worthwhile. Animals never stayed at the depot for long; sometimes it was five days before the transient grouping of animals was broken up and sold to slaughterhouses across the southeast, and sometimes just one day. Anderson demonstrated that even a single day was long enough for one sick animal to infect the entire temporarily assembled herd. Pooling animals for profit had unintentionally created a new means of disseminating disease—one that could spread an infection originating on one small farm over hundreds of miles.

Anderson's alertness to the way that changes in farming altered the occurrence of human disease affected how he worked at the national laboratory, where he supervised analysis of disease specimens taken from animals and humans from across England. The samples allowed his team to draw relationships between cases of illness even when they were separated by long distances. Those analyses let them see that antibiotic resistance was rising across England—and the rise was most visible in organisms that originated in animals and crossed to humans on food. In 1961, 3 percent of the *Salmonella* strains that came into the lab carried some form of drug resistance. By 1963, the proportion was 21 percent.

By 1965, it was 61 percent, and in one strain of *Salmonella,* 100 percent of samples—every single one that came into the laboratory in 1965, from a person or a cow—carried drug resistance in some form.

What troubled Anderson the most was that some of the strains from animals—which sometimes caused illness and sometimes resided quiescently in their guts and exited in their manure—possessed molecular defenses against four, five, or six unrelated families of drugs. It was extraordinary in an organism originating in an animal, and he could not make sense at first of how it had happened. Resistance was supposed to occur when a bacterium was exposed to antibiotics, but it did not make sense that any single cow would have been given that many different rounds of drugs.

While Anderson was puzzling it out, a piece of research that would reframe the problem of resistance was making its way around the globe. The source was a Japanese researcher named Tsutomu Watanabe who had been studying an outbreak of *Shigella,* a foodborne illness that causes fever and diarrhea just as salmonella does. He discovered that the bacterium was resistant to both a drug that had been used to treat it and also a drug its victim had never received. This made no sense at first. Then Watanabe and his colleagues determined that the snippets of genetic code that confer resistance were not locked inside bacteria, bound into chromosomes so that they passed by inheritance only to the organisms' descendants. Instead, the genetic material could break free of one bacterium and migrate into another, transferring the molecular protection of resistance from an organism that had already experienced a drug into one that had not yet. They dubbed the snippets "R-factors."

The implication, instantly perceived by worried scientists, was that R-factors could migrate randomly through the bacterial world without leaving any mark of their passage. They would allow any bacterium to accumulate resistance to multiple antibiotics, stacking up the R-factors like trading cards stashed in a box.

The R-factor hypothesis gave Anderson a means to understand what his lab was discovering. Virulent *Salmonella* infections were becoming more common in people and in livestock; there were 10 times more cattle deaths from salmonella in the 1960s than there had been a decade before. Because of that threat to their livelihood, farmers were dosing their cattle with more antibiotics—not growth promoters (which in England were legal only for chickens and pigs) but preventive antibiotics to keep diseases at bay. One prominent farmer was notorious for tying little bags of antibiotics around the necks of his just-weaned calves when he sold them. It guaranteed they would get a final protective dose as they arrived at the sales depot—and though this was not his intention, it guaranteed they would pass resistant bacteria to the other calves in the depot too. Year by year, larger proportions of the *Salmonella* found in England were not only drug resistant but also resistant to multiple families of antibiotics.

With all of that evidence in hand—the free use of antibiotics in meat animals, the increase in resistant disease in animals and people, and the new knowledge that resistance created in one bacterium could cross into another—Anderson accepted a few *E. coli* samples from the Middlesbrough outbreak into his lab for study. What he found confirmed his darkest concerns. The outbreak was not due to a single strain of *E. coli;* he found several just in his small selection. But the resistance patterns in the different strains were identical and protected against seven different antibiotic types.

It was unlikely that each infection had been exposed to all of those drugs and evolved its own resistance to each of them. It was even less likely that three different strains would possess identical arrays of resistance. Anderson concluded the resistance in each strain had migrated from other organisms. It was a living demonstration, halfway around the world, of the bacterial traffic that Watanabe had revealed in Japan. There, R-factors had been a laboratory exploration. In Yorkshire, they caused children to die.

At least some of the children who were sickened in Middlesbrough were infected while they were in the hospital; the bug was transferred from one child to another by the hospital's environment or equipment or by doctors or nurses. But Anderson was confident that the extraordinary amount of resistance in the organisms was not a hospital product. The *E. coli* strains were resistant to so many unrelated families of drugs—including ones that would not have been used as treatments for *E. coli*—that medical use alone could not have produced what he saw. That belief led him to two insights that would change how he, and later other scientists, perceived the peril in farm antibiotic use.

Anderson's first recognition was that Jukes and his cohort—the researchers who had been so enthusiastic about growth promoters in the 1950s—had been wrong about something fundamental: that the effect of antibiotics on livestock was a benefit that came with no cost attached. They knew that growth promoters would induce some resistance in the bacteria that reside in the gut, the benign organisms that assist with digestion and metabolism. But they assumed that the process had a built-in circuit breaker: When resistance in gut bacteria rose to some never-determined level, growth promoters would stop working. Thus, they never considered that resistance might move in animals' guts from benign bacteria to disease-causing ones, in the manner that Watanabe had demonstrated.

Beyond that, Jukes and his cohort never contemplated that those disease bacteria could exit animals' guts—in the manure they excreted or the meat they became—and carry resistance away with them. That was Anderson's second insight: animals and humans shared one bacterial world, and the resistance-bearing R-factors that Watanabe identified could swim anywhere within it. (Decades later, wildlife biologists would describe this insight with the title "One Health.") There was a cost to growth promoters that matched or exceeded their benefit: resistant illness occurring in people who had no connection to the farms where the dosed animals were raised. Long before anyone developed the molecular tools that could parse bacterial genetics, Anderson intuited that antibiotic resistance was a single epidemic, carried from animals to humans by the invisible conversations of organisms.

Reducing the risk that antibiotic-resistant illnesses posed to people, by reducing the use of antibiotics in animals, would become Anderson's crusade for the rest of his career. It would ignite a political movement in England, making the United Kingdom the first country to attempt to control the emerging threat. And it would pose difficult questions that other countries would have to confront as antibiotic use in livestock increased around the world.

CHAPTER 5

PROVING THE PROBLEM

THE GRAY CLAPBOARD HOUSE on the two-lane road in a western suburb of Boston looked, in fall 1974, the way you would expect a comfortable old Massachusetts house full of children to look. It was rambling and tall, made out of a house and a barn butted together. There were other barns out back, down a long gravel drive that stretched to a grove of trees: small sheds and one big building, 200 feet on the long side, painted an iconic barnyard red. There were a milk cow, a few horses, a couple of pigs, and chickens: white laying hens, some in a chalet-shaped coop and some skittering underfoot between the trees.

Inside the house, kids were everywhere, seemingly as much underfoot as the chickens were: Richard, Mary, and the twins Peter and Paul; Steve, Ronnie, and Mike; Christopher, Christine, and Lisa. Their parents, Richard and Joan Downing, were Catholic and had wanted a big family. Richard had done well in business, establishing credit-reporting firms around Boston. When the twins turned out to be their last biological children, he and Joan had committed to using their money to improve other kids' lives, so there were always extra children in the house, up to a dozen at a time. Some stayed a few months or years as fosters; others joined the family permanently through adoption. The Downings were sturdy people, no-nonsense but warm. Their house was mostly happy chaos, with kids racing around corners clutching sports equipment and homework, and dashing outside to feed the pigs and milk the cow.

There was just one note of stern discipline, in a notice stuck on the front of the refrigerator. It had been written by Mary, the oldest daughter, in the careful printing she was learning in college biology. It said, in block letters, "No Juice Until I Get Your Poop."

Inside the refrigerator, along with celery and cold cuts and the much desired juice, were a couple of brown paper lunch bags. All of the bags held the same thing: a cluster of long, clear tubes, tightly capped, each one holding swabs that looked like long Q-tips. Each of the swabs was stained at one end. The swabs in one bag came from the Downing kids. The swabs in another came from the neighbors, who dropped them off once a week. (They didn't wait for juice.) The swabs in the third had been swirled over the butts of chickens in the big barn at the back.

It looked like arrangements for a kid's jokey science fair exhibit, but this endeavor was deadly serious. The Downings had agreed to host an experiment, one that centered on the animals out back and involved their sprawling family and their neighborhood. The experiment was the first attempt to explore, in an organized, documented way, what Andy Anderson maintained he had observed:

that antibiotics given routinely to animals represented a threat to human health.

Actually, at first, it intended to disprove it. The sponsors of the study were not public health scientists like Anderson but industry: the Animal Health Institute, the trade group representing the companies that made and sold the antibiotics used on farms. It was a quarter-century now since Jukes's discovery, and antibiotics were a routine component of agriculture: 40 percent of the ones made in the United States were going into livestock, not to human patients. But it was also 10 years since Anderson had begun claiming a connection between farm antibiotic use and human illness, and growth promoters and preventive dosing were coming under increased public scrutiny. The animal drug industry was under pressure to prove its products were safe, and it had agreed to fund a study to demonstrate that.

The study would not go the way the industry hoped, and it would change the debate about agricultural antibiotic use for good.

—⁓—

FARM ANTIBIOTICS WERE being studied in rural Massachusetts because, in England, Anderson had pursued his conviction that R-factors—which allowed bacteria to trade resistance genes undetectably—made antibiotic use more dangerous than anyone had imagined. There was no way to halt the invisible traffic that made the trade possible. But if antibiotic use could be curbed, the threat of resistance spreading might be lessened.

Anderson's position at the top of one of the United Kingdom's public health laboratories guaranteed him the attention of the editors of the major English scientific journals, the *British Medical*

Journal and the *Lancet,* and he made use of it, publishing more than 20 scientific papers on the danger of transmissible resistance. But he also crafted an advocacy strategy that would look normal today but was unheard of then: He took his case directly to the public. He got to know the science editors of the major newspapers and popular science magazines, who all worked in one area of London just 10 miles from his lab office. He persuaded friendly journalists to place articles in the *Times* of London, the *Financial Times,* and the *Guardian,* and finagled television coverage on the BBC, which ran two of Britain's national networks (there were only three) and all of its regional ones. His smartest act was recruiting the support of Bernard Dixon, a microbiologist turned journalist. In 1967, Dixon wrote a scathing article for the national magazine *New Scientist.* He counted up the outbreaks and deaths that had been occurring and excoriated the "complacent" British government for not taking action. The article's title was a stark warning: "Antibiotics on the Farm: Major Threat to Human Health."

Dixon and Anderson had a right to their passion. Before the children died in Middlesbrough and before half the outbreaks that Anderson investigated had even occurred, the British government had attempted to examine farm antibiotic use and its possible consequences—but the effort was halfhearted and designed to fail. The government impaneled a committee of scientists to investigate but handed the chairmanship to possibly the most powerful representative of farming in England: James Turner, the chairman of a fertilizer conglomerate and president of the National Farmers' Union, who had just been given a peerage for services to British agriculture. The committee's conclusions, delivered in 1962, could have been predicted: Growth promoters posed no risk, there was no reason to regulate them, and farmers' right to use antibiotics ought to be expanded.

Immediately after that, Anderson began to document the movement of resistant *Salmonella* across the United Kingdom. Then, in 1964, the country was riveted by a searing exposé of factory farming, *Animal Machines,* written by an animal welfare campaigner named Ruth Harrison and full of photographs of chickens crowded in cages and veal calves chained in dark, windowless barns. The effect was as explosive in England as the publication of *Silent Spring* was in the United States—Rachel Carson wrote the foreword—and the book left Britons questioning where their agriculture was headed. And then in 1967, that agriculture suffered a horrific blow: a massive epidemic of foot-and-mouth disease, which forced the slaughter of almost half a million cattle, sheep, and pigs. Whole areas of the countryside were quarantined; farmers perched on their gates with rifles to keep out anyone who might track the fast-spreading diseases in on their boots. It was impossible to turn on the news without seeing smoking piles of carcasses being burned.

In the aftermath, it felt urgent to reexamine how English meat animals were being raised. Within a few months of the end of the epidemic, a new commission was created, headed this time by a scientist: Michael Swann, a molecular biologist who was also Vice Chancellor of Edinburgh University (and later the chairman of the BBC). Anderson, who was not selected for the Swann committee, campaigned ferociously for it to enact the antibiotic controls that Turner's group had refused. "I called almost three years ago for a 're-examination of the whole question of the use of antibiotics and other drugs in the rearing of livestock,'" Anderson wrote in the *British Medical Journal.* "The problem exists despite denials of its significance."

The Swann committee was thorough. It took testimony from 35 scientists and 58 representatives of government labs, pharmaceutical companies, and trade associations, and reviewed hundreds of pieces of research. It uncovered startling statistics: Farm animals in England were consuming, every year, 168,011 kilograms of antibiotics (370,400 pounds, or more than 185 tons), about two-thirds of what humans were receiving. Some of that was for treatment of sick animals, which no one disagreed with, but most of it was for promoting growth. Some of the drug dosing fell into a murky preventive category, similar to what the FDA had approved in the United States, though in England, the agriculture industry called preventive dosing "antistress." "'Stress' is best defined as that state in which the giving of antibiotics in dietary concentrations below therapeutic dose levels leads to an economic response," the committee noted, with what sounds like sarcasm.

Economics were at the heart of farm antibiotic use, they found. The group calculated that the extra income farming gained by using the drugs were at least one million British pounds per year and possibly as high as three million pounds; in 1969 dollars, that would have been $2.5 million to $7.5 million. That vast profit convinced the committee that agriculture would never curb antibiotic use on its own. The drugs would have to be regulated. In an 83-page report delivered in November 1969 that was measured, thorough, and surprisingly bold, the committee specified that growth promoters had to go.

"It is certain that the use of an antibiotic in animal feed produces large numbers of resistant organisms, including organisms with transferable resistance, and that these resistant organisms may be transmitted to man," the committee members wrote. "There are grave potential disadvantages for animal and human health in adding in this way to the pool of organisms which are resistant to the antibiotics of most value for the treatment of disease."

The committee asked, and answered, a question that no one in agriculture had yet confronted: Since all antibiotic use caused antibiotic resistance to emerge, when was that risk worth taking? Their answer was that using antibiotics to treat sick animals was worth the risk, but using them to boost profits was not. Their recommendations were thoughtful but also strict. They broke antibiotic use in agriculture into two categories: therapeutic ones given to treat disease and growth promoters given in feed. Then they took most of the drugs that farmers were currently administering in feed—all the penicillins, both tetracyclines, sulfa drugs, and tylosin (a veterinary antibiotic chemically related to the human-use drugs azithromycin and erythromycin)—and classified them as therapeutic. Henceforth, the committee said, those antibiotics ought to be available only if a veterinarian wrote a prescription. Farmers could still buy and use "feed" antibiotics on their own—but that category would be limited to drugs that were not used to treat disease, in humans or animals, out of fear that resistance would undermine them. This reduced farmers' options to a much smaller set of drugs, while taking away the ones that were most effective, but also most responsible for resistance that could harm human health.

The Swann committee's recommendations were instantly controversial. Parliament accepted them, and they became law in March 1971, though they would be argued over and fought against for years after. In the story of farm antibiotic use, the recommendations are notable not because they were perfect but because they were first. They represent something extraordinary: a forthright statement by a government that the power of antibiotics should be conserved and protected. The United States, where most of the miracle drugs were discovered and growth promoters were invented, could not ignore the Swann Report. Other countries were paying close attention to what the U.S. government might do.

—៣—

WHILE ANDERSON AND THEN the Swann committee had been combating the farm antibiotics problem, the United States had been making weak stabs at defining it. The National Academy of Sciences convened a meeting of 47 scientists in 1955. The attendees presented research on how antibiotics were being used in different farm animal species but only briefly raised—and immediately dismissed—whether that might pose a risk to humans. But by the time the Swann commission assembled 14 years later, researchers were feeling the first tickle of disquiet at how rapidly salmonella was spreading across the United States and how frequently the strains retrieved from human patients bore some drug resistance. Yet another committee that was asked to examine the problem made suggestions that had no political force behind them: "only truly low levels of various antibiotics should be used in feeds, in water and in feed ingredients"; "antibiotics should not be used routinely."

Publication of the Swann Report made it impossible to ignore the issue any longer. The FDA created a 16-person task force—10 federal scientists and 5 from universities and industry—which summoned scientists to Washington to give testimony, and also traveled around the country to hear from farmers and agricultural organizations. In 1972 the group delivered to the FDA a shorter version of the Swann Report, containing recommendations that were equally strict. They said older antibiotics that were being used as growth promoters should be put under veterinarians' control, and newer antibiotics with novel chemical structures should be kept from agricultural use until it could be proved there was no risk to human health. The task force's research made clear for the first time how rapidly agriculture was appropriating new drug formulas and, potentially, generating resistance to them. When Jukes recognized the growth-promoting effect in 1948, there had been barely a half-dozen families of antibiotics. Now there were 30, and 23 of them were being used in both livestock and people.

The task force recommendations ignited a political uproar. Almost half its members wrote a minority report saying they disagreed with what was supposed to be a consensus recommendation. Congressmen, livestock and pharma lobbyists, and animal science researchers from land-grant universities all bore down on the FDA to prevent the recommendations from being enacted. Under the onslaught, the agency announced a compromise. It would allow pharmaceutical companies and agricultural interests to bring forward their own research into growth promoters, resistant bacteria, and threats to human health. If they could not prove the products' safety, any antibiotic doses too small to treat animal diseases—which included both growth promoters and "preventive" dosing—would be banned after April 1975.

Jukes was furious on the industry's behalf. He had moved to the University of California, Berkeley at this point, but he was still defending his invention, and he blasted the FDA in a scientific journal for knuckling under to "a cult of food quackery whose high priests have moved into the intellectual vacuum caused by rejection of established values." There seemed to be a real chance the United States would follow the United Kingdom in banning most farm antibiotic use. The companies making the drugs scrambled to prove their products were safer than both governments thought. That brought the Animal Health Institute, then and now the veterinary pharma companies' powerful defender in Washington, to the door of Dr. Stuart B. Levy, a researcher in Boston.

Levy was 36 in 1974. He looked a lot like the actor Elliott Gould, famous from the 1970 antiwar movie *M.A.S.H.*: shorter and more slender than Gould's 6'3", but with the same heavy dark hair, straight eyebrows, and thick mustache framing a buoyant grin. He was the son of a family doctor from Delaware and had grown up accompanying his father on house calls and discussing cases afterward. He was a faculty member at Tufts University School of Medicine, in a part of Boston that is gentrified now but was cheap and seedy then, and he had taken a circuitous route to get there, studying first literature, then medicine, and then microbiology in Italy and France. Along the way he read Watanabe's first papers on R-factors and how they allowed organisms to become resistant to a drug they had never been exposed to. He was captivated. He persuaded Watanabe to accept him in an informal internship and relocated to Japan to work in his lab.

This was in the 1960s, at a time when the *New England Journal of Medicine* was warning, "It appears that unless drastic measures are taken very soon, physicians may find themselves back in the pre-antibiotic Middle Ages." Watanabe died of gastric cancer in the early 1970s, just as a new generation of scientists was renaming his R-factors with the term *plasmids*. Those younger researchers confirmed his discovery that genes conferring antibiotic resistance could stack up on plasmids and be carried from one bacterium to another as the plasmid moved. That would allow an organism to acquire resistance in advance of ever being exposed to a drug, while also allowing multiple types of resistance to spread. The process would go like this: Imagine that within a colony of bacteria, a few organisms harbored a plasmid containing the genes for resistance to drugs A, B, and C. Then imagine that just one of those drugs—B, let's speculate—was used against the colony. All the vulnerable bacteria would die, but the bacteria containing the plasmid would be protected. In surviving, they would preserve all of the resistance genes within the plasmid—not just B, but A and C as well—and make them available to be passed on to other bacteria, separately or together, horizontally in the same generation or vertically when the plasmid was inherited by daughter cells.

The possibility that resistance to an antibiotic could be amplified by the use of an *unrelated* drug was alarming. It threatened to make resistance much harder to track and combat. And Levy found it fascinating.

Levy's research specialty was the family of tetracycline drugs, the group to which Aureomycin belonged. He had identified the genes that conferred resistance to them and had pinned down the novel strategy by which that resistance operated. Some antibiotics killed bacteria by interfering with their cell walls, and resistance to those drugs kept the antibiotics from attacking the outside of the cell. Tetracycline, however, slipped inside the cell—and tetracycline resistance worked by forming minuscule pumps that squirted the drug back out of the cell like a bouncer ejecting an unruly customer. Tetracyclines were the original growth promoters; transmissible resistance made growth promoters' effect unpredictable; Levy was an expert in both. He was also a relatively new researcher, with a publication record that was not yet lengthy.

The Animal Health Institute found him and offered to fund a study on behalf of farm antibiotics.

That was why there were tubes of poop-stained sample swabs in the Downings' refrigerator. They were tools that would help Levy establish, or disprove, whether resistance could migrate through the environment, from animals that had received antibiotics, to animals and people who had not. Growth promoters' proponents hoped the answer would be no.

—〰—

LEVY DIDN'T KNOW the Downings, but he knew what he needed to make the proposed study possible: a place that looked like a farm but had not been functioning as one. He needed new animals that had never received antibiotics, a place to raise them where antibiotics had not been used in the recent past, and a group of animal handlers numerous enough to conduct the experiment and healthy enough to not be taking antibiotics themselves. For bonus points, the location needed to be close enough to his office that he and his staff could travel back and forth affordably. In the affluent suburbs of Boston, that was a complex order to fill. He was not sure even where to look, but he began to ask around.

Boston's rural exurbs are very unlike the city, but so many people commute in and out that they are more tightly tied than they look. The news of Levy's search for a place to conduct his study percolated through the medical community, and after a while, someone got in touch: a veterinarian responsible for the mice and other animals that Massachusetts General Hospital kept to do research. He lived in the tiny town of Sherborn, 20 miles southwest of Boston. His across-the-street neighbors were a relaxed, irreverent family; they had lots of kids; and they lived on a big parcel with a couple of barns that had once been an egg-sorting business. He offered to make an introduction.

Levy drove out to meet the Downings. He described what he envisioned: a temporary farm housing 300 chickens, to be maintained for at least a year. Richard Downing liked the impish, focused physician, and he liked the idea of contributing to knowledge and letting his kids watch an experiment up close. But he had grown up on a poultry farm, in the coastal town of Weymouth, and he knew Levy did not know how to accomplish what he wanted to do.

"I told him he was crazy—he had no idea what this would take," Downing recalls. "He'd have to build the pens, buy the feed, set up the watering system, put in heat, find someone to look after them, get someone to clean up. And he said I was right, and he hoped we could help."

The Downings accepted the challenge—for fun and out of curiosity, and because being unconventional had never worried them before. To manage the experiment, they nominated their oldest daughter, Mary. A sophomore at a local college, she was living at home to save money. She wanted to go to France after graduation, but with so many other kids in the household, spare funds were in short supply. She and her parents and Levy struck a deal. She would supervise the chickens, water and feed them, and collect all the data Levy would need, which, she learned, meant collecting poop, and not just from the birds. He offered to pay her $50 every week, about $250 now. She signed on.

In the big barn at the back of the property, Levy and 10 medical students, along with the Downings and their kids, built six wire pens, each equipped with gas heaters and independent food trays and water systems. Four were inside the barn, 50 feet apart; two more stood outside its thick timber walls. Then Levy went in search of chicks. To ensure there would be no contamination, nothing in the chickens' systems that would slant the study results, he bought them from a company that supplied "pathogen-free" eggs to laboratories. In July 1974, the day-old Leghorn chicks arrived in Sherborn and were stashed in one of the pens with a heat lamp and water and antibiotic-free food. When they were two months old, the experiment began. Levy divided the chicks into six batches, 50 per pen. At the local feed store, he bought two types of feed, one antibiotic free and another that was sold premixed with antibiotics. It contained oxytetracycline—the drug that had originally been named Terramycin, made by Lederle's rival Pfizer—in a ratio of 100 grams per ton. (It was marketed as a preventive dose rather than a growth promoter, because it was intended for egg-laying chickens; rapid weight gain was not a priority for them.) Half of the birds, in three of the six pens, got the drug-free feed. The other half got the tetracycline-laced mix.

Levy had several questions to answer. First, did the antibiotics in feed cause resistant bacteria to emerge or multiply in the chickens receiving it? Second, did that resistance cross from those chickens to the rest of the flock? And third and most crucial, could it make the leap from chickens to humans? They were the same questions that had troubled Anderson in the 1950s and propelled the Swann Report in the 1960s. But back then, researchers had perceived resistance, looked backward to where it might have come from, and made assumptions about how it arose and spread. This time, no assumptions would be necessary: Levy would be able to watch resistance emerge (or not) and spread (or fail to)—not in bacteria or rats in a lab, but in farm animals in a barn, and in a farm family as well.

But for the experiment to establish everything that Levy planned, he needed to recruit people beyond the Downings to participate. At their invitation, he drove out to meet the neighbors. Joan and Richard threw a barbecue and invited the five families—10 parents, 14 kids—who lived up and down the road. After the hamburgers and hot dogs and corn were handed around, the Downing boys rolled a washtub over to make a podium. Levy had discussed with the parents what he planned to say, and they had reassured him that it was best to be straightforward. Still, he felt a little nervous as he climbed up on the tub.

"We're asking you all to be part of an experiment," he told the guests. The parents made murmurs of interest and shushed their kids. He described the puzzle of antibiotic resistance, how the chickens might help solve it, and that the Downings had agreed to assist. Then he got to the hard part.

"We'd like you to donate something that you have to science," he said. The small crowd perked up with curiosity. He heard one woman say, "This is exciting!"

He took a deep breath. "Frankly," he said, "we need your shit."

The silence was deadly. Then three-year-old Lisa scrambled up, round-eyed. "You want our *poops?*" she squeaked.

That broke the ice. Everyone laughed, and they all agreed to help; none of the families backed out. And that in turn justified Mary's substantial pay for managing the experiment. Her job was not just feeding and watering the chickens and swabbing their butts every seven days; it was also persuading and bugging and nagging her siblings and the neighbor kids, and her parents and the neighbor parents, to do their sampling duty. Every week, the Downings' refrigerator shelf filled up with bags full of tubes, and then emptied again when Levy's staff picked them up.

The results came quickly. Samples taken before the experiment began showed that very few bacteria in the guts of the chickens, family, and neighbors possessed genetic defenses against tetracycline. That was to be expected, given the random roulette of mutation. But once the drug-laced feed was administered, those bacteria multiplied in the birds harboring them and spread to birds that had started out clean. The first changes appeared in 36 hours, and within two weeks, 90 percent of the chickens were pooping out resistant bugs. The dose of antibiotics in the feed had killed the gut bacteria vulnerable to the drug but did not harm the ones that were protected by minor mutations—and those resistant survivors thrived and multiplied in the living space left empty when the other bacteria were killed. Researchers including Anderson had assumed this was what occurred in animals given antibiotics, turning them into factories for producing resistant bacteria. But no one had measured it in the wild before, and no one had expected to see it happen so fast.

For a few weeks, the chickens getting drug-free feed remained clean of resistant bugs. Then things changed. First, the bacteria from the chickens getting the antibiotic-laced feed became resistant to multiple drugs—sulfas, chloramphenicol, streptomycin and neomycin, and two derivatives of penicillin—even though the feed contained only tetracycline. Then the multidrug-resistant bacteria appeared in the chickens that had never received that feed and had no contact with the birds that did. And soon after, the same multidrug-resistant bacteria showed up in the Downings' poop as well.

The resistance was carried by plasmids. Levy confirmed this by creating marker bacteria in his lab that he transferred into just four of the chickens and then retrieved from the others and from the Downings. But the route they traveled was not obvious. The groups of chickens were never let out of their pens or allowed to mingle. If any birds managed to escape, Mary took them out of the study to a coop on the other side of the property. When she fed and watered and tested the chickens, she moved through the barn in a precise pattern, going to the drug-free birds first and washing her hands and changing boots after each pen. And the Downings were not eating the experimental chickens (though they did have a giant barbecue once the study was done).

To the disappointment of his sponsors, Levy had demonstrated what they hoped to disprove. Even though the feed contained just tiny doses of antibiotics, those doses selected resistant bacteria—which not only flourished in the animals' systems but left the animals, moved through the farm's environment, and entered the systems of other animals and of humans in close proximity. (But not in any of the neighbors, who were the experiment's control group; the resistant bacteria did not spread to them.) The altered bacteria were an untrackable form of pollution. And because they could accumulate resistance genes invisibly, they were an unpredictable threat as well.

There was one footnote to what Levy had found, and for years it would influence efforts to control farm antibiotic use: The Downings had not gotten sick. There are many strains of *E. coli*, and the one that resided in the chickens' guts and crossed to their owners was not a disease-causing one. Instead, it was a commensal, one of the range of benign bacteria that occupy the gut and pervade the world without causing illness. On the scientists' side, this did not diminish the risk; it only made the bacterial traffic more complicated. But it would allow those who chose not to believe in the threat to downplay the danger.

—⚏—

LEVY PUBLISHED HIS DESCRIPTION of what happened on the Downings' farm in September 1976. In April 1977, the new commissioner of the FDA—Donald Kennedy, in office just two weeks—stood up at a meeting and dropped a bombshell: The U.S. government would follow the British example and ban growth promoters from American agriculture.

Kennedy was 46, a slender man in big glasses who vibrated with impatient energy. He was a biologist, had a Ph.D. from Harvard, and had become a department chair at Stanford University when he was only 34. Stanford had loaned him to the White House, and he had been working there part time since the beginning of 1976, helping the administration of President Gerald Ford set up the new Office of Science and Technology Policy. Late in 1976, little-known former Georgia governor Jimmy Carter squeaked past Ford in a close election, positioning himself as an outsider who could cleanse the country of the scandal of Watergate and the lingering taint of the Vietnam War. Carter brought a crew of earnest young reformers with him. The FDA was high on their agenda. The agency had just been hauled before Congress by Senator Edward Kennedy for accepting fraudulent data from pharmaceutical manufacturers, and it was embroiled in a fight over the cancer-causing potential of saccharine, the only artificial sweetener sold in the country and a huge source of income for soda companies. The FDA needed a leader who could speak up for science and owed no favors to the Washington power structure. Donald Kennedy fit. The White House expected him to be bold.

But maybe not as bold as he turned out to be. On his first visit to the FDA's National Advisory Food and Drug Committee, a group of scientists and industry representatives who for months had been fractiously debating farm antibiotic use, Kennedy made it clear the discussion was over. In a short statement, he announced that his department would immediately ban any growth promoter use of penicillin and the tetracyclines, and it would also ban preventive use as soon as researchers could identify compounds that farmers could use instead. Henceforth, he added, the drugs would be used in animal feed only if a veterinarian wrote a prescription for them.

"The benefit of using these drugs routinely as over-the-counter products to help animals grow faster or in prophylactic programs does not outweigh the potential risks posed to people," Kennedy said. "Although we can point to no specific instance in which human disease is more difficult to treat because drug resistance has arisen from an animal source, it is likely that such problems could have gone unnoticed."

Kennedy was a scientist, accustomed to weighing evidence; in the language he spoke, *potential* and *likely* had weight. But the politicians who were his audience did not receive what he said as proof. They heard it as a supposition, and also as a threat, to an enormous industry with outsize political influence. By this point, almost every food animal raised in the United States was receiving antibiotics sometime in its life: close to 100 percent of the chickens and turkeys, 90 percent of veal calves and swine, 60 percent of cattle. Within weeks, Kennedy was hauled before Congress and vilified, beginning a series of hostile hearings that would go on for months.

His proposal to take away growth promoters—and place all antibiotic use under the control of veterinarians—was "totally unworkable," according to the American National Cattlemen's Association. It "could jeopardize nearly any product employed in animal agriculture," the Animal Health Institute fumed. "There are no widespread epidemics of untreatable disease among humans because of antibiotics in livestock feeds," insisted a representative of American Cyanamid, the makers of Aureomycin. The Southeastern Poultry and Egg Commission said forgoing the drugs would cost egg producers $4.25 million. The Arkansas Poultry Federation claimed the loss of growth promotion would require farmers to grow 45 million more pounds of corn for feed, along with 23 million more pounds of soybeans. A congresswoman from Nebraska, appearing for the feedlot industry, charged that the move would force Americans to spend two billion dollars more for food. Practically spitting in derision, the director of the Northeastern Poultry Producers Council snarled, "The Commissioner's logic would eventually lead to a government edict to abort all pregnant women because every aspect of life—every minute of every day—poses some potential risk to health."

Kennedy and his staff ignored the condemnation. On August 30 and October 21, the FDA filed two lengthy documents in the *Federal Register* describing their case against growth promoters: one for penicillin, one for the tetracyclines. Technically the documents were "notices of opportunity for a hearing," inviting the drugs' manufacturers to request an appearance to defend their drugs. But in practical terms, they were legal briefs: precisely footnoted arguments that built the case against growth promoters, piece by piece. And like any good brief, they made clear who the defendants were. For each drug category, the notices listed the products at stake. There were 26 containing penicillin and 31 for the tetracyclines, and they came from almost all the major pharmaceutical and animal feed manufacturers in America: American Cyanamid, Cortex Chemicals SPA, Dale Alley Company, Dawes Laboratories Inc., Diamond Shamrock Corporation, Eli Lilly

Company, Falstaff Brewing Corporation, Hoffman-La Roche Inc., Merck & Company, National Oats Company, Pfizer Inc., Rachelle Laboratories, Ralston Purina Company, E. R. Squibb & Sons Inc., Texas Nutrition & Service Company, and the Thompson-Hayward Chemical Company of Kansas City.

Although the scientific evidence was substantial, the economic might of those companies, and all those companies' customers, outweighed the science. A message was passed from Congress up to the White House and then down to Kennedy: The hearings would not be allowed to happen. Representative Jamie Whitten, a Mississippi Democrat who since 1949 had chaired the body where the FDA's budget originated—the House Appropriations Subcommittee on Agriculture and Rural Development, the fount of farm pork-barrel funding—had vowed to hold that budget hostage if Kennedy proceeded. If he did, it would put at risk other reforms the Carter administration had planned.

The White House brokered a deal. Whitten's agricultural contacts were insisting there were gaps in the evidence; he wanted to see them filled. If Kennedy abandoned the attempt to ban growth promoters, Whitten would not kidnap the FDA's budget. In fact, he would add some extra money, enough to allow the FDA to do the study his allies were asking for. In 1978, he approved enough funds to allow the National Academy of Sciences to study the public health impact of growth promoters. The appropriation allowed enough money for three years of research.

Coincidentally, Kennedy was due back at Stanford within two years, at the end of the leave it had granted him to work in government.

It might have seemed that Whitten was just kicking the can down the road, until the arrival of Kennedy's successor at the FDA.

But the old Southerner had been a politician for as long as Kennedy had been alive, and he was craftier than that. Whitten added a rider to the next appropriations bill: Until a study answered the question of public health harm to his personal satisfaction, the FDA would not be allowed to act on any antibiotic bans. There would be dozens of studies over the years and eventually hundreds. None would be good enough. Whitten renewed the rider every year until he retired in 1995.

Whitten's obstinacy on behalf of agriculture cemented the security of farm antibiotic use in the United States. England and, later, other European countries arrowed off on a different path, defined by what a philosopher would call the precautionary principle: that preventing a harm is more important than waiting for all the evidence. The United States, being more empirical and less interested in protective regulation, let growth promoters and preventive dosing flourish unregulated. More and more evidence—which is to say, larger, deadlier outbreaks—would accumulate for decades before anyone found the courage to try to control farm antibiotics again.

PART 2

How Chicken Became Dangerous

CHAPTER 6

EPIDEMICS AS EVIDENCE

THE DIFFICULT REALITY OF the difference between the U.K. and U.S. attitude toward farm antibiotic use was this: Judged on the scale of America, the resistant foodborne outbreaks that persuaded England to change course were not very big. In the 1970s, the U.K. population hovered at 56 million; in 1977, the year of Kennedy's unsuccessful attempt at regulation, the United States housed 220 million. In a place so much more populous and extensive, an outbreak that originated in farm antibiotics would have to be much bigger than the English ones to attract attention or change policymakers' minds. The media barely covered Kennedy's attempt at reform, and in the years just afterward, world attention would fix on two more visible epidemics. First there was smallpox, history's most deadly disease: It was declared eradicated, wiped from the planet, in May 1980. And then there was AIDS, the first signs of which were identified just 13 months later, in June 1981, when the CDC and several private physicians discovered an odd form of pneumonia among gay men in Los Angeles.

Smallpox had killed millions. HIV would soon begin to rack up its own worldwide toll. Against that backdrop, the possibility that antibiotic resistance could be spreading through the food supply must have seemed unimportant. But a few scientists who were alert to the danger understood that the way to prevent resistance from becoming a significant epidemic would be to identify its outbreaks when they were still small.

Scott Holmberg had a vivid sense of how large an epidemic could get. After graduating from Harvard, he had volunteered to join the Peace Corps and was sent to Ethiopia, working alongside teams of vaccinators who were fanning out across Africa and Asia to bring an end to smallpox. He had been an English major, but those hot, dusty years tracking cases through tiny villages convinced him that fighting outbreaks was what he wanted to do with his life. He pivoted toward that, attending medical school at Columbia and doing his medical residency at Brown University, where he began work just as the AIDS epidemic began to sicken men in California and New York. In summer 1982, Holmberg relocated to Atlanta, to the CDC. The Epidemic Intelligence Service, the elite rapid-reaction corps that trained Reimert Ravenholt in the 1950s, had grown into a powerhouse of disease detection. Physicians just out of residency and researchers who had just finished their Ph.D. competed for the scarce slots, about 80 each year, that would put them inside outbreaks all over the world. It was part of the training to be sent out without any warning. EIS officers, as they were called, joked that they could leave their office to eat lunch in the bare-bones cafeteria and come back to find an airplane ticket on their desk.

Holmberg had been in the program a little over six months when he got a call from his boss—at home, on a Saturday—in February 1983. The Minnesota Department of Health was worried about some cases of salmonella. The victims had a strain called *Salmonella* Newport, which was uncommon in the Midwest, and they were very ill. The Minnesota department had a reputation among public health people; it was so well funded and well staffed, thanks to a generous state legislature, that it functioned like a mini-CDC. When it asked for help, that was a sign that something interesting was happening. The next day, Holmberg headed north.

He stepped out of the airport into cold so deep it sucked his breath away and was scooped up by Michael Osterholm, an intense, ruddy-faced Midwesterner who had just become the state's epidemiologist. Holmberg knew about Osterholm already; everyone in U.S. public health did. He was one of only a few leaders in the national hierarchy who had not come up through the CDC—he had gone to the health department straight from a small Iowa college and earned two master's degrees and his doctorate while working there—and he had a reputation for being forceful, skeptical of authority, and willing to slice through red tape. Osterholm had already noticed that most of the victims had taken antibiotics just one or two days before their salmonella symptoms started. He was worried the illnesses might have been caused by contaminated drugs, and was concerned that he might have to order batches of antibiotics to be yanked from shelves in all of Minnesota's pharmacies if necessary.

Osterholm was a few weeks shy of his 30th birthday, Holmberg had just turned 33, and both were eager to get to work. Though it was a Sunday night, the two dove into studying what was known about the 10 victims. They had relatively little in common. They ranged in age from 8 to 43; eight of them lived in Minneapolis and neighboring St. Paul; and eight out of 10, including the 8-year-old, were female. Most of them had had some sort of upper respiratory infection, bronchitis, sore throat, or an ear infection; that was why they had gotten the antibiotics that preceded their developing *Salmonella* infections. One was still hospitalized.

The details the two epidemiologists pored over that night ruled out one cause and raised another. They realized that contamination could not be the problem because the victims had not been given the same drugs: Some had taken amoxicillin, some penicillin—dispensed by different pharmacies—and one victim no antibiotics at all. But something did link the cases. The *Salmonella* infecting all of them was antibiotic resistant, in an identical pattern across all 10 patients—to the drugs in the penicillin family that they had taken and also to tetracyclines, which none of them had received.

The question was where the tetracycline resistance had come from. It was the kind of signal that the CDC was teaching Holmberg to watch for, one that warned this small outbreak might herald a much larger problem.

—m—

INVESTIGATING A FOODBORNE outbreak takes painstaking work. Everyone eats food, of course, and just about everyone eats many kinds of food every day. So narrowing down which food might carry illness requires conducting long interviews, and confirming recollections against packages that might still be in refrigerators or receipts that might be lingering in a wallet.

Holmberg dove into interviewing the victims. To sort out what was different between their illnesses and any other foodborne outbreak, he also talked to local people who had had salmonella before this outbreak began. He met both groups, visited their houses, and looked in their refrigerators and cupboards. Everyone was eating the same things: milk, eggs, beef, chicken. There was no food that set the outbreak victims apart, nothing that they had eaten and the control group had not. In frustration, he headed back to Atlanta, wondering if he could cast a wider net. He thought first that if the unusual strain of *Salmonella* had occurred elsewhere in the country, another health department might have reported it. And then it occurred to him that since *Salmonella* originates in animals, the resistant strain might have appeared in them as well.

He called the National Veterinary Services Laboratory in Ames, Iowa, the animal diseases' equivalent of the CDC, and asked for any *S.* Newport strains that had been taken from livestock in the past two years. They sent him 91 bacterial samples, nine from the Midwest and the rest from across the country. As part of his CDC training, Holmberg had learned how to perform genetic fingerprinting of plasmids, so he took the isolates to one of the CDC's labs to see what the strains might contain. Out of all of them, just one contained a plasmid, holding all the same resistance factors, that was identical to what Osterholm's lab had found in the Twin Cities patients. It came from a calf that had died on a farm in South Dakota, hundreds of miles from Minneapolis. This was a tantalizing clue, but it was only interesting, not significant—until Holmberg checked with South Dakota's state epidemiologist, Kenneth Senger, and discovered that the state had also recorded four human cases with an identical resistance pattern. Three of the four sick people had fallen ill the previous year, before any of the Minnesota patients got sick. One of the four was the farmer who had owned the dead calf.

In any disease investigation, there is a moment when random pieces of data begin to look related, like a picture emerging from the surface of a partially assembled jigsaw. Holmberg headed to the airport again. Senger met him in South Dakota to help him find the farm.

The three people who had been ill the previous year (whose names were never made public, for privacy) belonged to the same extended family, who all lived southwest of Sioux Falls. There was the man whose calf had died, a 33-year-old dairy farmer; a 29-year-old woman who was his cousin and lived on another farm nearby; and her 3-year-old daughter. They had all gotten sick in December 1982, with the same progression of illness as the Minnesota cases: bronchitis or a sore throat, and then an antibiotic prescription, and then a sudden onset of fever, cramps, and vicious diarrhea.

The cousins did not live on the same farm; their properties were several miles apart, and they only saw each other at church and to wave to on the road. But as they talked to Holmberg and Senger, a closer connection emerged. They had each gotten a gift of meat from a mutual uncle who ran a beef cattle feedlot that shared a property line with the 33-year-old's dairy farm. The uncle was known for his generosity: At least once a year, he would choose a good cow, kill and butcher it, and share the steaks and trim among the extended family. As he told the epidemiologists about his uncle, the dairy farmer remembered: There had been a strange episode the previous year involving the feedlot. One of his dairy calves had gotten through a gap in the shared fence and wandered through the beef cattle. A few days after he got the baby back, his entire dairy herd had endured a quick sharp outbreak of a virulent diarrhea, and several calves had died. The illness was so unusual— he had won awards in the past for how well he ran his farm—that he had asked the state veterinarian to check all of his animals. That was how his calf's sample had gotten to the national lab and then had been passed to Holmberg.

Holmberg and Senger went to meet the uncle. He told them there had been 105 head of cattle in his feedlot when the dairy farmer's calf had wandered over, but they had been sold months ago, in January 1983. But, he said, he had difficulty believing his cows could have been the source of any illness. He showed them the source of his belief in his farm shed: bags of Aureomycin, chlortetracycline, that he had bought at the local feed store and fed to his cows to keep illnesses away.

Holmberg felt another piece of the puzzle click into place. He was pursuing a resistant infection. He had found it first in one set of sick people, and one of the things they ate was beef. Then he found the infection in another set of sick people, who also ate beef, from a single, identifiable farm. The same resistant infection had occurred in other cattle that had wandered onto the farm. And now the owner of that farm was showing them where the resistance might have originated: in the antibiotics he scooped into his own cattle's feed, about 100 grams per ton.

Holmberg had tracked the chain of infection back to its origin; now he had to trace it forward, to see if it reached all the way to Minneapolis. The uncle's cows had been trucked from South Dakota into the southwest corner of Minnesota to be slaughtered—and, unusual in 1983, the slaughterhouse manager was an early adopter who used a personal computer to keep track of his sales. He searched his spreadsheet and reported that 59 of the

105 had been sent to a packinghouse in Nebraska. There they had been broken down into prime cuts for butchers—ribs, briskets, sirloins—and all the leftover bits had been boxed into giant lots and sold to wholesalers. The boxes containing scraps from the uncle's cows had been resold through the Midwest; they became part of a load of 40,000 pounds of beef put together at a plant in Minneapolis and part of another load of 30,000 pounds at a similar business in central Iowa. Both plants subdivided the loads into smaller lots and then sold those to supermarkets that ground their own hamburger. When Holmberg went back to the first set of interviews he had done with the Minnesota outbreak victims, he found the final link in the chain. The supermarkets that bought beef from the Minneapolis plant for grinding—beef that included scraps from the cows from the South Dakota herd—were the same ones where the Twin Cities salmonella victims had shopped.

The one remaining question was why all the cases were so separated in time. The farm family's 29-year-old cousin and her daughter had fallen ill in December 1982 and the dairy farm owner in February 1983. The Twin Cities victims' dates of illness stretched from mid-January 1983 into the second week of February, and the last known case occurred in May. Holmberg and Osterholm spent extra time studying the patients' records, trying to understand what happened. Then they spotted the connection: At some point, different for each of them, every victim had developed some unrelated problem: respiratory infections, for the most part, common in the winter chill. They went to doctors and got prescriptions—and those antibiotics had killed both their intended target, the chest infections, and also an unintended one, the diverse bacterial flora in the victims' guts. The resistant *Salmonella*, lurking quiescently in their intestines, had exploded in the empty space and multiplied to a level that made the victims ill.

Tracing that process left Holmberg with an uneasy sense of how silently antibiotic resistance could travel. The story of the last known victim sadly underlined that. There was a fourth case in South Dakota, a 69-year-old man, the only person killed by the outbreak strain. He was infected early, in the same span of time when the 29-year-old and her daughter fell ill, before the uncle's cows were slaughtered and their meat started its journey through wholesalers and middlemen and into supermarkets. He could not have purchased any beef that would have infected him; he had no connection to the farm family, and no way to have received a private gift from their herd. In fact, during most of the time the outbreak had been burgeoning, he had been in a hospital in Sioux Falls, recovering from a disastrous farm accident. His injuries were so severe that he had undergone an emergency colonoscopy, and his spleen and part of his large intestine had been removed.

Holmberg realized he had already heard the name of that hospital. The 29-year-old cousin, the first to fall ill in mid-December, had been admitted there too.

He called the nurse who ran the hospital's colonoscopy suite. Check your records, he asked her. Was there any chance both people had undergone procedures on the same day? The nurse checked the day's schedule and said no. He thought about that, and about hospital protocols, and asked her, just in case, to check other days as well. His hunch was correct. The woman had been the last patient on one working day. The man had the first appointment on the day after. The long, flexible scope that had been snaked up the backsides of both of them ought to have been thoroughly disinfected between uses, but some portion of the *Salmonella* must have survived to contaminate the injured man's intestines. It had burgeoned in his guts, leaked into his blood vessels, blew up into fatal blood poisoning, and killed him. After he died, hospital pathologists found *Salmonella*—later proved to be the resistant outbreak strain—in his intestines, lungs, and blood.

The resistant foodborne illness that had migrated silently from cows to people had become a hospital-caused infection, passed unknowingly from one patient to the next. There was no way to know whether that had happened elsewhere or when it might happen again. Out of thousands of people who might have eaten a burger contaminated with a scrap of beef from the uncle's herd, only 18 had been identified, and only because an alert health department with good detection systems had noticed, and a federally funded epidemiologist had agreed to spend months on the case. Those 18 people had been sickened by meat from just one farm—but farms all over the country were using antibiotics as freely as the South Dakota farmer did. In 1984, when Holmberg published his description of the outbreak, American farmers spent $270 million on antibiotics for livestock, buying about half of all the antibiotics made in the country that year.

After the investigation was over, Holmberg tried to define the magnitude of the risk that farm antibiotics were creating. He pulled all of the CDC's reports of salmonella outbreaks, as far back as 1971 and up to 1983, the last full year for which the agency had data. Agriculture was still contending that resistance arose only from misuse of antibiotics in human medicine, but in the records, Holmberg could see that two-thirds of cases of resistant salmonella traced back to animals. In some cases, the link was eating a contaminated food from an animal source: raw milk, ice cream, roast beef. But in others, the resistant illness passed from an animal to a human, and then went from person to person. In one especially poignant example, a young pregnant woman caught resistant salmonella from a calf that had been fed antibiotics. She passed the infection to her baby at birth, and then, while the baby lay in the hospital nursery, nurses unknowingly picked up the infection and transferred it to other newborns.

Hundreds of people had been made sick by salmonella that had been made resistant by farm antibiotics, but in twos and threes, clusters that no one would have considered to be related until Holmberg perceived the link. It was an epidemic wider and more dangerous than anyone had imagined. His data showed that people who caught the farm-related resistant strains, compared with strains that still responded to antibiotics, were 21 times more likely to die.

—✠—

KENNEDY'S ATTEMPT TO BAN growth promoters six years earlier had failed because politicians, and their agricultural patrons, refused to accept a connection between farm antibiotic use, foodborne bacteria, and human illness. The midwestern investigation made it impossible to deny the link. The journal *Science* called the findings a "smoking gun." The *Washington Post* put the study on the front page, and within a few months, the House of Representatives hauled FDA officials into an unfriendly hearing. Levy's experiment 10 years earlier had shown that resistance could travel from animals to humans, but that was in a small, controlled experiment on a single property. Holmberg, Osterholm, and Senger had demonstrated that resistance could travel in that same manner through free-living animals and humans, across much longer distances than the length of the Downings' farm. And unlike what had happened with the fortunate Downings, they had shown that farm-created resistance could make people sick, and that it could kill.

It reframed the debate and inspired the first citizen activism against farm antibiotics. The nonprofit Natural Resources Defense Council (NRDC), based in New York, formally petitioned Margaret Heckler, the secretary of health and human services, asking her to immediately suspend subtherapeutic penicillin and tetracycline use in livestock. "On the basis of recent scientific evidence it can be established that hundreds of human deaths each year . . . could result from subtherapeutic use," NRDC argued. The group said that continuing to allow the drugs to be used would constitute an "imminent hazard," meaning it met a regulatory standard that would allow the secretary to remove drugs from the market without holding a hearing. Heckler rejected the petition and refused to act, ignoring not just Holmberg's investigations but also two massive new studies—ones funded by the money that Congressman Whitten had authorized when he claimed Kennedy had not done enough research.

The issue was harder to squelch this time. In a blistering report released on the last day of 1985, the Committee on Government Operations of the House of Representatives excoriated the FDA for neglecting to regulate how drugs used in livestock were distributed and applied. The committee said that up to 90 percent of 20,000 drugs recently approved by the agency had never been evaluated for safety or even for effectiveness. It published evidence from a whistle-blower deep within the FDA that the agency knew 4,000 of those new drugs might have been causing "significant adverse effects" in animals and people. It charged that the FDA had not set up adequate processes for determining whether drugs used in food animals posed a risk to humans, and it exposed that meat, milk, and eggs often contained drug residues, both cancer-causing compounds and antibiotics: nitrofurans, used in poultry; carbadox, used in young pigs; and chloramphenicol, formulated in the 1940s and banned in humans for causing fatal blood disorders.

The committee report said that animal drug manufacturers were openly defying the FDA, making new formulas and releasing them onto the market without testing or approval, and continuing to sell drugs—including more than 200 forms of sulfa antibiotics—that the FDA already had told them to take off the market. And it gave a glimpse of the free-for-all marketplace that had developed around farm drugs. The committee had sent undercover investigators on road trips through Iowa, home ground of hog agriculture. They reported back that, walking into businesses as strangers with no paperwork, they had been able to buy animal drugs that the FDA considered dangerous and were supposed to be restricted to prescriptions—including antibiotics—at half of the warehouses and almost all of the veterinary clinics they tried.

The congressional researchers were not motivated by concern over antibiotic resistance; they were focused on manufacturers' defying FDA rules. But their blockbuster report showed how hard it would to enact reforms: because the drugs were so profitable, and because they were sold in such a distributed manner. The report also hinted at something that would soon become a much bigger issue. Agriculture was not only using old, inexpensive antibiotics. It was also scooping newer and more powerful drugs, intended to combat resistant infections in humans, onto farms as well.

—ᴍ—

THE SUCCESS OF HOLMBERG'S Minnesota investigation made Osterholm realize how useful it would be to have an Epidemic Intelligence Service officer around all the time. He made a deal with the CDC: He would create a permanent job in his department, and each year, one member of the rapid-reaction corps would have the option to train in the Twin Cities instead of Atlanta. The slot was immediately popular, and EIS officers competed to occupy it. Osterholm's people had a record of solving outbreaks so big or difficult they went into textbooks: 200,000 cases of salmonella caused by one of the world's largest ice cream plants, some of the earliest cases of MRSA outside hospitals, and the first occurrence of a mysterious, long-lasting gastric upset that they named for the town where it surfaced: "Brainerd diarrhea."

In exchange for the excitement of working on outbreaks, the EIS officers who came to Minnesota were required to perform a big data-focused project, something that would be published in a medical journal. And so it was that in the late 1990s, a veterinarian named Kirk E. Smith sat in the health department office, wondering what could be extracted from a sheaf of reports that the state had been collecting for almost 20 years.

Smith came from a farming family—his parents and grandparents had owned small farms in North Dakota—and with degrees in wildlife biology, veterinary medicine, and public health, he was alert to the bacterial traffic between animals and humans. The database he was looking at was a deep dive into that traffic. In 1979, the state legislature had required that Minnesota doctors report to the health department every case of foodborne illness that they thought might be campylobacter. *Campylobacter* is not an organism that most people think about, because it often causes just mild symptoms. But it was (and still is) the most common cause of foodborne illness in the United States. In addition to the usual fever and nausea, it can cause heart infection, meningitis, miscarriage, blood poisoning, long-lasting arthritis and kidney damage, and a rare polio-like paralysis called Guillain-Barré syndrome. In children, the elderly, and AIDS and cancer patients and others with compromised immune systems, *Campylobacter* infection can be deadly, killing about 100 people each year.

But Smith had more in mind than just the public health burden of an underappreciated illness. He suspected that the 17 years' worth of reports in the department's database contained evidence that *Campylobacter* was becoming a new kind of threat. The organism has a unique quirk: It is extremely common on chicken. And in just the past few years, there had been a significant change in chicken production: the adoption of a new family of antibiotics called fluoroquinolones.

Fluoroquinolones arrived on the market in the 1980s; they were an accident of lab chemistry, a by-product of an attempt to improve malaria drugs. (The name comes from the ancient malaria remedy quinine, the same compound that makes tonic water bitter.) Fluoroquinolones, especially the best-known version, ciprofloxacin, were a big advance and the best-selling antibiotics of their time, because they had few side effects and worked against a wide variety of illnesses: pneumonias, urinary tract infections, sexually transmitted diseases, bone and joint infections.

They also cured very serious infections with *Campylobacter* and *Salmonella,* the kind that required hospital care. And they seemed to be something that medicine would be able to rely on for a long time, because they were based on a new molecule that had been synthesized in a lab instead of being extracted from mold or a soil organism. Resistance ought to be slow to develop, because bacteria had never encountered the drugs before.

But the fluoroquinolones were maybe too successful, because agriculture, especially poultry production, clamored to be allowed to use them too. In the mid-1990s, the FDA licensed two versions of fluoroquinolones, sarafloxacin hydrochloride and enrofloxacin, for farm use: not as growth promoters but to prevent and treat diseases occurring in the close confinement of chicken and turkey houses. Smith thought the health department's records would show what happened next. The lab analyses the department had been doing for almost two decades, to see whether a foodborne illness was caused by *Campylobacter* or some other organism, relied on a compound called nalidixic acid. Nalidixic acid happened to be a chemical relative of the fluoroquinolones. So if Smith wanted to know whether *Campylobacter* in Minnesota were still sensitive to fluoroquinolones or becoming resistant, he would not have to do fresh lab work. He would just have to do a database search.

Ciprofloxacin reached the market in 1986 and sarafloxacin in 1994, so Smith decided to start looking in 1992, to understand what happened before and after the farm versions of fluoroquinolones arrived. In the first reports he reviewed, he could see a small amount of fluoroquinolone resistance; that made sense, since ciprofloxacin had been on the market a while. But in 1995, resistance jumped upward; in 1997, after enrofloxacin entered the market, it jerked up again. By 1998, the proportion of Minnesota *Campylobacter* samples resistant to the fluoroquinolones was 10.2 percent, almost eight times what it had been six years before.

That was a provocative coincidence, but not proof: The increase in resistance he was seeing could have come from either the human drug or the animal ones. To prove the problem was food, Smith and some state government colleagues fanned out across the Twin Cities, bought 91 packages of raw chicken at 16 markets, and tested them. Eighty of the 91 packages harbored *Campylobacter*, one out of every five organisms was fluoroquinolone resistant, and the resistance matched genetically with the human isolates in the state database. That left no doubt: Rising rates of resistant *Campylobacter* in Minnesota were coming from chicken.

But it was not just a Minnesota problem. The chicken Smith and his colleagues bought was labeled with 15 brands and came from nine states. Smith's Atlanta colleagues did a nationwide version of his database search using the CDC's records and found the same trend: By 1997, 13 percent of human *Campylobacter* infections had become resistant to fluoroquinolones; by 2001, 19 percent. In Georgia, the leading state for producing broiler chickens, 26 percent were resistant.

It was not solely an American problem either. When Smith took another look at the state database of infection reports, he spotted an unexpected detail: Year by year, resistance increased in a smooth upward curve. But month by month, the line was jagged, peaking higher in January and then falling back. Something was happening every January that caused Minnesotans to pick up many more resistant foodborne infections than at other times of year. As a lifelong resident of the Upper Midwest, Smith guessed immediately what that was: winter vacations, a desperate flight from the deepest dark and cold. Mexico was warm, close, and cheap, and a huge user of fluoroquinolones. Between 1990 and 1997, according to UN data, the amount of chicken meat produced in Mexico had doubled, and sales of fluoroquinolones for poultry use had quadrupled. The resistant bacteria those drugs generated were returning to the United States in the guts of unknowing tourists.

Mexico wasn't the only source. In the Netherlands, a place where human antibiotics were prescribed conservatively, there was no resistance to the quinolones in 1982. Then enrofloxacin was introduced to agriculture in 1987; within two years, 14 percent of *Campylobacter* on chicken and 11 percent in human patients was fluoroquinolone resistant. In Spain, fluoroquinolone resistance went from nonexistent to present in one-third of all the *Campylobacter* sampled. In England, enrofloxacin was approved only in 1994, and within a year, more than 4 percent of *Campylobacter* infections in British people were resistant to fluoroquinolones, a trend that was accelerated by the million tons of chicken imported each year from European countries that were already using the drugs.

Resistance was rising so rapidly because fluoroquinolones were being used so freely; they had been intended to solve a human medical problem, but they had become an agricultural bonanza. In 1998 the World Health Organization determined that about 120 metric tons of fluoroquinolones were being used in animals in the United States and Europe, up from zero just a few years before. The Swann Report and the ban on growth promoters had not prevented it because fluoroquinolones were not growth promoters. Resistance was rising from the legally allowed use of antibiotics to prevent and treat disease. And it was rising everywhere at once—not from a single outbreak originating on a single farm, the kind that Anderson first revealed in England and Holmberg uncovered in the Midwest, but across the world.

—∭—

WE TEND TO THINK of foodborne illnesses as being all the same thing; we even call bouts by the same name, food poisoning, no matter which organism caused them. But there are crucial differences. *Campylobacter* might be the most frequent cause of foodborne illness in the United States, but *Salmonella* is the more dangerous one. Each year in the United States, it causes up to 160,000 doctor's office visits, 16,000 hospitalizations, and about 600 deaths, six times what *Campylobacter* does. The number of deaths would be higher, but in severe salmonella cases, antibiotics are literally lifesaving, and the drug responsible for pulling people back from the brink is ciprofloxacin, which most people know as Cipro. In the late 1990s, roughly 300,000 Americans received Cipro for acute diarrhea (the main symptom of *Salmonella* and other foodborne organisms) every year.

In the mid-1980s, a novel strain of *Salmonella*—formal name, *Salmonella* Typhimurium DT104—had started spreading in England. One out of every three people who caught it ended up in the hospital, compared with one in 100 for most *Salmonella* strains. It seemed to be transmitted by a variety of raw meats, it could be found in a range of farm animals, and it was already very drug resistant to five separate families of antibiotics. Only Cipro still worked. But as soon as enrofloxacin was licensed in the United States in 1994, fluoroquinolone resistance appeared in DT104's defenses as well.

DT104 spread rapidly worldwide and soon arrived stateside. It sickened 19 kids at an elementary school in the tiny farm town of Manley, Nebraska, in October 1996, and at least 110 people near San Francisco in February 1997. Raw milk, and cheese made from it, were the only products that linked the outbreaks. In May 1997, a small family dairy farm, Heyer Hills Farm in Vermont, was devastated by the strain. The calves were sick first, and then members of the Heyer family, who drank their herd's milk: first a five-year-old boy, Nicholas Heyer; then his grandmother Marjorie, the farm owner; then six other relatives; and finally Cynthia Hawley, Marjorie's daughter and Nicholas's aunt, who almost died. Physicians at a small rural hospital pumped her full of antibiotics. Nothing worked. Their veterinarian saved her. He had sent a sample from the calves to Cornell University's College of Veterinary Medicine, and their lab results alerted Hawley's doctors that the strain was multidrug resistant but that a fluoroquinolone ought to work. It did. Hawley survived. Thirteen of the 147 cows died.

DT104 exploded. It zoomed from nonexistent in the United States before 1980 to more than one-third of all the CDC's *Salmonella* Typhimurium samples by 1996. The agency estimated there might be 340,000 cases in the country, and the health of all those people who caught it depended on whether one single drug, ciprofloxacin, still worked. The USDA could not match the CDC's depth of knowledge, because it had only just launched a system that checked animals for *Salmonella*. But in an emergency assessment cobbled together at the end of 1997, it said it was receiving reports of the organism in cattle from every part of the country.

A virulent, highly resistant illness was permeating world agriculture, killing people in other countries, advancing across the

United States, and threatening a crucial remaining drug. As reluctant as U.S. regulators had been to move against agricultural antibiotics, they had no choice but to act.

—⁂—

ON OCTOBER 31, 2000, the FDA took a long-delayed step. In a Notice of Opportunity for a Hearing published in the *Federal Register*—the same tool Kennedy had tried to use 23 years earlier—the agency announced that it intended to withdraw fluoroquinolone use from poultry in the United States.

This was bold, even given the international alarm over losing ciprofloxacin. The drug and agricultural lobbies were even more powerful than they had been in 1977. But the context had changed. Congressman Whitten, who had repeatedly blocked FDA action by threatening congressional reprisals, had retired six years earlier. Jukes, tireless denier of any downside to agricultural antibiotics, had been dead for a year. Smith's Minnesota study had been top news nationwide, and the advance of DT104 was an international story. In addition to Britain and Denmark, the multidrug-resistant strain was rising in France and Germany and causing a deadly outbreak in Ireland. The outbreak in Vermont had been on the cover of *U.S. News & World Report,* with a tragic close-up of the Heyers' dead calf. The World Wide Web was only 10 years old, and people were still getting most of their news from print; images on newsstands, passed by millions of commuters and shoppers, could still drive national conversation.

And one more in the series of studies that Congress kept commissioning to reexamine Levy's findings, two decades after Whitten insisted more research was needed, had endorsed what everyone except agriculture already accepted: Farm antibiotic use was creating a hazard to human health. "There is no doubt that the passage of antibiotic-resistant bacteria from food animals to humans occurs," the National Research Council, part of the National Academy of Sciences, wrote in July 1998 in a 253-page report that had been six years in the making. "A demonstrable link can be found between the use of antibiotics in food animals, development of resistant microorganisms in those animals, and . . . spread of pathogens to humans."

Still, the FDA built its case carefully. As in the case of growth promoters two decades earlier, the *Federal Register* notice—technically, a copy of the FDA docket behind the proposal—read like a legal brief footnoted with science journal citations. The FDA estimated that in a year, 1.24 billion pounds of retail chicken, the meat most likely to carry *Campylobacter*, was harboring a resistant form of the organism; 190,421 U.S. residents were falling ill from it, and 11,477 of them had been endangered by their infection's resistance to the drug that would have made them well. But despite the size of that risk, the agency could not simply cancel the license it had previously granted for the drug. It was required to offer the manufacturers a chance to defend their licenses by proving their drugs were not dangerous. Abbott, the manufacturer of saraflox-acin, declined and took the drug out of the U.S. market. But Bayer, the maker of enrofloxacin (which it sold under the name Baytril) decided to fight—and fight, and fight.

The company's defiance triggered an administrative law process, a within-government equivalent of a courtroom trial, with the FDA as the plaintiff and Bayer as defendant. (The Animal Health Institute, the trade group representing the drug manufacturers, signed on as a codefendant.) For two years, the company called expert witnesses, put researchers through depositions under oath, and filed 32 separate Freedom of Information Act requests. When the case went to trial in April 2003, Bayer summoned all the researchers it had deposed to Washington, to be interrogated a second time in the FDA's courtroom.

Osterholm and Smith were among the witnesses whose presence and documents were demanded, and Osterholm, now the founder and director of the Center for Infectious Disease Research and Policy at the University of Minnesota and a prominent policy advisor on epidemics and bioterror, still seethes at what felt like intimidation. "We had to provide all kinds of information to them about what we did and who we were," he remembers. "It was very personal, and very litigious. It was hard to challenge our data, because it was so straightforward. But the frustrating part was that it took so long." In March 2004, the administrative law judge found for the FDA. The company and the AHI appealed, triggering a fresh round of briefs and evidence gathering. But the final authority, the FDA's commissioner, upheld the decision. After threatening to continue the battle by appealing to the federal courts, Bayer folded. In September 2005, Baytril lost its license. It was the only animal drug ever forced off the U.S. market for generating resistance that threatened human health.

It was a victory, no question. But it would be the only one for a while. The battle revealed what the FDA would be up against—in intransigence, in publicity, in being forced to spend funds and time—if it tried to control any other farm antibiotics. The process revealed something else too: Until the evidence gathering began, the FDA literally had not known the quantities of antibiotics being used in animals in the United States. It possessed no data of its own and had to rely on limited figures tendered by industry, which were supplied by the AHI out of its member companies' proprietary information. The realization that the FDA did not have enough information to make policy spurred a socially committed research organization, the Union of Concerned Scientists (UCS), founded in 1968 at MIT in opposition to the Vietnam War, to fill the gap. In January 2001, the group released a devastating analysis of agriculture's antibiotic consumption. Human medicine was using three million pounds of antibiotics per year, they said. But according to government records, farm censuses, and sophisticated calculations, agriculture was consuming 24.6 million pounds, more than eight times as much, just for growth promotion and prevention.

The report's jaw-dropping math was instantly challenged by industry. The AHI maintained that all farm antibiotic use—growth promoters, prevention, and treatment of animal diseases—accounted for just 17.8 million pounds and growth promoters for only 3 million of those. But the UCS's documentation was detailed. It calculated that every year, cattle were receiving 3.7 million pounds of antibiotics, more than humans got; hogs were getting 10.4 million pounds; and chickens, the animals that were most numerous but lived the shortest amount of time, were receiving 10.5 million pounds. If those doses had been given in human medicine, they would have been considered outlaw, because none of them were being used to treat disease, only for growth promotion and prevention. In the cost-benefit analysis of whether to risk provoking antibiotic resistance, they represented no benefit, only cost.

CHAPTER 7

THE TRIUMPH OF
THE HYBRIDS

THE REVELATION THAT AGRICULTURE was feeding live-stock millions of pounds of antibiotics each year felt shocking. People who had never had a reason to imagine the lives of farm animals before they turned into meat were moved to wonder: What had happened in agriculture that required such vast deployment of drugs? The answer, as with so much in farm antibiotic use, is best explained by chicken.

Before Thomas Jukes performed his experiment or Jesse Jewell began driving feed trucks around north Georgia, there were more than six million farms in the United States, compared with two million now. They were mostly small properties, growing a mix of crops and animals, and they almost all raised chickens. Which type of chicken was a complicated question, because there were so many to choose from. The January 1921 issue of the *American Poultry Journal* carried six pages of small-type classified ads featuring dozens of varieties from hundreds of breeders nationwide: Single-Comb Anconas, Silver Wyandottes, Brown Leghorns, Black Langshans, Light Brahmas, Sicilian Buttercups, Golden Campines, White-Laced Red Cornish, Silver-Gray Dorkings, Silver-Spangled Hamburgs, Mottled Houdans, Mahogany Orloffs, White Minorcas, Speckled Sussex. Most farms kept small flocks, from a few birds up to about 200, and for most of them, the point of chickens was eggs; birds were sold for meat only when hens were spent or when chicks hatched out male. Farmers chose the variety they raised based on what other farmers in their area preferred—breeds that had adapted well to whatever wet or dry, windy or humid conditions prevailed where they lived—or because they were persuaded by boastful ads that talked up the egg production of breeds that won medals at state and national poultry expositions.

Chicken production's first step on the road to innovation, and the expansion that followed it, came in 1923 with the first electrically heated incubator. That freed farmers from having to choose and maintain breeding stock, and also from losing the months in a hen's short productive life when she would be hatching her eggs instead of laying more. Now they could outsource those tasks to a new tier of the industry, thousands of hatcheries shipping newborn chicks by mail. But the breeds those chicks came from were still selected for maximum egg production, not for any eating pleasure they might offer after their laying years ended. Choosing birds for how many eggs they could lay was a smart strategy through the privations of the Great Depression and the restrictions of World War II: It maximized the protein you could get from a bird without sacrificing the bird herself. But after the war, when beef and pork emerged from rationing, eggs seemed dull by comparison, and laying hens' lack of tasty muscle made them an insufficient alternative. People had willingly cut their meat eating for a long time, to support the war. Now they wanted to indulge.

One smart retailer saw the problem coming. Howard C. Pierce, the poultry research director for the A&P Food Stores supermarket chain, told a November 1944 poultry meeting in Canada that someone needed to develop a sumptuous chicken, a bird with a breast like a turkey's. By the next summer, his wish ignited an extraordinary undertaking: the Chicken of Tomorrow contest, organized by the USDA, with the backing of A&P and the support of every major poultry and egg organization in the country, all aimed at breeding a better chicken.

The effort was massive. The contest had 55 national organizers—scientists and bureaucrats loaned from government agencies, producer organizations, and land-grant colleges—and hundreds of volunteers in 44 states. (That was out of 48; Alaska and Hawaii had not been added yet.) It began with state contests in 1946, progressed to regional judging in 1947, and ended with a national competition, held at the University of Delaware's Agricultural Experiment Station, in 1948.

What they aimed to achieve was droolingly described in the *Saturday Evening Post* in 1947, after the contest was two-thirds through: "one bird chunky enough for the whole family—a chicken with breast meat so thick you can carve it into steaks, with drumsticks that contain a minimum of bone buried in layers of juicy dark meat, all costing less instead of more." Anyone who wanted to compete—and they ranged from small farmers to large, established companies—was granted one year to devise and breed a bird that possessed the sturdy, meaty qualities the contest was hoping for. If they reached that goal, they then had to prove their bird was reproducible, by breeding enough birds in enough generations to last through a three-year beauty pageant.

This was a significant challenge. Creating better poultry varieties had been a goal for decades, but maintaining reliable crosses had been challenging. Farmers distrusted crosses, worrying they would be sickly and not breed true, so most of the aspirants to the Chicken of Tomorrow contest competed by tuning up pure breeds that they were already raising. In the final stage of the contest, only eight of the 40 contestants entered birds crossbred from the historic standard breeds.

By March 1948, all 40 breeders, plus six more in case any were eliminated, shipped 720 eggs each to a hatchery on Maryland's Eastern Shore. The shipments came from 25 states and were loaded onto trains according to a precise timetable so that each arrived at the right hour to go into incubators. The batches were coded so that only a few people knew their identity and put into one hatching pen per breeder, dark chickens next to white in case any got loose and tumbled into the pen next door. Once the eggs hatched, 410 chicks—400 for judging and 10 extra in case any came to harm—were picked at random from each batch of 500 and driven to new purpose-built barns.

The chicks were allowed to grow for 12 weeks and two days and then were killed, defeathered, weighed, and chilled, as if they were going to be sold. Out of each breeder's batch, 50 were picked for judging. That meant the judges were looking at 2,000 birds, and evaluating each of them on 18 criteria, from body structure and skin color to how early they had developed feathers and how efficiently they converted feed to muscle. On June 24, 1948, the judges announced their results, on a stage adorned with boxes of chicken carcasses from each of the contestant's batches, and frozen cross-sections of the top-scoring birds. The first runner-up was Henry Saglio, the teenage son of Italian immigrant farmers in Connecticut, who had bred his family's pure line of White Plymouth Rocks into a muscular, meaty bird. The winner was Charles Vantress from California, who had crafted a red-feathered hybrid out of the New Hampshire, the most popular meat bird among East Coast growers, and a California strain of Cornish.

That evening, the contest celebrated the breeders' achievements with a parade through Georgetown, Delaware, with floats depicting the phases of Delmarva's poultry industry and a smiling, waving Festival Broiler Queen perched on top of a car. It celebrated not just the new birds but the new economy their developers hoped to create: a time when the Chicken of Tomorrow would be the dominant meat on farms and in markets, cheaper than beef, more docile than hogs, desired on its own behalf and not as a cast-off carcass after egg laying. A second version of the contest three years later brought that day closer: Vantress won again, with another crossbreed, once again displacing a purebred bird. He would go on to establish one of the industry's top hatchery companies, rivaled only by Saglio's family company, Arbor Acres, which abandoned its winning purebred White Rock for a hybridized broiler in 1959. That same year, Vantress hybrids served as the sires of 60 percent of broilers across the country. The sturdy, free-living, weather-tolerant purebreds that had dominated American barnyards for almost 100 years vanished from commercial use.

—⁂—

THE WINNERS OF THE CHICKEN of Tomorrow contests did more than create new birds; when they transformed chickens, they recreated the chicken industry too. The earliest attempts at hybrids had been single crosses between two breeds: mother from one variety, father from another. But to ensure they could reliably reproduce the characteristics on which they were building businesses, breeders turned to creating complex crosses. The intricacy of the family trees they constructed ensured that the birds could not be reproduced outside the companies that bred them. If a farmer who bought the new hybrids tried to mate them on his own property, the birds would not breed true. Previously, broiler farmers had bought chicks from hatcheries mostly for efficiency, but now they had no option. Raising hybrid birds became like growing hybrid soybeans or corn: It required returning to the company to start each new crop. In a remarkably short span of time, the open-source birds that had populated millions of farm yards and back gardens for thousands of years became an ingredient in proprietary intellectual property. Simply by the mechanics of genetics, without even the assistance of patents, the patrimony of purebreds vanished behind the restrictions of trade secrets.

The tight hold that companies exerted on that property became more significant as the number of companies holding it diminished. The narrowing of the market was probably a natural consequence of the cost of sustaining breeding programs. Keeping the traits in a modern broiler balanced and consistent requires constant breeding of parent, grandparent, and great-grandparent flocks that may total hundreds of thousands of birds. Isolating and adding a new trait, without losing or unbalancing anything the birds already possessed, could take years. The time and expense not only discouraged other firms from entering the business— after 1960, no company that had not been a Chicken of Tomorrow competitor broke into the business—it also forced rapid consolidation. By 2013, just three companies owned the genetics of almost all of the billions of broilers produced every year worldwide: Cobb-Vantress; Aviagen, which included Arbor Acres; and, in Europe, Groupe Grimaud.

It wasn't only businesses that became consolidated. As the Chicken of Tomorrow contest receded into the past, the variety of birds that existed before the competition faded too. Rangy or sturdy, maternal or feisty, white or brown or barred, all of those qualities were subsumed into the meatiness the contest elevated above all other qualities. The birds entered into the contest were allowed 86 days to reach the weight at which they were slaughtered, and at the end of that time, Vantress's birds were three and a half pounds. (That was a full pound more than the average weight when the USDA began keeping statistics in 1925—one reason the winning birds were such an achievement.) Now, the average weight at slaughter is six pounds, and chickens get there in an average of 47 days. That extraordinary change is due to growth promoters, but it is also due to the breeding companies' constantly selecting among their flocks for birds that pack on muscle while eating less. In 1945, it required four pounds of feed to put one pound of flesh on a broiler, a ratio the industry calls "feed conversion." Now it takes less than two pounds.

The birds that emerge from modern breeding programs—the Cornish Cross that Vantress developed and similar ones produced by the other two companies under different names—look nothing like the old purebreds. For one thing, they are always white; companies realized early that light-feathered chickens appear cleaner and more appealing after defeathering than ones with dark pinfeathers still in the skin. For another, they are disproportionate: They have been bred to bulk up the most in the breast muscles, which are the white meat that American eaters prefer. The breasts on today's broilers are twice the size of those in the purebred birds that predated them and can account for one-fifth of a bird's whole body. Those muscles develop faster than the bones and tendons that support them—six weeks, a modern broiler's age at slaughter, would be barely preteen in a purebred chicken—and the weight unbalances birds' bodies. A fast-growth broiler has the teetering instability of an olive propped up on two toothpicks. Broilers can develop areas of dead or hardened muscle in their breasts and fluid accumulations in their abdomens, both signs their circulatory systems cannot keep up with bringing oxygen to their muscles and carrying metabolic waste away. They suffer from distorted leg bones and difficulty walking, and scientists who raise them for research have found that when given a choice between regular feed and one laced with painkillers, lame chickens choose the drugged feed. The more difficulty they have moving, the more drug they will choose to eat.

A purebred chicken will peck, scratch, flap onto a perch, and even fly for short distances. But the descendants of the Chicken of Tomorrow are not bred to perch or flap, and the barns that were developed to hold them, known as houses, give them little opportunity to try. They walk, in shoals with thousands of other chickens, or they sit. When houses are not well maintained, the litter that birds sit on becomes saturated with ammonia from their waste, burning the skin of their feet and hocks, the backward knee joint halfway up a chicken's leg. The confined conditions and their own passivity make them more vulnerable to disease—a major reason why, after growth promoters became so successful, farmers sought to use millions of pounds of preventive doses of antibiotics as well.

If all the chickens of the world were only Cornish Cross, there would be no way to know how the chickens of yesterday looked or behaved, how they ate or flew or fought off diseases or mated or cared for their chicks. But in a few scattered corners, they survive undiluted. They hold the genetics of the birds that existed before the industrial hybrids, secrets that might someday be deployed to create a better chicken once again.

—◊◊◊—

MARQUETTE, KANSAS, population about 640, is a place where no one ends up by accident. It lies almost dead center in the state, surrounded by flat plains, gentle hills, and trees that were planted 100 years ago to break the biting wind that whips across the grasses. No one drives through; Denver is six hours to the west and Kansas City three hours east, but Interstate 70, which links them, lies 30 miles north. No one can ride through either—the Union Pacific and BNSF railroads bracket it north and south—and no one who cared to try would float far on the nearby Smoky Hill River, so oxbowed it looks like a toddler's scribble. To get to Marquette requires intention, and to get to Good Shepherd Poultry Ranch, which lies just outside it, requires daylight and a paper map and the trust of a child in a fairy tale: beyond the edges of cities, beyond the forests, beyond the hills, lies a treasure.

Frank Reese, Good Shepherd's owner and sole employee, chose the location 25 years ago to safeguard a treasure that had been entrusted to him: dozens of varieties of chickens and turkeys that were once the backbone of small farms—birds that could feed themselves outdoors, find their own places to roost, and fight off disease without assistance. The battered tractors, the porch of the weathered Victorian farmhouse, and the ground outside the metal-sided barns are covered with them: black and white, russet and bronze, barred and streaked in silver and cream, flapping up to fence rails, skittering under the farm truck, and pressing around Reese's ankles as he wades through.

"Rhode Island Red," Reese counted off as chickens burbled past his feet. "Blue Andalusian. Silver-Laced Wyandotte. White-Laced Red Cornish. New Hampshire. Black Spanish. Single-Comb Ancona. Rose-Comb White Leghorn." He paused, and seemed to be counting. "There might be just 50 of those left in the world."

Reese, who is almost 70, is a lean man, with corded muscles from heaving heavy bales into trucks and big hands chapped red. His head is big too, round and wide above his ears, tapering to a full mustache over a narrow jaw. He wears his hair cropped to fuzz; over his heavy hooded jacket, it gives him a monkish air. He spotted a black and white hen sheltering under a feeder, fluffed-out feathers striated like ripples on a pond, and grinned. It transformed him.

"She's a Barred Plymouth Rock," he said, scooping her into the crook of his arm. "I've had them 52 years."

Every bird on Reese's parcel of prairie was hatched there, from an egg that was laid there, from parents that were hatched and raised there before them. The farm is a living archive of history and genetics, preserved because the birds bring him joy—and also because he believes, in defiance of the trends of decades, that the poultry industry erred in sacrificing them, and will someday need them again.

Reese was born the year of the first Chicken of Tomorrow contest, to a family that landed in Pennsylvania in 1680, migrated through Illinois, and arrived in Kansas just after the Civil War, chasing a federal promise that anyone who was willing to work the land could have 160 acres to homestead for free. The Reeses had been farmers before they came west, and they were farmers after, down to Frank's parents. His folks ran a mixed property— beef and dairy cattle, hogs, and chickens, turkeys, ducks, and geese—and as the third of four children, too small to milk cows or venture safely into the sow pens, Frank was given the birds as his chores. He fed the chickens and collected eggs, and he herded the turkeys from the barn out into the fields, where they pecked for insects and flapped onto fence posts and tree branches.

From his earliest memories, Reese was transfixed by them. In first grade, told to write an essay about a pet, he penned "Me and My Turkeys." When his father took their Herefords to the American Royal, a giant late-autumn livestock show in Kansas City, he would sneak away to the poultry section, tugging at the elbows of the breeders who ruled over the barns. His family were locally famous for their Barred Rocks, and by age seven, he was winning ribbons at county fairs for chickens he had bred.

The Plains states were turkey country—so much so that farmers used to herd their turkeys to distant markets, thousands at a time, on cattle drive–like "turkey walks"—and the fairs were the turkey breeders' kingdom. Winning offered more than bragging rights; selling a champion that carried the best characteristics of a bloodline could earn more than a thousand dollars. Exactly what constituted a champion had been codified almost 100 years earlier in the American Poultry Association's *Standard of Perfection,* the bible of unique qualities that distinguished one breed from another. "Standard" sounds like faint praise, but to the poultry old guard, "standard-bred" was the highest approbation. The men and a few women who ruled the turkey pavilions of the state fair had been earning it since before Reese was born.

"There weren't youth shows back then," he recalled. "Whether you were 14 or 84, you competed in the same division. So I would be lucky if I got fifth place, because the big guys always beat me: Rolla Henry, who bred Bronze turkeys, and Norman Kardosh, who bred Narragansetts. Well, I got tired of that, so when I was 14 I took one of our trucks and I drove myself 50 miles over to Abilene to see Sadie Lloyd, who had been breeding turkeys so long, she had showed birds at fairs with those men's mothers. I told her, 'Sadie, I want to beat Norman and Rolla,' and she cackled and said, 'We'll show them.' And that year I won."

The breeders were a prickly, exacting group: bachelors and spinsters, for the most part, people who poured the love and attention they would have given spouses and children into preserving poultry bloodlines that were already vanishing. They must have seen something of themselves in the avid, jug-eared farm kid, and they began teaching him how to tell an average bird from one worthy of carrying a variety into another generation. Developing a eye for color and stature was essential; some aspects of the standards depended on measurement—the length of a bird's neck, the weight of an egg—but most judgments of conformance relied on familiarity and skill.

Sensing Reese's promise, Lloyd sold him some of her Bourbon Reds, a mahogany turkey with white primary feathers and a white ruff of tail. Golda Miller sent him her Jersey Giants, meat chickens that dated back to the 1880s and could grow to 13 pounds. Ralph Sturgeon, a legendary breeder, gave him the Barred Plymouth Rocks that Reese still treasures. Kardosh, who lived hours away in tiny Alton, Kansas, became his chief mentor, schooling him in the history of the eight turkey lines recognized in the Standard.

But Reese didn't envision a life as a poultry hermit. He left Kansas, first for the army and then for nursing school, where he became a registered nurse anesthetist. He settled outside San Antonio, and though he kept his chickens and turkeys going, they were a private pleasure, not a cause. In the late 1980s, his mother asked him to return to Kansas; she wanted him nearby, and the small local hospital needed an anesthetist. Reese loaded a van with his chickens and turkeys and drove 700 miles north to his former, and future, life. Local wisdom said the best site for a turkey farm was on a slope, to let waste drain away, and not too close to water, because predators would come to drink. He found the 160-acre farm that became Good Shepherd near the top of a low hill that slopes two miles down to the Smoky Hill River, between Marquette to the west and the Swedish-settled town of Lindsborg to the east.

Not long after, a friend called him. Tommy Reece, no relation, was a small-scale chicken raiser too, in the remote hill country west of San Antonio, and for years his passion had been Indian Game Cornish, a compact, muscular bird with tortoiseshell feathers, related to the birds Vantress used to make his Cornish Cross. Reece was dying. "'He said to me, 'Save my Cornish,' and I promised to try," Reese told me. "He sent me two dozen eggs, and out of the two dozen, three hatched."

So many of Reese's mentors were gone. Kardosh, who trained him, was the last. In 2003, Kardosh summoned his former acolyte to a central Kansas hospital. He was 76 and knew he had not long left. He bequeathed his turkey bloodlines to Frank, begging him to keep them going. Crying too, Reese promised that he would be their caretaker and not let the birds die out.

Without ever intending to, Reese had become the guardian of dozens of historic lines of poultry, birds the industry considered so irrelevant that no one else noticed when they were about to be lost. In the past, there would have been a generation of farmers holding the bloodlines of poultry in trust. Now it seemed possible that there would be only him.

—ɱ—

REESE'S CHICKENS AND TURKEYS ought to be valuable. They preserve the genetic sources of sturdy immune systems that require no antibiotics, balanced bodies that allow them to run and flap, instincts that allow them to find their own food and teach their chicks to do the same. They are utterly unlike commercial broilers, and also unlike the hybrid Broad-Breasted White turkey, developed in the 1960s and now the staple of every commercial turkey company, which is so unbalanced by its overgrown breast muscles that it cannot get into position to mate and has to be artificially inseminated.

They grow slowly, as did all birds before the hybrids arrived. Reese's chickens take 16 weeks to reach what would be slaughter weight, compared to six weeks for modern broilers. The turkeys take six months and would live to five years if allowed. But to keep the bloodlines true, it is necessary to keep breeding the birds, and their long lives and ability to mate naturally left Frank with an ever-expanding flock. He began selling eggs for hatching and chicks and turkey poults for other farmers to raise, but he was strict in how he sent them out: never by mail, only to people who would come to the farm or pay a driver to deliver them.

He realized that to keep his farm going, he would have to sell birds for meat, but that was more complicated than it sounded. The Chicken of Tomorrow contest had not only pushed the indus-

try toward confinable, reproducible hybrids; it also taught consumers, over decades, to prefer the birds it created, with big wings
and breasts that are fine-textured and pale. The meat of Reese's
pecking, perching birds reflects their long, exercise-filled lives: It
is lean, dark, and deep flavored—something a chef might showcase
for adventurous customers but not packaged supermarket fare.
Even getting his birds to chefs or supermarkets posed problems.
The geography that keeps Good Shepherd safe from development
worked against him: Restaurants prefer to receive chicken fresh,
but the kinds of restaurants that could persuade customers to try
a heritage bird were so far away that Reese would have to ship his
birds frozen. That was if he could get the birds killed and processed
at all. He needed a slaughterhouse that was close by and USDA
certified, with equipment that would fit his nonstandard birds and
a processing schedule that could accept small lots arriving irreg-
ularly. Yet small, independent slaughterhouses have been shutting
across the United States, the aftereffect of the consolidation that
subsumed small farms into corporations.

Rarity, distance, and difficult processing all funneled down to
price: Reese had to make a case that his birds were worth what he
would need to charge. He found online merchants to help him.
For the turkeys, there was Heritage Foods USA, a spin-off of the
American arm of the international slow food movement, which
enshrines threatened heritage varieties in an "Ark of Taste." Emmer
& Co., a start-up focused on creating a market for old breeds,
began marketing his chickens. For once-a-year turkeys, consumers were willing to spend more than $10 per pound, even though
that pushed the price of a holiday bird into hundreds of dollars.
But chicken met price resistance at half of that. "The industry can
produce a baby turkey for 90 cents, maybe a dollar," Reese told

me. "It costs me seven to eight dollars to produce the same bird." The turkeys subsidized his chickens. He estimated he would need to sell 1,500 chickens every month to break even, but when I met him in 2013, he was managing to sell barely 2,700 a year.

The irony was that Reese would prefer never to kill a bird at all. He did it to thin and perfect the flock and because it was his only way to raise the funds he needed to keep Good Shepherd Ranch viable. On a day that I visited him, he sat on the concrete pad beneath a feed bin and watched his birds mill around him. A cold wind was rising, but the setting sun glinted on their bronze feathers and bright eyes. "I would love to stop killing them," he said, softly. "And just have a preserve, and save them."

—✺—

THE HYBRIDS THAT BEGAN as the Chickens of Tomorrow did not require the living space that the standardbreds did. The urge to fly or perch had been bred out of them, and automated feed and water systems spared them from having to hunt and peck. Since birds had no use for the extra living space, poultry production took it away, packing more and more chickens into barns. At the same time, automation and economies of scale made it possible to make barns larger, and to add more barns to properties without increasing a farm's amount of land. From the 1960s to the 2000s, the average chicken barn—"house" is the industry term—grew from less than 13,000 square feet to three times that size, windowless sheds as long as two football fields. Over the decades, thanks to growth promoters, genetics, and tuned-up feed recipes, birds grew more quickly and came and went in shorter and shorter cycles. In the 1950s, most broiler farms produced fewer than 100,000 birds in a year; by 2006, the average was 600,000 birds.

So many birds in such huge groupings increased not just the possibility of illness in a flock (and the need for antibiotics to keep them well), but fumes and gases, sucked out of the houses by fans taller than a person. And flies. And most of all, manure. A barn holding 20,000 broilers—small by today's standards—will produce 150 tons of "litter," a spongy mix of droppings, shed feathers, spilled feed, and exhausted bedding, in a year.

"The lights, the smell, the noise of the fans, the flies," Lisa Inzerillo said. "I can't open my windows. I can't hang clothes out to dry. I've lost the ability to sit out on my porch at night."

Inzerillo is a flight attendant married to an emergency room doctor, but she is also the latest owner of a farm that first belonged to her great-great-grandfather. The 66-acre property is a wide swath of hay and woodland, centered on a trim white farmhouse and a pond, north of the town of Princess Anne, Maryland, on the Delmarva Peninsula. Inzerillo grew up walking the fields with her grandfather and riding on her dad's tractor, and like any other native of the birthplace of broiler growing, she expected her neighbors to keep chickens: one or two houses, no more than a few thousand birds at a time, as part of a mixed farm of other animals and row crops.

When she and her husband, Joe, moved back to the farm in 2010, they discovered how outdated that expectation was. Chicken operations in Delmarva were becoming what the U.S. Department of Agriculture calls "no-land" farms, with mega-houses built as close as possible to the legal setback of 200 feet from a property line. A property just north of their farm held 12 of the longest houses; to the south, there were 31, clustered into several abutting fields owned by a family that did not live on the property. Whenever the wind shifted, the Inzerillos got a blast of ammonia and fumes.

"My neighbor has just become asthmatic," Joe said. He nodded at his wife. "She has sinusitis. I've had bronchitis. Forty-five years I've worked in emergency rooms, and I've never been sick until now."

When the couple and their neighbors learned that one of the nonresident farm owners planned an array of the longest chicken houses directly across from their farm, they organized into a neighborhood group and confronted the Somerset County Board of Commissioners. Over local industry's protests, they won an incremental improvement, an increase in the setback to 400 feet—but in payback, the board allowed farms that had already been granted permits to proceed. Their battle encouraged neighborhoods in the other counties of the Eastern Shore to push back against metastasizing chicken farms, but it also made clear that Delmarva faces an unsolvable problem: too much manure for the land to absorb. By the time the Inzerillos moved to Lisa's family farm, poultry in Delmarva were pooping out 1.5 billion pounds of manure per year.

Farmers have always used manure for fertilizer: It is abundant, cheap, and endlessly renewed. Poultry litter in particular is high in nitrogen and phosphorus, the minerals that make up fertilizer. But it contains about the same amount of both, and crops do not take them up at the same rate—so when litter is applied to fields, most of the phosphorus it contains goes unused. And when a lot of litter is applied or heaped in open fields or under shed roofs for storage, those extra nutrients wash off into streams and sink into aquifers. In Delmarva, which is laced with streams called "branches" and fringed with bays and inlets, storm water and groundwater don't travel far before they reach the coast. For decades, the Chesapeake Bay to the west and Delaware Bay to the east have struggled with nutrient overloads. The nutrients encourage algae to grow; when it dies and decays, it blocks the sun and robs the water of dissolved oxygen. Without oxygen, fish and seafood, including the Chesapeake's iconic blue crabs, do not survive.

Water quality was supposed to be protected by the Clean Water Act, passed in 1972, but that legislation primarily dealt with "point sources" of pollution, such as a pipe jutting out of an industrial plant. It had much less power to regulate the complexity of contamination that arises from thousands of piles of dry litter held on farms or spread on corn and soybean fields, or from the storm runoff and groundwater that carry the litter's contents away. For decades afterward, fresh federal and state regulations struggled to catch up to that oversight. The regulations created permit systems that set limits on how much manure could be spread on the ground or held on a property, but activist groups often found those requirements were ignored or defied.

Meanwhile, battles over manure disposal pitted neighbors against neighbors. The combatants were not always predictable: New arrivals were as likely to be immigrants wanting to farm as they were retirees looking for a piece of countryside, and the long-time residents might be small property owners like the Inzerillos or big producers looking to expand. But the fights over who was responsible sparked lawsuits and poisoned relationships within small towns.

The Chesapeake Bay Foundation estimates that farm runoff accounts for up to half of the excess nutrients entering the bay. Federal agencies and the University of Maryland have demonstrated what happens next: dead zones defined by fish kills—rafts of dead fish, suffocated by the lack of dissolved oxygen, that wash up in branches and on beaches—and also by critically depressed populations of delicious, economically important shad, striped bass, oysters, and blue crabs. Harder to track, but just as important, are the antibiotic-resistant bacteria that also move out from farms—not just in foodborne organisms but via manure, airborne dust, and flies as well.

As Levy's research hinted in the 1970s, manure is the source of much of the resistant bacteria that spread from farming. When animals are slaughtered, some of the contents of their intestines may splash onto the meat they are becoming as they are cut apart. But while animals are being raised, the bacteria in their guts—and any unmetabolized antibiotics that their bodies did not absorb—pass out of them into chicken house litter, or the vast pits or ponds of liquid manure on pig farms and cattle feedlots. When that manure is disseminated through the environment—either deliberately, by spreading it on crop fields, or accidentally via rain or pond overflow or a leak in the liner of a storage pit—the bacteria it contains spread too. Researchers have found resistant bacteria in the soil around chicken farms, in groundwater under hog farms, and in dust borne away from an intensive farm by the wind. The trucks that bear chickens from farm to slaughter, stacked up in towers of wire cages, stream a plume of resistant bacteria behind them that can contaminate cars on the same road. Scientists have found resistant bacteria being carried away by flies from chicken farms in Delaware and Maryland and hog farms in Kansas and North Carolina.

Sometimes people are the vehicle that moves resistant bacteria away from farms. Workers on pig farms in Iowa and in hog-killing plants in North Carolina are more likely than their neighbors to carry strains of MRSA, drug-resistant staph, that are resistant to multiple families of drugs. One group of poultry farm workers in Delmarva was 32 times more likely to be carrying *E. coli* resistant to gentamicin, an antibiotic injected into the eggs that broilers hatch from, than their neighbors who did not work on farms.

All of those accidental exportations cause the environment outside farms to be loaded not just with resistant bacteria but with genes that confer resistance. As in a library packed with books, bacteria can acquire the genes and the plasmids that house them and pass them on to other organisms. At the end of that chain of acquisition and transmission lie people who have never entered the premises of a farm and have no reason to imagine a farm-related health problem would ever come their way. Thus, in Pennsylvania, thousands of people using primary-care medical clinics were more likely to have a MRSA infection if they lived near fields that had been sprayed with hog manure. And in Iowa, military veterans who lived within a mile of an industrial pig farm were twice as likely as their distant neighbors to be carrying drug-resistant staph.

The paths that resistant bacteria and resistance plasmids take into the environment are complex. Manure from farms washes into coastal waters, but so does effluent from pharmaceutical manufacturing plants. Hospital sewage can transport resistant bacteria out of the building; water treatment systems are designed to take fecal bacteria out of sewage, but they do not capture resistance genes. As a result, streams and lakes and surface waters are laced with resistant bacteria. Wildlife pick up those bacteria; so do fish. And seagulls and other water birds carry them across oceans.

Recently researchers have proposed that it is not just resistant bacteria and their broken-away genes that are a peril, but the still active antibiotics that flow into the environment from animals given them (and from humans who have taken antibiotics as well). Possibly one-quarter or less of any antibiotic dose gets used up in the body; the rest exits in waste, enters sewage systems, and travels on from there. Those intact antibiotics could push resistant bacteria to evolve further, in unpredictable ways—but the larger problem may be the effect the drugs exert on people who consume them unknowingly. Some researchers suspect those diluted doses could act in humans as growth promoters did in animals, and as corroboration, they point to the near-simultaneous arrival of antibiotics and beginning of the modern epidemics of obesity and diabetes.

Those occurrences might be only coincidences, happening simultaneously but with no causative connection. But provocative experiments in mice, by Dr. Martin Blaser and associates of the NYU Langone Medical Center, show that antibiotic dosing early in their lives alters the balance of bacteria in their guts and affects gene activity in ways that lead to weight gain and change how immune systems develop. If that were true in humans also, the obvious culprits would be the episodic antibiotics that kids get for ear infections and other childhood problems; the less obvious might be their constant exposure, or their mothers', through the environment as well.

Like the movement of bacteria through groundwater or the wind, the flow of resistance genes away from farms, through the environment and toward people—first farmworkers, and then those with no connection to farming—is invisible and subtle. It takes research, and significant resources, to unravel how the vast economy of modern chicken pumps organisms and resistance through those complex pathways. Unless, of course, there is a dramatic outbreak, such as the one that scooped up Rick Schiller in the autumn of 2013.

CHAPTER 8

THE COST OF CONTAMINATION

RICK SCHILLER HAD NO WAY of knowing, while he was coming down from his tooth-chattering fever, that a vast nationwide network was already pursuing what had made him so ill. The pattern-recognition alert in the CDC's PulseNet program, which compares the DNA fingerprints of organisms taken from foodborne illness patients across the country, had flagged a match between *Salmonella* samples uploaded from different cities in California. In another office in the CDC's complex bureaucracy, senior epidemiologist Laura Gieraltowski inherited the case. She alerted the California health department, which had already noticed the same match. The department's investigators were already interviewing people who were sick, and though it was still early, they had a sense the problem might lie in chicken.

That discovery was not as significant as it might seem, because so many people eat chicken so often. It is America's favorite protein, after all, and it shows up as such in the surveys that disease detectives conduct during outbreaks: At least three people out of every five will say they ate chicken in the past week. But those results include all kinds of chicken: frozen nuggets, fast-food sandwiches, bar wings, buckets of fried stuff, and supermarket rotisserie birds, as well as chicken cooked at home. To find the culprit food, investigators would need to sift the signal of a particular chicken product, contaminated with this particular strain of *Salmonella*, out of the background noise of everyone eating chicken all the time.

The program they were using was designed to help them do that. PulseNet arose from the realization that the systems that had once helped epidemiologists untangle foodborne outbreaks—the slaughterhouse spreadsheets that let Holmberg pursue one farm's beef, the mandatory reports to the state that allowed Smith to uncover resistant *Campylobacter*—could not cope with foods that were distributed over long distances. The CDC learned that from a tragedy, an outbreak that forever altered understanding of how grave a foodborne illness epidemic can be.

In November 1992, children across a swath of western states—Washington, California, Idaho, and Nevada—abruptly began falling seriously ill, with cramps and bloody diarrhea, terrifying their parents into emergency dashes to hospitals. Some of the children developed hemolytic uremic syndrome, known as HUS, a life-threatening complication in which red blood cells destroyed by infection clog the kidneys, backing toxins up into the bloodstream, and raising blood pressure and destroying kidney function. The cause was a strain of *E. coli*, designated O157:H7, that possessed an unusual ability to make a cell-destroying toxin. By February, 726 people, most of them younger than 10 years old, had fallen ill. Four children died.

The link was hamburgers sold by the fast-food chain Jack in the Box, which had locations all over the West. The burgers had been contaminated with the *E. coli* strain at processing plants that made them for the chain. The reach of the outbreak, involving 73 restaurants, signaled a new problem in food safety. Since the 1970s, the number of plants producing food—including meat, poultry, cheese, milk, and milled grains—had been dropping as smaller companies consolidated into fewer, larger ones. In the egg business, to take one example, 85 percent of the farms that existed in 1969 were gone by 1992, and each farm had grown much larger, from hundreds of hens to thousands and in some cases millions. Foods were originating in fewer places than before and being sold into networks that took them to many more points of distribution; when they were contaminated, they could cause cases over thousands of miles. Not long before the Jack in the Box outbreak, tomatoes from a single packing house in the South had made people ill in four states, and cantaloupes from Central America that were served on salad bars had sickened people in 28 states.

The way that the CDC had always learned about outbreaks, from reports sent from a doctor or lab to a state health department and then to the agency, did not suit this new reality. It took federal investigators 39 days to understand that the hundreds of cases of illness being reported in 1993 were linked. That required receiving reports of illnesses, interviewing families, gathering medical records, obtaining bacterial samples, conducting lab work, and looking for a common source. If it could have moved faster, hundreds of illnesses—and millions in legal claims afterward—could have been forestalled.

The new technology of pulsed-field gel electrophoresis, developed just nine years before, offered a way to move faster. The bar codes it rendered, black and white and so graphically simple they could be sent and received even over a slow Internet connection, were a perfect match for the infrastructure of the time, when email was still a novelty and bandwidth was precious. It was an expensive undertaking to set the machinery up in a state health department lab and train personnel to use it, but within 10 years, it was a routine feature of public health investigations, cutting out months of investigative time and saving lives.

—m—

WITH THE INTERVIEW RESULTS and the PulseNet matches in hand, Gieraltowski had evidence that some chicken, somewhere, was making people sick. But given the new complexity of food distribution, how to find the source? Another government surveillance system, also founded in the wake of the Jack in the Box outbreak, provided a clue. NARMS, for the National Antimicrobial Resistance Monitoring System, measures the presence of resistant bacteria in animals, meat, and humans. The CDC, USDA, and FDA all feed data into it. The CDC tabulates reports from public health laboratories of patients with certain infections. The USDA checks animals on farms and at slaughterhouses for the same resistant bacteria. And to see whether resistant infections are moving through the food system, the FDA sends out teams that buy meat for sale in supermarkets, bring it back to the agency's labs, and test it.

NARMS is indicative, not comprehensive: The CDC tests for only five types of foodborne bacteria; the USDA struggles to get enough access to live animals; and in the year that Schiller fell ill, the meat the FDA uses—chicken parts, ground turkey, ground beef, and pork chops—came from only 14 states. So it was a piece of luck that when Gieraltowski's counterpart looked through the NARMS records for any *Salmonella* Heidelberg of the pattern that PulseNet had highlighted in California, just one popped up. It had been collected in California, in the right time period, from a package of chicken.

The chicken had been produced by a company called Foster Farms. This was not much help to the investigation, because Foster Farms dominates the market for chicken in California. The company is family owned, local, long-standing—2013, when the epidemic was burgeoning, was its 74th year in business—and well known from amusing commercials in which puppets portraying badly behaved broilers aspire to be good enough to be sold by Foster Farms, and fail.*

The database hit allowed the CDC to go public. On October 8, 2013, four months after the first PulseNet signal and a week after Schiller's emergency, the agency made its first announcement that an outbreak was occurring and that it believed it knew the origin: "Consumption of Foster Farms brand chicken is the likely source of this outbreak." They based that assertion on the connections the three agencies had traced among the evidence they had gathered. Labs at the CDC and in the states had pulled seven different genetic fingerprints from victims, indicating seven closely related strains of *Salmonella*. The FDA had gone back into the NARMS database and discovered that four of the seven patterns had been found on five pieces of Foster Farms chicken. The USDA's Food Safety and Inspection Service had gone

---

* Foster Farms declined to be interviewed. For a statement supplied by Ira Brill, its director of marketing services, please see note on page 337.

into Foster Farms plants, swabbed the machinery, and taken samples of chicken from it. They found four of the outbreak strains in three facilities.

When Schiller got his phone call, just after coming home from the hospital, investigator Ada Yue asked him the same questions her colleagues had been asking everyone else who was ill: where he ate when he went out; where he did his grocery shopping; what foods he usually preferred to buy. He told her he could not remember what he had bought, or where. It had been too long and he had been too sick; the trauma had wiped the details from his mind. But he always shopped with a debit card, and it occurred to him there might be something in his bank's online records. He pulled up his account, and back in September, he spotted an entry for FoodMaxx, a local supermarket chain. He asked Loan Tran, his fiancée, if she remembered what they bought.

Of course, she said. She'd been going out of town, and she had baked some chicken so there would be something in the refrigerator in case he forgot to feed himself: thighs, because he preferred dark meat, with barbecue sauce because that was her special way. She remembered he'd eaten them too, because they were all gone when she got back from her long weekend. But, she reminded him, she had bought two packages and stashed the second one in the freezer. She'd been planning on defrosting it soon.

Schiller called the investigator back. Then he rushed home and checked the freezer. Tran had frozen the chicken in its original packaging, and Schiller recognized it at once. It was three pounds of thighs in a golden Styrofoam tray, covered with tight clear wrapping bearing a printed blue ribbon, a yolk-yellow label, and a crowing white rooster. He took a picture and texted it. The next day, a worker from the health department showed up to collect the chicken and take it to the lab. She gave him a receipt.

—∿—

WEEK BY WEEK and then month by month, the Foster Farms outbreak expanded. Three days after its first announcement, the CDC added 39 sick people to the tally, in three new states and Puerto Rico. A week later, there were 21 more victims. By the end of the month, 362 people were sick in 21 states. In 14 percent of them, the bacteria had surged out from their guts into their blood, causing the overwhelming infection and inflammation that had felled Schiller. That was almost three times the rate of bloodstream infections that was expected in a salmonella outbreak

By the middle of November, the count of sick people was 389 in 23 states. By Christmas, it was 416, and half of the bacteria the CDC had been able to analyze were drug resistant. By the middle of January 2014, the toll was 430—but no new cases had been reported for a while, and officials began to hope the outbreak might be over. They were wrong. By the end of February, there were another 51 cases; by the end of March, an additional 43. At the end of June, a year after PulseNet first rang the alarm, there were 621 sick people in 29 states and Puerto Rico. More than a third had been sick enough to be hospitalized.

To end an outbreak caused by food, it is necessary to do two things: stop people from eating the contaminated food already in their possession, and keep them from buying more. Federal and state agencies had done what they could to alert consumers that they might have contaminated chicken in their home kitchens: They held press conferences and published bulletins that urged people to check their refrigerators and freezers. But none of the agencies could prevent more chicken from being shipped to supermarkets. Their hands were tied by the laws governing food safety.

The Jack in the Box epidemic had sparked a nationwide activist movement, led by grieving mothers outraged that a food as iconic as a hamburger could have made their kids so sick. They besieged Washington, demanding reform, and won a significant change. From the time that Upton Sinclair exposed the disgusting details of turn-of-the-century meatpacking in *The Jungle* in 1906, food safety regulation had focused on preventing things from being carelessly added to meat: broken glass, metal shards, chemicals. By law, those were "adulterants," illegal in any amount; finding them gave the USDA the power to force a company to remove a batch of meat from the marketplace. But until the Jack in the Box outbreak, no one had ever thought a microbe could be an adulterant, in the sense of something that should not be there, and also in the sense of something so lethal that it required immediate action. In 1994, Michael Taylor, the administrator of the USDA's Food Safety and Inspection Service, announced that *E. coli* O157:H7 would become the only foodborne pathogen with that zero-tolerance designation. Later the agency would give the same designation to six other strains of *E. coli* that made cell-destroying toxins in the way O157 did.

But the USDA refused to recognize that any other foodborne organism could be deadly enough to be an adulterant—so there was no federal power to force meat to be recalled because it was spreading *Salmonella*. It was up to Foster Farms to volunteer, and it did not elect to do that for more than a year.

The inaction was galling to advocates who knew that the company had had outbreaks before and that regulators were aware of them. In June 2012, cases of *Salmonella,* bearing a PFGE pattern that would resurface a year later as one of the seven outbreak strains in California, had begun appearing in Washington and Oregon. By the time the outbreak ended in July 2013, there were 134 victims in 13 states. Like the California-centered outbreak that would suck in Schiller, there was multidrug resistance—to gentamicin, streptomycin, and sulfa drugs; and to amoxicillin, ampicillin, and ceftriaxone, also known as the human drug Rocephin, a cephalosporin that is the crucial treatment for children who develop serious *Salmonella* infections. Most of the victims in this earlier outbreak lived in the Pacific Northwest, and the cause was finally traced in the end to a Foster Farms slaughter plant in Kelso, Washington.

And, in fact, there had been even earlier outbreaks than that. In 2009, there had been 22 sick people in Oregon and two matching strains of *Salmonella* in 12 samples of chicken. In 2004, there had been 22 cases of illness, noticed first in Oregon and then linked by PulseNet comparisons to cases in Washington, California, Ohio, Hawaii, and Kansas; and six discoveries of matching *Salmonella* strains in Foster Farms chicken. In that first outbreak, the states complained to the USDA. The investigators sent to a Foster Farms plant found six *Salmonella* strains on the equipment, all matching the strains in the patients and the chicken meat. None of those outbreaks led the company to announce a recall.

For most of the outbreak that made Schiller sick, there was no recall either. The investigators at the federal agencies and in the various states where people were sick faced a classic public health problem. They had plenty of evidence that illuminated the epidemiology of the outbreak: the same *Salmonella* strains in processing plants, in victims, and in meat analyzed in government labs. But nothing connected those pieces of evidence in a way that proved the plant had caused the person's illness and that the meat had been the vehicle between them. The agencies needed an unbroken chain of evidence: a package bearing intact labeling indicating where it had been processed and sold, meat inside the package containing the outbreak strain, a person who had eaten from that same package, and an infection in that person caused by the same strain. It was a high bar to meet.

They met it in July 2014. Investigators found a 10-year-old girl in California whose parents had bought Foster Farms chicken on March 16. She ate it April 29 and fell ill on May 5. When the USDA learned of her illness on June 23 and visited her house, the rest of the package of boneless chicken breasts was still in the freezer, and the labels that were still attached identified both the supermarket where it was purchased and the plant where the chicken inside had been processed. With that proof, Foster Farms bowed to the need to declare a recall. But it recalled only the rest of the batch of chicken that the culprit package had come from: chickens slaughtered at three plants in California between March 7 and March 13, 2014. That represented less than a week's worth of production, in an outbreak that had been spreading for 16 months.

Even so, the recall demonstrated how far meat can travel from the place where it was slaughtered and packaged. In a bulletin, the USDA advised: "These products were shipped to Costco, FoodMaxx, Kroger, Safeway and other retail stores and distribution centers in Alaska, Arizona, California, Hawaii, Idaho, Kansas, Nevada, Oklahoma, Oregon, Utah and Washington." The single-spaced list of recalled chicken products—drums, wings, thighs, leg quarters, breast tenders, "seasoned splits," "double-bag fryers," and more—was four pages long.

—⟋⟋⟍—

IF FOSTER FARMS seemed recalcitrant—and food safety advocates, trying to protect more people from getting sick, certainly thought it was—the company thought its slow response was justified. Like every other major meat processor in the United States, the company has its own internal food safety program, which tests chickens emerging from its slaughter and processing lines. Because *Salmonella* is not a zero-tolerance organism, USDA standards allow the bacterium to be present in low numbers, and the company touted that its rates were actually lower than the federal standard, down in some instances to zero.

But if the tests were accurate—and there was no suggestion in any investigation that the company had cheated—how did *Salmonella* strains appear inside packages of meat? At the CDC—which is free of the USDA's built-in conflict of policing meat producers while supporting their production—researchers thought they knew the answer. The *Salmonella* had been there all along, but it was masked by a testing regimen that no longer accounted for American preferences in poultry or the way plants process it to meet that demand.

Dr. Robert Tauxe, the head of the CDC's foodborne disease efforts, explained it this way. The steps of chicken slaughter are killing, by shocking the birds into unconsciousness and slitting blood vessels in their necks; scalding, dipping intact dead birds into hot water to loosen the feathers; defeathering, to expose their bare skin; gutting, to remove their entrails plus heads and feet; and finally chilling, to get the carcasses quickly from the temperature of a living body down to 40 degrees Fahrenheit, where bacteria are less likely to grow. Tauxe had persuaded processors to let him watch the process, and it was the defeathering— whirling the just-killed, just-scalded birds in giant drums lined with flexible rubber fingers—that worried him the most. Chickens entered the slaughter line mucky with litter from their barns and with poop from the thousands of other birds packed around them as they traveled to the slaughterhouse.

"All those rubber fingers are covered in manure," Tauxe said. "They're massaging the skin to get the feathers out of the follicles they grow in, and whatever's on those fingers slams into that pore. And then they hit them with cold water or cold air, to chill them, and the pores close right up."

The gutted carcasses were tested for *Salmonella*—and cleared— after the chilling bath cooled them down. At that point, Tauxe theorized, the follicles in the skin were clenched shut from the shock of cold. But if chicken passed through more steps before it was packaged and sent out, it would inevitably warm up slightly, and that would allow the follicles to relax enough for bacteria to leak out again. And almost all chicken processed in the United States does move through more steps, because Americans no longer buy much whole chicken: four-fifths of the raw chicken sold in the United States is bought already cut up into parts.

Tauxe is so well respected that the USDA agreed to test his idea when they went into the Foster Farms plants that had been linked to the outbreak. In addition to the whole carcasses that the company always tested, they also collected chicken parts after the birds had been portioned farther down the cutting line. Very few of the whole birds carried *Salmonella*—but 25 percent of the parts and 50 percent of ground chicken did.

Where was it coming from? Many things can potentially pass *Salmonella* to chickens, and almost all of those things occur on a farm: contaminated feed, rodents getting into the chicken house, wild birds slipping through a gap in a door. But Foster Farms had experienced *S.* Heidelberg in four processing plants in two states, and all of those plants were fed by different farms. It made more sense that *Salmonella* was entering the supply chain of chicken upstream of all of those locations—possibly before the chickens that were becoming meat had even been hatched.

One little-known, far-away episode held a possible explanation. In 2011, Denmark's agricultural watchdogs noticed that *E. coli* resistant to cephalosporin antibiotics was rapidly increasing in Danish broilers—but Danish chicken producers had not used those drugs for at least a decade and used very few antibiotics overall. To find the source, Yvonne Agersø, a microbiologist at the Technical University of Denmark, sleuthed through the birth records of generations of chickens. Most of the broilers' parents had been imported from Sweden, but Sweden did not allow cephalosporins in chickens either, and so resistance could not have arisen there. However, the parents of those parent birds—the grandparents and great-grandparents that made up the breeding company's production pyramid—had been hatched in Scotland, and Scotland allowed cephalosporin use. Cephalosporin resistance had passed in *E. coli* from one generation to another, lingering even without drugs to sustain it, contaminating flocks that everyone assumed were antibiotic free. It was possible that *Salmonella* behaved in the same way; the organism can be transmitted through generations, from a hen to her eggs as they form in her oviduct. It would have taken only a few grandparent or great-grandparent hens to seed a hatchery's entire production pyramid with drug resistance.

—〜〜—

THE DEMONSTRATION THAT *Salmonella* could lie concealed in chicken parts jolted Foster Farms into action. The company embarked on an aggressive overhaul, investing $75 million in reforming its barn operations, processing, and microbiological testing. By the end of 2014, it was able to say that it had forced the occurrence of *Salmonella* in packages of parts down to 5 percent of the ones it sampled. In the wake of the outbreak, the USDA had set new rules, requiring that *Salmonella* could be present on no more than 15 percent of chicken parts. Foster Farms' success at reducing *Salmonella* below what the USDA required moved it from being the source of an outbreak to being a model for the rest of the industry.

But since the USDA had not set the acceptable level at zero—because *Salmonella* was still not a zero-tolerance organism in the way the toxin-producing *E. coli* strains are—it also meant that regulators were continuing to leave consumers vulnerable to *Salmonella* infection.

By August 2014, there had been no new cases bearing the outbreak strains for a month. The federal agencies considered the outbreak to be over. The victims did not have that luxury. Schiller's aftermath stretched for months. "Before the salmonella, I'd never been in the hospital, except once for stitches when I cut myself on something," he told me in February 2016. "But since then, I've been in nonstop." In his recovery, he had become part of a second epidemic, of people who suffer the aftereffects of foodborne illness throughout their lives.

This is a shadow epidemic, difficult to detect in the United States. The surveillance systems that uncover foodborne outbreaks here aim to find victims when they are acutely ill. Discovering whether people have lingering symptoms afterward requires maintaining contact for months or years, and neither federal surveillance nor health care records are set up to encourage that. But in other countries, ones with single-payer health systems, patients' records flow into centralized repositories, which make it possible to discern years after the fact that people who were part of the same outbreak in the past are experiencing similar problems now.

Thus, in 2008, researchers in Sweden reported that people who had survived infections with *Salmonella, Campylobacter,* or a third foodborne organism called *Yersinia* had higher-than-normal rates of aortic aneurysms, ulcerative colitis, or reactive arthritis (a type that develops after infections, not from aging or overuse). In 2010, scientists in western Australia discovered that children and teens who contracted foodborne illness were 50 percent more likely to be hospitalized for ulcerative colitis or Crohn's disease than people born in the same place and time who had not suffered the initial infections. And five years after a 2005 outbreak of salmonella in Spain, two-thirds of the victims developed muscle and joint problems, nearly three times more frequently than people who lived in the same area but had not been affected by the outbreak.

Those studies were done retrospectively, which makes it possible to question their findings; perhaps something else happened to the victims between when their foodborne illness happened and when researchers found them. But another study, done in Canada, established the connection in a way that could not be challenged: by identifying victims while they were enduring an outbreak and monitoring them as years passed.

This research had its roots in a small town's agony. In May 2000, drinking water in the farming community of Walkerton, Ontario, became contaminated with *Campylobacter* and also *E. coli* O157:H7, the toxin-forming strain that caused the Jack in the Box outbreak, after heavy rains washed manure from cow pastures into its aquifer. Almost half the town's population—more than 2,300 people—fell ill with fever and diarrhea. In 2002, the Ontario provincial government asked researchers to track the victims' experience afterward, and in 2010, they reported what they had found. People who developed acute *Campylobacter* or *E. coli* infections during the outbreak were 33 percent more likely to have high blood pressure, 210 percent more likely to have heart attacks or strokes, and 340 percent more likely to have kidney damage that might lead in time to dialysis or a kidney transplant.

Schiller's experience mirrored theirs. He was left with arthritis in his right knee, fatigue, and damage to his digestive system. Three months after his *Salmonella* infection, he had an episode of diverticulitis, in which weak spots in the wall of the large intestine collect waste and become inflamed; in December 2014, more than a year after the attack, he underwent surgery to excise a section of his bowel. In the year following that, he had five excruciating obstructions in which scar tissue snarled his intestines together. "I go to the emergency room, they admit me, and I'm there three or five days," he said.

In January 2016, Schiller suddenly developed abdominal pain, and when he went to the hospital again, doctors discovered that a weak spot on his colon had ruptured and was spilling fecal material into his abdomen. Emergency surgery followed.

The first time we talked, he had 21 staples in his abdomen holding his surgical wound together, and a colostomy bag that channeled his waste outside his body had been installed in his abdominal wall.

After he recovered from the acute attack, Schiller found a lawyer: Bill Marler, probably the best-known attorney representing foodborne illness victims in the United States, who began his career representing the children sickened and killed by Jack in the Box. Marler won him a settlement from Foster Farms, not huge, but enough to pay for his medical care and his time away from work. Still, the money could not change how the infection had altered his life. "I've had nothing but problems since that *Salmonella* thing," Schiller told me. "I've been struggling pretty hard."

CHAPTER 9

THE UNPREDICTED DANGER

THE FOSTER FARMS OUTBREAK that changed Schiller's life racked up 634 known victims, by the CDC's count, and the earlier outbreaks linked to two of the company's slaughterhouses accounted for 182 more. There were probably more victims than that, because in foodborne outbreaks, there almost always are: People fall ill but do not see a doctor, or they get care but never have a sample analyzed. The known numbers alone made the Foster Farms outbreak one of the largest recorded of resistant foodborne illness. But it is dwarfed by a larger, longer-lasting epidemic, passed by food and created by farm antibiotic use, that has been moving unnoticed across the world.

The first signals of that epidemic emerged in 1999, in one of the most common problems in medicine: urinary tract infections. UTIs, as they are known, are a problem that almost no one takes seriously, except for the people who experience them. Women get more UTIs than men, and young women more than older—so in the implicit hierarchy of medical problems, UTIs are doubly disadvantaged, for being a women's health issue and for arising from the impression that young women are having too much sex. (UTIs are so strongly linked to heterosexual intercourse that they used to be called "honeymoon cystitis," from the idea that couples who were virgins when they married would overdo it on their wedding night.) UTIs are overwhelmingly caused by *E. coli*, and medicine historically interpreted them as accidents, caused by bacteria from the gut slipping out of a woman's colon and being wiped or pushed the short distance to her urethra. The assumption within that understanding was that each woman who developed a UTI was the source of her own infection, and every woman's infection was unrelated to any other.

At the University of California, Berkeley, Amee Manges, an epidemiologist working on her Ph.D., was interested in whether those longstanding assumptions were correct. She was studying UTI bacteria to see whether sex partners might transmit them to each other, and she had recruited a group of female students and their boyfriends to give her regular stool samples to illuminate the bacterial traffic between them. A university campus full of young people is a great setting to study UTIs, because there are likely to be many. But at the time Manges was working, something odd was happening on the sprawling, historic campus across the bay from San Francisco. Women who had been diagnosed and given prescriptions at the university's health service and treated in the usual manner were coming back complaining that their UTIs had recurred. *E. coli* cultured from the women's urine were resistant to a combination antibiotic called trimethoprim-sulfamethoxazole, commonly known as Bactrim.

This was a problem, because Bactrim was the first drug on the list of preapproved antibiotics recommended for UTIs. The women's infections had not recurred. They had never gone away because the drug given for them had not worked. It was odd to see so many UTIs in one clinic reacting to one drug in the same manner, when UTIs were supposed to be unique and random.

Some of the women complaining of the recurrent UTIs were part of Manges's study. The next time she went to the health service, which was holding their stool samples for her, she scooped up their urine samples as well. She isolated the *E. coli* in the women's samples and did the prep work to analyze them by PFGE, the DNA-fingerprinting technology that the CDC would later use to untangle the Foster Farms epidemic. When the assay was done, she clicked off the timer, pulled out the gel sheet, and walked it over to the ultraviolet viewer that would allow her to see what the fingerprints looked like. If the women's infections had anything in common, the patterns in the gel would resemble each other.

Manges clicked the viewer on, focused on the sheet, and blinked: Half of the isolates possessed an identical pattern. The UTIs at Berkeley were not random. They were an outbreak. And she would need to find out where they were coming from.

—⚎—

AN OUTBREAK OF UTIs—many similar cases, occurring at the same time, caused by the same organism with the same pattern of antibiotic resistance—was almost unheard of. As Manges would discover, this had been documented just once before, in England, and there too, it had seemed to arise out of nowhere.

Thirteen years earlier, in December 1986, a physician and a microbiologist who were colleagues at St. Thomas' Hospital in London had written a short letter to the medical journal the *Lancet.* They said that in just the past month, their hospital had treated 60 patients infected with a seldom-seen type of *E. coli.* The patients were all from the neighborhood, the borough of Lambeth; of all ages, from one year old to 97; and all very sick with bloodstream infections. Normally the kidneys serve as filters, taking metabolic waste products out of the blood and sending them downstream to the bladder. But because so much blood passes through them, the kidneys also can serve as an accidental back door that allows bacteria to flow upstream into the blood. That had happened to the St. Thomas's patients, and as the *E. coli* spread through their bodies, it also caused pneumonia, meningitis, and infections of the valves of the heart. Lab analysis of the *E. coli* showed that it did not respond to six different families of antibiotics, including the two types that make up Bactrim.

The letter was meant to sound the alarm, and also to solicit information about any similar outbreaks. Ten weeks later—quite fast, in the era before online publishing—the *Lancet* printed a reply from microbiologists at Queen Mary's Hospital in Roehampton, several miles west. In the past few months, they had seen eight instances of infections caused by the same strain as the ones in Lambeth: starting as a UTI, not responding to antibiotics, and then spreading upward. Seven of the eight cases were elderly women; the eighth was a 16-year-old who had just given birth. Two of the eight died.

More than a year later, the researchers who wrote the first letter reported back. The outbreak that had originally felled 60 patients had roared through their hospital and was still going. It had sickened 385 patients with urinary tract and kidney infections; 34 with bloodstream infections; and 19 who had infections in which the resistant *E. coli* strain had landed unpredictably in their lungs, ears, and eyes. With so many people ill, they had an abundance of samples to analyze, and they had identified the source of the problem. It was a plasmid carried by the *E. coli,* which had accumulated genes for an array of resistance—up to 10 different antibiotics in some cases. But they had something else to report as well: Despite appearances, the hundreds of cases at St. Thomas were not a hospital outbreak—that is, the patients had not developed their infections after being admitted. In all but a few instances, they had been carrying the highly resistant *E. coli* when they arrived. Whatever was making them sick was present in the everyday life of two London neighborhoods.

All outbreaks, even ones that feel random when they are happening, follow some kind of pattern. Epidemiologists draw it as a curve, a plot of the onset of cases against the dates when they occurred. An epidemic curve tells an outbreak's history, but it also can suggest what the cause might be, even if the organism responsible has not been identified yet. The curve of a communicable disease such as flu, in which each victim incubates an illness and then passes it to others, is a long, shallow climb. The curve of an outbreak in which many people are exposed to the same thing at the same time is steep.

The curve of the Lambeth outbreak was steep: The *E. coli* infections had bloomed suddenly, built up quickly, and then trailed off. Manges recognized the pattern right away when she read the *Lancet* articles years later. It matched the curve of the Berkeley cases and reinforced that what was happening at Berkeley was an outbreak, even though that went against the medical understanding of UTIs.

And it suggested where she should look for a cause, because it was also the shape taken by outbreaks caused by food.

—ᴍ—

THERE WAS NOTHING NEW, of course, in the idea that bacteria carried on food could cause many illnesses at once—but UTIs had never been counted among those illnesses. The *E. coli* that cause urinary tract infections had always been assumed to be permanent benign colonizers of the gut, unlike the diarrhea-producing or toxin-forming types that trigger food poisoning. At about the time that Manges started her studies, though, researchers began proposing that the types of *E. coli* causing UTIs were an entire third category of that organism. Microbiologists dubbed them "ExPEC," for "extraintestinal pathogenic *E. coli.*"

The UTI-related strains possessed additional genes that let them attach to tissues outside the gut and protect themselves from the immune system's attack. They were capable of migrating to the abdominal cavity; the lining of bones; and the lungs, brain, and spine. Previously, anyone looking at infections in those body sites would have considered them unrelated, because they occurred in different organs. But recognizing that the illnesses all stemmed from this one grouping of *E. coli*, researchers began to reframe all the infections caused by the ExPEC strains as a single phenomenon—an imaginative shift that recast them as less like foodborne illness and more like tuberculosis, which is understood to be the same disease whether it develops in the lungs or bones or brain.

That redefinition allowed researchers to recognize that the ExPEC problem was enormous. Two scientists in particular, Dr. James R. Johnson of the University of Minnesota and Dr. Thomas

A. Russo of the State University of New York at Buffalo, mined existing public health data to try to define it. They calculated in 2003 that there were six million to eight million UTIs in the United States every year, and ExPECs caused almost all of them. There were 250,000 cases of kidney infection, 100,000 of which sent people to the hospital; ExPECs caused 90 percent of those. ExPECs were linked to thousands of cases of diverticulitis, appendicitis, peritonitis, pelvic inflammatory disease, gall bladder inflammation, and pneumonia, and more than a hundred cases of meningitis in infants. Among cases of sepsis, one of the leading causes of death in the United States, ExPECs might be causing 40,000 deaths every year.

The staggering toll trailed a staggering bill. The researchers calculated that those sepsis deaths alone might cost the United States almost $3 billion in health care spending; surgical infections, $252 million; pneumonia cases, $133 million. Even the lowly UTIs, they estimated, required so many doctor's office visits—and repeat trips for prescriptions and lost work time waiting for symptoms to subside—that they were probably costing the United States $1 billion each year.

Compiling those stunning numbers prompted the question: Why had no one recognized earlier that ExPEC was such a threat? It was the increase in antibiotic resistance that made the problem visible, thanks to Manges, who drew the links between illnesses that would otherwise have seemed unrelated. When she compared the urine samples from Berkeley students with ones given by women students with UTIs at the health services at the University of Michigan and University of Minnesota, she again found high rates of resistant infections. At both schools, 40 percent of the resistant infections were caused by the same strain of *E. coli* that had caused the Berkeley UTIs. The campuses were not unique in that: Resistance in UTIs had been rising across the United States. During the 1990s, the proportion of infections resistant to Bactrim doubled from 9 to 18 percent, and began to rise also for antibiotics that were second choices when Bactrim did not work. But medicine was slow to recognize the trend because the rising rates were masked by the way UTIs are diagnosed.

Medical encounters for possible UTIs usually worked like this: A patient, almost always a woman, arrived complaining of burning when peeing or pain over her kidneys. She gave a urine sample, and the office performed a rapid test to confirm infection, then wrote a prescription using a list of antibiotics recommended in advance by one of several medical societies. The rapid tests were almost always dipsticks that are dunked in a urine sample; they are reliable, inexpensive, and, most of all, quick, taking minutes instead of the 24 hours required to grow a bacterial culture. Achieving a quick test result meant a woman could begin taking antibiotics more quickly, but that innovation came at a cost: If there were no grown-out bacteria, there was nothing to test for drug resistance. Thus, when a woman's symptoms recurred and she returned to her doctor, it was assumed that she had gotten infected a second time. There was no earlier culture that could demonstrate that the infection from the first visit had never gone away.

So many women went back to gynecologists complaining of recurrent UTI symptoms that primary care physicians began talking about it at annual medical conventions. Hospital specialists who treated kidney and bloodstream infections began comparing notes on how many were occurring in healthy young women. The main professional organization for physicians specializing in infections struggled with how to respond. It told its members in 2011, 12 years after Manges's first study, to take stock of local resistance data—though in many places, such data had never been gathered. During that span of time, though, some other researchers took seriously her intuition regarding the source of the problem and began to investigate whether ExPECs were carried on food.

—ɱ—

JIM JOHNSON, the Minnesota professor who first predicted how many illnesses ExPEC strains would cause and how much they would cost, had an unusual perspective on the problem. He treats infectious diseases at Minneapolis's VA hospital, where most of the patients are men and most are also much older than the students Manges had studied. His patients too had UTIs, though for different reasons than women did: from an enlarged prostate that kept a man's bladder from emptying or from an immune system faltering from age. They also had ExPEC infections elsewhere in their bodies, proof that the resistant *E. coli* problem was more widespread and more complex than illnesses only in young women. Johnson had a long research interest in *E. coli;* he had helped Manges and her Berkeley adviser, Lee W. Riley, analyze the first resistant strains she had found, and he shared their sense that the problem resembled a foodborne outbreak.

But almost no data existed to back up that intuition. Though the multiagency federal NARMS project had been established three years before the Berkeley outbreak began, its nationwide searches looked for only one *E. coli* strain, the toxin-making O157:H7 variety, and not the newly recognized ExPECs. If Johnson wanted to define their incidence, he would have to gather his own data. Fortunately, a model existed to illustrate how to do that: the investigation that Smith and Osterholm had just carried out at the Minnesota health department, not far from Johnson's office.

Since Cipro was a crucial alternative drug in UTI treatment, Johnson decided to research whether the fluoroquinolone resistance that Smith and Osterholm had documented in *Campylobacter* on chicken was also appearing in *E. coli*. He compiled a list of grocery stores around Minneapolis: 24 stores, independents and members of national and local chains, in 17 cities and towns. Every two weeks between April and December 2000, he and colleagues from the Minnesota Department of Agriculture randomly selected stores and bought 10 packages of precut chicken from whichever brands were sold at those locations. They did that for seven months, accumulating 169 packages to analyze. When they found *E. coli,* they analyzed it for quinolone resistance and then sorted the resistant ones to determine their type: benign commensals; diarrhea-producing foodborne illness strains; or ExPECs, with their unique genes for attachment and virulence.

They found even more resistant ExPECs than they had anticipated: two-fifths of all the *E. coli* strains they found in the chicken packages were resistant to quinolones, and one out of every five resistant strains was an ExPEC. In other words, almost 10 percent of grocery store chicken was carrying *E. coli* that could cause these newly recognized severe infections, and was resistant to one of the main drugs used to cure it.

The result showed that Manges and Johnson were on the right track: In these chicken samples, resistant ExPECs were a foodborne organism. But there would have to be much more work to confirm the link between food and ExPEC infection. In 2005, Johnson and his collaborators reported that they analyzed 346 samples of foods that they bought around the Twin Cities during the course of a year—beef, pork, chicken, turkey, and produce—to see which was most likely to carry ExPECs and whether the strains were resistant. They found 12 resistant strains, all on turkey, possessing defenses against 10 different human-use antibiotics. Four of the 10 ExPECs that they found on the turkey samples matched *E. coli* strains that were already known to infect humans.

In a second study published that same year, they scrutinized 1,648 food samples, including 195 that were chicken and turkey. Simply assessing which food type carried *E. coli,* one of the first steps in their analysis, began defining the problem: Only 9 percent of fruits, vegetables, and seafood carried the organism, but 69 percent of beef and pork did and 92 percent of the chicken and turkey. As a second step, they asked which of the *E. coli* strains they had found were resistant. Those were most likely to come from the poultry samples; some of them were multidrug resistant to as many as five separate classes of antibiotics. In their third step, they examined the resistant *E. coli* they had identified, looking again for the molecular evidence of ExPEC. Compared to other meats, poultry was twice as likely to carry ExPEC strains.

To determine whether poultry was transmitting those strains to eaters, the team recruited the cooperation of four hospitals in towns near the Twin Cities. They asked 622 patients in the hospitals to donate samples of their feces immediately after being admitted, because what was in their intestines at that point would have been carried in from the outside world. At the same time, they bought even more meat: 40 packages of chicken in each of the four towns over the course of a year, as well as 40 chickens bought directly from farmers who raised their birds without antibiotics. They identified *E. coli* strains on the chicken meat, sorted out which of them were resistant, and separated out the ExPECs from other resistant strains. They retrieved resistant *E. coli* from the patients' fecal samples as well. Then, using molecular methods that were more precise than what Manges had had access to years earlier, they explored whether the resistant and drug-susceptible *E. coli* from the patients, and the resistant and drug-susceptible *E. coli* from the chickens, were alike in any way.

The resistant and sensitive strains in the humans' guts were not related; the resistant human strains matched the strains from the chicken meat. Manges's intuition years before had been correct. The vast epidemic of resistant ExPEC was originating in food.

Johnson's hypothesis that poultry was the source of human *E. coli* infections caught the attention of other researchers, and corroborative findings poured in from other parts of the globe. In 2006, a research team in Jamaica uncovered similar patterns of resistance in *E. coli* in chickens and in local people. The same year, researchers in Spain found resistant *E. coli* in hospital patients that closely matched chicken strains. Between 2009 and 2014, so did researchers in Idaho, Italy, Germany, the Czech Republic, Finland, and several provinces in Canada.

In 2010, a large study in Denmark demonstrated close matches between multidrug-resistant *E. coli* causing UTIs in people, and multidrug-resistant strains carried on chicken meat and also by chickens about to be slaughtered. Because human antibiotic use is so low in Denmark, the researchers were also able to show that medical prescriptions were not to blame for the resistant strains in people. In 2011, yet another study drew on data from 11 countries in Europe to show that in *E. coli* recovered from humans and from poultry, the patterns of resistance to four different antibiotic families were essentially the same.

After finishing her doctorate at Berkeley, Manges returned to Canada, becoming faculty first at McGill University and then at the University of British Columbia. Working with other researchers at those schools and from Canadian federal agencies, she showed that women who ate chicken frequently were most likely to contract an antibiotic-resistant UTI, that *E. coli* strains retrieved from humans and from retail chicken were closely related genetically, and that poultry meat in Canada carried multidrug-resistant ExPEC strains.

In the nearly two decades since Manges first began checking on the Berkeley students' infections, there were almost 50 separate studies reporting strong similarities between chicken meat and the resistant *E. coli* it carried, and resistant ExPEC infections caused by the same or similar bugs. Scientists usually choose conservative language but those studies used words such as *indistinguishable* and *identical* to describe the human ExPEC epidemic and its apparent poultry cause.

At the same time, the ExPEC problem demonstrated how difficult it can be to move from solid scientific evidence to policy change. It was not possible to trace any of the infections to one poultry producer, as federal investigators had done for Schiller. By the time a woman experienced an ExPEC UTI or a more serious complication, the meat and the package that carried it were long gone. No one could say how long: It might be weeks or months, and in that time, depending on where she dined or how she shopped, a woman might eat meat processed by dozens of companies. Any one serving might have been the culprit, or several, or many.

The cause of an infection, and the resistance it carried, could not be traced to any single farmer or meat production company—

and therefore, no producer or processor was likely to volunteer to relinquish the antibiotics that made resistant ExPECs so dire. The thousands of illnesses, and the millions of dollars of health care spending and lost productivity they caused, were a problem that no one was willing to own—or to take seriously, because there was always another antibiotic to use. And then, in 2015, it became clear that perhaps there was not.

—⟋⟍—

IN THE SUMMER OF 2015, Timothy Walsh, a microbiologist from Cardiff University in Wales, was finishing a research visit to China. Walsh is a specialist in the genetics of resistance, and he had been helping Chinese scientists explore how resistant bacteria were emerging in the country that has become the largest producer, and consumer, of antibiotics on the planet.

On the way to the airport, one of his colleagues shifted uncomfortably in his seat. "Tim," he said, "I need to tell you something."

Walsh, mentally already halfway aboard his flight, looked up. His friend looked down.

"We found something," his friend said. "We found colistin resistance, and it is transmissible."

Walsh felt his jaw drop. To anyone outside the circle of researchers who study resistant bacteria, the statement would have been opaque. But to those who understood, it was portentous. Colistin was not just an antibiotic; it is the last antibiotic, the one that still works while others have fallen to resistance. Few were better equipped to understand that than Walsh, who had inadvertently become globally famous while battling for recognition of the seriousness of new resistant bacteria. That battle had brought him to China, and to the car ride where he learned an antibiotic apocalypse might be close.

Walsh, an Australian now in his 50s, is square-jawed, irreverent, and intense. He grew up on the island of Tasmania, relocated to Great Britain to do his doctorate, and stayed there. By 2006, he was a full professor and department chair at Cardiff, and a leader in an informal but tight network of researchers concerned that resistance was becoming a global peril. In the middle of that year, Walsh got an email from another researcher in the network, who was taking care of a 59-year-old man of South Asian heritage who lived about 100 miles from Stockholm. The patient was not in good shape: He had lived with diabetes for many years and had suffered several strokes, and now was in the hospital with deep bedsores. The other researcher, Christian Giske of the Karolinska Institute, was less worried about the lesions than about a strain of bacteria he had found in the man's urine. It was *Klebsiella pneumonia,* a common hospital pathogen, but it seemed unusually resistant, and Giske wanted Walsh to unpack it.

Walsh's analysis revealed the *Klebsiella* did not respond to any of the antibiotics that would have been first choices to treat it, and also harbored a never-before-seen gene that conferred resistance against a last-resort category called carbapenems. Carbapenems are used only in medicine, not in agriculture. They have become a mainstay of medicine because they still worked when other older drugs did not; they were precious. This new gene undermining that effectiveness was alarming, the more so because it appeared to have traveled around the world. The man had probably picked up the *Klebsiella* that harbored it during a recent hospitalization in India on a visit home.

Adhering to microbiological conventions, Walsh named the new gene for its place of origin: NDM-1, for New Delhi metallo-beta-lactamase. He and Giske and others announced the discovery in December 2009. And then all hell broke loose. NDM resistance was uncovered in patients in the United Kingdom, and then in the United States and throughout Europe. The discovery became entangled in the economics of medical tourism and the politics of national pride; Walsh was excoriated in the Indian press and denounced in its parliament. When he tried to go to India to do more research, pursuing a concern that NDM-bearing bacteria were spreading not just in hospitals but in tap water and puddles, his visa was denied. (He recruited a British TV cameraman, in India on assignment, to smuggle out water samples, and analyzed them and proved his fears were correct.)

By the end of 2012, varieties of the NDM gene had migrated into 42 species of bacteria and had been found almost a thousand times in 55 countries. In the West, it prompted the first government responses to antibiotic resistance since the Swann Report. In March 2013, the chief medical officer of the United Kingdom, Sally Davies, warned that advancing resistance posed "a catastrophic threat" that ought to be considered as serious a national emergency as terrorism. In September 2013, the CDC released its first ever "threat report" on resistant bacteria and called carbapenem-resistant organisms "nightmare" bacteria. "We will soon be in a post-antibiotic era," said Dr. Tom Frieden, CDC's then director. "For some patients and for some microbes, we are already there."

What saved NDM from being an immediate emergency was that bacteria bearing the gene remained vulnerable to one or two antibiotics that had not yet fallen to resistance. Chiefly, they were sensitive to colistin, an old drug so crude in formula and destructive to the kidneys that it had been pushed to the back of the medicine cabinet; no one would have prescribed it as long as more effective drugs were available. The arrival of NDM turned colistin from neglected to valuable. In 2012, the World Health Organization put the drug on its warning list of critically important antibiotics, ones that the world needed protected at all costs.

That warning was already too late. In 2015, researchers who drilled into European Union records discovered that most of the animal species raised for food in EU countries were receiving more than one million pounds of colistin per year, and resistance was already emerging. Their work got almost no attention. Colistin use was legal, despite the bans on growth promoter antibiotics that followed on the Swann Report, because it was being used preventively.

It was being used that way in China as well. And Walsh's Chinese colleagues had just uncovered evidence that colistin resistance in that country was not just emerging; it was an established genetic resident of agriculture, leaking out into the human world. The Chinese researchers told him they had first found the gene two years earlier, in a project they were running to search for antibiotic resistance in farm animals. They had verified it was on a plasmid, then determined the plasmid could move freely between bacterial species; they dubbed the new gene MCR-1, for the first instance of mobile colistin resistance. Then they had searched for MCR in animals, meat, and people, in multiple places across a 1,000-mile swath, and repeatedly found it. Colistin was not part of China's national formulary; that is, it was not used in human patients there. So there had been no signal of the drug's not working for human patients that would have sounded an early warning of the new resistance being present.

Walsh and his Chinese colleagues released their findings in November 2015. The uproar that followed dwarfed the furor over NDM six years earlier, because what was threatened this time was not the almost-last drug, but the very last. New antibiotics that might take colistin's place in medicine were in development at a few pharmaceutical companies, but some were years from reaching the market; others that were closer to being ready to sell had been made in limited quantities, with very high prices to ensure conservative use. There would not be enough of either category to supplant colistin if the last drug lost its effectiveness everywhere.

By 2016, MCR had been found in animals and humans in more than 30 countries. That does not mean that, in all of those countries, bacterial infections had become completely untreatable. In some places, including the United States, bacteria bearing the gene were not causing infections at all; they were accidental bystanders in the colon or bladder, found by chance when a patient was checked for something else. In others—China, as well as Argentina, Denmark, Germany, and the Netherlands—bacteria carrying MCR were causing serious infections. Fortunately, some still responded to other antibiotics. As researchers had established, resistance genes stack up on plasmids by random chance, like gamblers receiving cards in poker. But colistin resistance was the ace, the card that beat all others. If it landed in a cell that possessed the equivalent of two of a kind, that bacterium would still be sensitive to some antibiotics. But if it completed the bacterial equivalent of a royal flush, possessing resistance to colistin and to carbapenems and to cephalosporins and to penicillins, that patient would have few options left. The uncountable trillions of bacteria that coat the world birth new generations every 20 minutes. That is an infinite number of hands of poker and of plasmids transporting genes invisibly, like an ace passed under the table to perfect a poker hand.

The discovery of MCR was a reckoning. NDM had forced medicine to take an uncomfortable look at its own long-standing, substantial misuse of antibiotics. But MCR compelled the recognition that it was not really possible any longer to divide antibiotics arbitrarily into "medical" and "agricultural" and imagine that bacteria would respect the difference. There was only a single bacterial world, the shared province of the microbes that live on and in humans and animals and water and soil; any harm or alteration to any of them would ripple through an entire global ecosystem. It was time to treat antibiotics as agents of unintended consequences that could help, or could harm, the health of the whole planet. In a few places in the world, that had been happening already—and what regulators and farmers did there would provide a model for the United States to study as it began to move toward reform.

PART 3

How Chicken Changed

CHAPTER 10

THE VALUE
OF SMALL

THE ROAST CHICKENS that were heaped on the silvery platter, divided into wings and legs and thighs and quartered breasts, were gorgeous: deep brown, flecked with pepper, crisp skinned and aromatic, and resting in a pool of juice. Bernard Tauzia, who raised the birds on his farm in southwest France, started to hand them to me, and hesitated.

"Do you understand," he asked in French, "that our chickens are different from American chickens? We like our chickens strong-flavored, chewy. We think of yours as . . ." He stopped for a minute, sorting through words in his head. "Soft," he decided. "We think of yours as soft."

It was lunchtime in the town of Campagne in the Landes, a triangular *département* tucked into France's Atlantic coast one province above the Spanish border. We had spent the morning walking the property that Tauzia, a lean and fresh-faced man with iron gray hair, owns and runs with his wife, Marie Odile, and his brother-in-law and son. We had watched as thousands of Tauzia's *poulets jaunes des Landes*—rust-feathered and long-legged, with balanced bodies and upright heads—flew up to branches and ran between trees in the pine forest that covers part of his farm. The birds had more energy and covered more ground than the white-feathered, big-breasted hybrids of most American chicken production, so it was no wonder their flesh was different too: not dry or tough, but bouncy and leaking juices with every chew. It was sweet from the corn they had eaten, grown on another part of the property; herbal from the forest plants, and mineral and faintly metallic from the blood that had coursed through their muscles.

In the United States, a chicken like Tauzia's would be unusual and expensive: difficult to find and, as Frank Reese's experience proved, difficult to buy. In France, it is routine. Tauzia raises his chickens under a government-sponsored program, the *Label Rouge* (Red Label), which verifies that farmers follow a strict set of standards. It permits only slow-growing chicken breeds, requires that birds be allowed outdoors from the start of working hours until twilight, and restricts the size of the houses that chickens are confined in overnight, as well as the number of houses on a farm and how much land is preserved on which the birds can roam. Crucially, Label Rouge poultry standards forbid antibiotics—not only for fear of resistant bacteria, but because the drugs are the precondition for industrial production. The birds raised under the program cost, on average, twice what conventional chickens do— and yet that turns out to be no market barrier: Label Rouge birds account for more than three-fifths of the whole chickens bought in France.

"Eaters want Label Rouge; they know that it guarantees good welfare, and a good taste," Tauzia said. He flicked his eyes down to the table in front of me, and grinned. "When they eat it, they don't leave anything on their plates."

—⟋⟍—

THE LABEL ROUGE PROGRAM was born from the same uneasiness over intensification that in England led Anderson to sleuth out the sources of drug resistance and inspired the antibiotic controls in the Swann Report. In France, that discomfort emerged not because of the peril of resistant infections, but because a way of life was threatened, along with a cherished food. The postwar pressure to grow more protein was acute in Europe, leading to the import of fast-growing Chicken of Tomorrow varieties that gained weight on low-quality protein. Industrial poultry farms began to spread in Brittany on the northwestern coast; but though people welcomed the abundance and low prices, they talked disapprovingly of *poulet au goût de poisson,* chickens whose flesh tasted of the ground-up trash fish they had been fed.

At the other end of the country, the Landes was struggling— not just from the physical and economic devastation of the war, which had killed more than one percent of France's population, but from a disaster that marred the fragile recovery: the largest forest fire in modern Europe, in August 1949. It had burned almost 200 square miles of trees and destroyed the historic industry in the Landes of tapping pine trees for resin, the raw material of turpentine. More than 10,000 tappers lost their livelihoods. The Landes had little to offer to replace those jobs and income, but a few astute farmers recognized an opportunity within the crisis. They proposed shaping a new industry around the province's heritage chicken, the *poulet jaune,* descended from birds brought over the border 12 centuries earlier by the soldiers of Moorish Spain.

Raising the poulet had always been a second crop for local grain farmers; the birds were named *jaune,* yellow, for the golden tinge that corn kernels pecked from the fields imparted to their skin. They were distinctive, with a ring of bare flesh on their necks just below their heads, sturdy from roaming into the edges of the forests, and savory from exercise and age. In 1959, a coalition of local farmers formed the first poultry producers' association in France, the Syndicat de Défense du Poulet Jaune des Landes. In 1965, they asked the government to certify not just the provincial origin of the poulet jaune—something it had already done for another chicken variety, the blue-legged *poulet de Bresse* from near the Alps—but also the uniqueness of the supply chain that produced them: the size of the farms, the historic design of the sheds, the recipe for their feed, and the age at which they were slaughtered.

The result was the first Label Rouge certification, which has since been extended to 29 additional poultry production groups around the country, which all raise slow-growing, outdoor-hardy chicken breeds. Each combination of breed and producer group has been granted an indication *géographique protegée,* a European Union seal that attests they come from a legally defined region and that also implies a guarantee of *terroir,* a flavor specific to that location. Inspections and third-party verifications are built into the process, including audits of taste and texture by experts and consumer tasting panels.

The inspections include checks for food safety, and the results validate the healthy-diet, outdoor-raising process. The rate of

Salmonella in Label Rouge birds is 3 percent. In modern intensively raised French chickens, it is 70 percent. And because the birds receive no routine antibiotics—they get drugs only if a flock is proven to be sick—the *Salmonella* is not drug resistant. There has never been a major outbreak of foodborne illness caused by Label Rouge birds nor any antibiotic-resistant infections.

—⚭—

WHILE WE WALKED HIS FARM, a property of not quite 300 acres, Tauzia explained that avoiding antibiotics for growth promotion and prevention—that is, for any use other than to cure birds that were visibly sick and diagnosed by a veterinarian—was crucial to Label Rouge production. But shunning antibiotics was one of several steps toward a goal of preserving small farms and local farm economies. It was September, and the stalks of the three kinds of corn he grows—feed corn, sweet corn, and another variety that he sells for industrial starch extraction—were rustling and beginning to yellow across from his house. The original syndicate that created the program in 1965 broke and re-formed over decades into a succession of agricultural cooperatives. Tauzia was a vice president of the largest organization to emerge from that evolution, Maïsadour, a portmanteau of the French for "corn" and the name of the river, the Adour, that bisects the Landes. Tauzia descends from farmers, but the earlier generations of his family only rented property. He bought his land in 1996. It was corn, he said, that allowed it to be successful.

"Label Rouge is a concept based on territory," he explained as we walked down a grassy lane to a field where chickens were housed. "Our forebears wanted to protect production that was grounded in our locality, that would bring value to what we do here. It was based on two things: on preserving the slow-growing chickens and on developing the economy of corn, because that is what we grow." Under the rules, 80 percent of the feed the chickens are given must be corn; it is supplemented by the cooperative with legumes and plant proteins, and then supplemented again by the chickens when they eat plants and insects in the fields. Tauzia raises 48,000 chickens and guinea fowl, which is large for a Landes farm; many produce no more than 12,000, half the population of a single American chicken house. He grows all the corn they eat.

The lane ended in a small field bounded by pines that held a row of small sheds: wooden sided and metal roofed and about 30 feet long. The walls ended at the bottom in horizontal hinged panels. Tauzia flipped them up and locked them open; the chickens, red feathered with an undercoat of white, surged out toward the trees. They moved with purpose but without nervousness, an arrow of feathers a thousand birds wide. They would spend the entire day there, until the Tauzias locked them up again at dusk. Under the rules, Landes poultry are raised *en toute liberté*, with no restriction on how far they can roam, but they naturally come back to the sheds when dark falls. Once the flock was taken to slaughter—they would not be ready for at least 81 days, twice the American standard, and depending on market conditions, might go to 92—Tauzia would sanitize the sheds and then drag them with a tractor to another field, to let the vegetation recover.

The sheds are tiny, measuring no more than 650 square feet. (Tauzia called them *cabanes*, huts.) The design is older than anyone remembers. The huts were always built from what was available in the forest, but since the pine trees benefited the pitch tappers more when they were upright and growing than when they were cut down for lumber, the amount of wood used was conservative. Over the years, the inexpensive structures have become less of a necessity enforced by the landscape and more of an advantage: something that makes it possible for young people to enter farming. "With this system of little buildings, you don't have to invest much," Tauzia said. "You have to work a lot, of course. But even if you don't have money, you can make a start."

Maïsadour farmers always have a second crop besides their birds, to diversify their income, but it is not always corn. The second crop for Jean-Marc Durroux and Anne Marie Labarbe, whose property is 30 miles east of Tauzia's farm, is grapes: They are producers of Armagnac, the style of brandy unique to southwest France. (Better-known Cognac comes from farther north on the Atlantic coast.) Durroux buys feed from the co-op, but as a supplement, he lets the chickens in among his vines.

"They adore grapes," he told me in French, as he watched a flock that had just been released from a hut flap uphill into a vineyard where the grapes had just been cut. "We have to keep them out sometimes, because they will eat everything. But after harvest or in the winter, they clean things up for us, and they enrich the soil with their droppings, so that we don't have to add other manure.

"It's fundamental to our culture to have two productions on the farm, a plant and an animal," he said. "It's not first and second, but complementary, and it helps us balance the risks of bad weather or disease."

Durroux is a blocky, cheerful man in his 60s, whose family has lived on the same property since 1921. When I followed him through his morning chores, he walked me into one of the small chicken houses, so I could watch him dispense the day's feed ration through a giant tube threaded off the back of a farm truck. Before we ducked inside, he handed me a surgical mask; the process was noisy and dusty, and it looked repetitive and tiring, far from the automation built into the giant chicken houses on American farms. But it was also an opportunity for him to spend much more time among his birds than an American producer might.

"This kind of production, there are few people who want to do it," he shouted over the rumble and whoosh of the feed going through the hose. "But it is very rewarding. Producing quality is a showcase of our poultry, but also of our self-worth. We were for the ecology long before we were asked to be."

—⁂—

LABEL ROUGE'S REFUSAL from the start to allow farm antibiotic use is notable—not just because it predates the U.K.'s Swann Report, published in 1969, but also because the Label Rouge farmers and their government supporters managed to maintain their program even as the ambitious, high-profile British recommendations failed. The reformers who made up the Swann Commission—and Anderson and the media allies who agitated for its success—expected that the regulations inspired by the report would force changes in livestock raising. The thinking went like this: With growth promoters removed from the market and antibiotics related to human drugs restricted to the control of veterinarians, packing animals into ever more densely crowded farms would have to stop.

That was optimistic. It was also naive. Almost as soon as the regulations went into effect, agriculture found ways to cheat around them. "There is in fact little indication that the overall sale of antibiotics for veterinary use has decreased as a consequence of the Swann Report," A. H. Linton, an influential professor at the University of Bristol, warned in a veterinary journal in 1977, "and it is probable that farm animals are receiving just as much antibiotics at the present time ostensibly for different purposes." Farmers were obeying the letter of the new regulations; they were getting the newly restricted antibiotics through prescriptions written by veterinarians. But so many prescriptions were being written that *more* antibiotics were used on British farms in the wake of the Swann Report than before. Every year after it was published in 1969, farm use of drugs that were important to human medicine rose by 15 percent, though the amounts should have been falling. Two years before the report, 41 tons of antibiotics went into livestock in the United Kingdom; six years afterward, 80 tons did.

With the drugs flowing freely, food-related drug-resistant infections burgeoned. In 1980, the *British Medical Journal* furiously demanded in an editorial: "Why Has Swann Failed?" The journal accused farm media—local small newspapers and trade magazines—of encouraging farmers to break the new rules by buying drugs from a countryside black market that let them avoid veterinarians' control. It was an early sign that simply banning growth promoters would not be enough to solve the problem of drug resistance emerging from agriculture. Within a few years, it was as if the Swann Report had never been published.

It is perplexing that the United Kingdom allowed this to happen, because, in human medicine, it had already seen how quickly antibiotics could lose their power. After all, England was the home of penicillin and then had suffered some of the worst early outbreaks of penicillin-resistant staph. Thus, it became the place where the pharma company Beecham Laboratories developed the semisynthetic substitute methicillin—which it boasted in the *British Quarterly Journal of Medicine* in 1960 was "effective against all staphylococci"—and also the place where the first cases of MRSA occurred just one year after the drug's debut.

The undermining of methicillin illustrated how difficult protecting the few remaining antibiotics would be. To stand in for it, medicine turned to a little-used antibiotic, vancomycin, that had retained its potency because doctors had been unwilling to risk its side effects. Between 1980 and 2000, worldwide use of the drug increased 100-fold. Resistance emerged, of course, especially in enterococci, bacteria that inhabit the bowel but can contaminate hospital surroundings. Vancomycin-resistant enterococci (VRE) arose around the world so very rapidly that researchers wondered whether medical use alone could explain it. Was something else compromising the effectiveness of the precious drug?

Something was, in agriculture. European livestock production was using huge amounts of the antibiotic avoparcin, a chemical relative of vancomycin in a family of drugs known as glycopeptides. Vancomycin was licensed only for use in humans, avoparcin only for animals; yet their molecular similarity was close enough to cause a resistance problem that no one had thought to look for. Which antibiotics were safe to allow and how usage could be policed threatened to become an enormous struggle—except for farmers such as the Label Rouge producers, who had never adopted antibiotic use at all.

—⚭—

IT HAD BEEN EASY, watching Tauzia's birds stream into a stand of young trees, to imagine that Label Rouge poultry is an agrarian fantasy come to life. That ignores the hidden technology that holds up the historic system. The feed belching through the pipe Durroux held had been compounded by a researcher with a doctorate in poultry science to complement how nutrients from forest plants and insects changed with the seasons. The cooperative studied the chickens' behavior when they ate, recording not just their energy and weight gain but whether they appeared to enjoy what they were eating. Each of the production groups under the Label Rouge umbrella bought, built, or allied with local versions of businesses that supported their farmers in getting a product to market: grain buyers, nutrition divisions, slaughter and packing houses. Every Label Rouge chicken for sale in a supermarket bears numerical codes that link it backward through each of those businesses: from the date it was shipped, to the date it was slaughtered, to the slaughterhouse that processed it, and back to the flock it came from and the identity of the farmer who grew it. In a supermarket outside Paris, I flipped open a box of eggs that were individually stamped with the name and address of the farm they came from and the date they were laid and snapped an image of a QR code inside the lid. A web page with the farmer's address and biography and a little video portrait of his farm popped up on my phone.

"Only we do that," Pascal Vaugarny said, looking over my shoulder. "Usually you only know the date of lay with eggs from hens that are caged. Our hens are *en liberté,* but our farmers know when their eggs were laid, because they are never far from them."

Vaugarny works for the Fermiers de Loué, the second cooperative to receive Label Rouge certification after the association that became Maïsadour convinced the government to create it. He had volunteered to be my guide, and he was well suited for it; his father,

Raymond Vaugarny, had been one of Loué's founders in the 1960s. The two groups maintain a mostly friendly rivalry—they jointly own a breeding company, Sasso, that maintains the genetics of the required slow-growth birds—but Loué is larger than its historic rival. One out of every four Label Rouge chickens eaten in France is raised by a Loué farmer, compared with one in ten for Maïsadour, and the methods of production—within the bounds of the Label Rouge rules—are different as well. Loué chickens are a different breed from the yellow-skinned Landes birds; they are *blancs*, white, because their feed is based on wheat instead of corn. (Before I left Maïsadour, Tauzia had said, shrugging: "In the southwest, a white chicken looks to us like an ill chicken. But further north, that is what they prefer.") And instead of roaming in forests, they are let out daily onto large fenced fields and return at night to sheds that are larger than the Landes huts, though still smaller than an American poultry house.

"We call them Loué buildings, because they were created by us," Vaugarny said, whipping his car around a traffic circle. "They are metal, 50 meters long, 9 meters wide, 400 meters squared. They hold 4,400 birds, and you can only have four at a time. If your farm is big enough, you can have four each in two places. But only three flocks per year." I did the math in my notebook, bracing myself against the car door. A Loué barn would hold only one-fifth of the birds in an American barn, and each of the birds would have one-third more space than an American broiler does. What a Loué farm produced in an entire year could be tucked into a corner of an average American poultry property, with thousands of square feet to spare.

Yet Loué was obviously competitive in its market—it is the dominant cooperative among all the Label Rouge producers—and obviously well financed. After the supermarket, Vaugarny had taken me through the clattering packaging plant, as pristine as an operating room, where eggs were washed and boxed in batches so the operators could label them by source. Then we had toured the slaughterhouse, where eviscerated birds tagged with the farms they came from and the feed they had been given—*bio,* organic; *sans OGM,* without GMOs—were perched on racks and shuttled through a series of refrigerated rooms. Air chilling was more labor intensive, Vaugarny indicated, but it maintained flavor better, and reduced the opportunity for foodborne organisms to spread.

Vaugarny was driving me to meet the cooperative's leadership in its headquarters outside the medieval city of Le Mans—the place with the 24-hour car race—130 miles southwest of Paris. The sleek, energy-conserving office building was very different from rustic Maïsadour, stenciled on the outside with sayings: "Good taste shared, good taste assured, good taste rediscovered—we share the same ideas." Indoors, there were posters of winking, ironic ads comparing the Loué chickens favorably against industrial ones: a slender swimmer side-eyeing a bulked-up bodybuilder and a healthy-looking policeman strolling down a street while another dozen cops crammed into a van so tiny it looked like a clown car. (In addition to meaning "chicken," *poulet* is slang for "police.")

The conference room was lined with old black and white photos of Loué's annual farmers' convocations. Vaugarny pointed out his father in one of them, mustached and solemn, wearing wire-rimmed glasses and a '70s checked jacket. The meetings recorded in the yellowing photographs still take place; the more than 1,000 farmers who jointly own and govern the company meet each year to decide on its direction and goals.

"Our ambition is not to grow in order to grow," Alain Allinant, a burly, round-headed chicken farmer who is Loué's president, told me in French when Vaugarny took me to meet him. "We want consumers, when they have a choice between standard poultry and Label Rouge, to choose the label; and between other Label Rouge poultry and Loué, to choose Loué. But we would never put more poultry in the market then the market demanded. If the market wants 30 million chickens, we will supply 30 million—not 31 million."

This was so different from the relentless expansion of conventional poultry, sustained by antibiotics and propelled by rapid growth—of birds, but also of companies—that it was hard to imagine Label Rouge being part of the same industry. The way Allinant explained it, everyone involved with their chicken—the farmers, the cooperatives, the supermarkets, the shoppers—agreed that low prices and high profits were not the highest objective. It was important to keep income flowing, he agreed. But it was equally important to protect smaller farms' independence and to sustain the rural economy that supports farming. The rules that kept farms medium sized ensured no one would be overwhelmed by debt. The limits on flock size kept birds healthy; the insistence on outdoor raising preserved a place in the market for the slow-growing breeds that conventional poultry had abandoned.

The ultimate endorsement was that people kept buying their chickens, even though Label Rouge birds cost more than industrial chickens that are imported from big operations in eastern Europe or Brazil. Though French eating patterns are changing—stores and consumers are asking for poultry parts and boneless cuts, and Loué had just introduced seasoned chickens presealed into a *sac cuisson,* an oven-roaster bag—people remained willing to pay the premium. And farmers kept applying to be part of the cooperative; each year, a commission of Loué's members interviews applicants to be sure their properties, their finances, and, especially, their motivation are a good match.

The ultimate goal, Allinant explained, was balance: among the animals, on the land, in the market. That guaranteed quality, which included the quality of the farmers' lives. "The point is to maintain farms that are the right size for a family," he said. (He used the verb *conserver,* which can mean "to keep," "to defend," or "to treasure.") "A husband, wife, maybe their children, with two, three, maybe four poultry barns. A hundred hectares (250 acres), that's about right. I have seen how it works in the United States, where farmers or their wives have two or even three jobs, but that is very rare for us. This size pays for itself."

The Label Rouge farms had achieved a model of production with minimal antibiotic use that maintained their heritage, honored animal welfare, and fed a market that wanted their products. But while they could boast that they produced the dominant chicken in France, that assertion provoked more questions. Were they successful because France is a small market? Or because French consumers are uniquely interested in the quality of their food, enough to pay a premium for it? A persistent critique of antibiotic-free meat is that even the largest assemblage of small farms will be inadequate to feed the world's population—what is often referred to as the "coming nine billion"—and that only intensive industrial production can meet the world's protein needs. Another model, not far from Label Rouge, demonstrated that livestock can be raised at industrial scale without routine antibiotics to prop them up.

CHAPTER 11

CHOOSING COOPERATION

YOU CAN SEE A LONG WAY in the southeastern Netherlands. The land is as flat as if it had been scraped by a ruler. The farm fields are squared off and geometric, fitted together like tiles and bordered by canals. The trees are sequestered into small copses between the properties; the hedges lining the fields are low. The roads are narrow, and in the villages, they meet in small traffic circles. The landscape feels like a well-tended park, thoughtfully arranged and tightly managed, without wildness save for the geese honking and arrowing overhead.

From a distance, winking in and out of view between houses and hedges, Hoeve de Hulsdonk looks like any farm anywhere: small brick house, big metal barns, churned-up mud where the tractors roll through the gates. It takes getting up close to see that it possesses the same openness as the landscape that surrounds it. A public bike trail runs through the property, and there are public picnic tables between the trail and the main swine barn. Huge horizontal windows built into the barn's sides show off the piglets and their mothers to anyone who might want to see them, and a glassed-in viewing parlor is in the barn's attic, accessible by broad metal stairs and a door that is always unlocked during daylight. In the parlor are tables and chairs, restrooms, coffee and water, and posters of pigs. And on a mild, sunny winter morning, there was a farmer, Gerbert Oosterlaken, wearing a "Pigs Are Cool" T-shirt.

"Most people have never seen a pig farm from the inside, and so all kinds of stories go around," he said. "I think we need to show our neighbors that we have nothing to hide."

What is not visible on this sizable, intensive industrialized property is any sign of routine antibiotic use: no labels on feed bags, no buckets of powdered drugs, no bottles of solutions ready to be drawn up into syringes. That is because, at Hoeve de Huls-donk, almost no antibiotics are in use. This is unusual in global agriculture yet utterly normal in the Netherlands, thanks to strict national standards that the government and farmers mutually agreed on in 2010. Oosterlaken was one of the leaders of that effort. He is a tall man, lean, with close-cropped hair and thick rimless glasses. He pauses often, but when he speaks, words tumble out as though he has a lot to say. He pivoted on the metal landing, grinding mud from the edges of his outdoor boots, and opened the door to the viewing parlor. The hoots of the sows and squeals of their piglets came up from the floor below.

"We decided that animal health, and human health, would be our priority," he said over his shoulder. "I don't need to take antibiotics every day. There's no reason my pigs should either."

OOSTERLAKEN GREW UP where he now farms: outside Beers, a pretty village of fewer than 2,000 people, 10 miles from the German border. His father was a pig farmer; the parents of his wife, Antoinette, were too. From the time Oosterlaken graduated from agricultural school in 1982, conventional, intensive pig farming, with large numbers of animals and routine antibiotic use, was the focus of his life. When his father retired, Oosterlaken and his brother divided the family farm between them, and to guarantee a steady income, Antoinette took a part-time job at a local bank. They raised two daughters and a son and built up the farm, secure that family knowledge and their own hard work had taught them what they needed to raise and protect their herd. And then a series of emergencies showed them how incomplete their preparations were.

In 1997, a deadly disease, dubbed classical swine fever—old-timers called it "hog cholera"—crossed the border from Germany and tore through farms in the Netherlands. The disease spread so rapidly that no one could trace what was carrying it. It might have been workers or salesmen, or trucks coming and going from markets and rendering plants, or feed, or manure, or boar semen used in artificial insemination, or the wind. The only way to stop it from roaring onward was to kill all the pigs that carried it and any pigs that seemed at risk of catching it. The Dutch government interdicted most of the southeastern Netherlands, decreeing the destruction of almost 11 million pigs.

The fever never came to Hoeve de Hulsdonk, but it did infect another farm in the vicinity—and under the draconian methods deemed necessary to contain the epidemic, that was enough to doom the Oosterlakens' herd. On a morning in 1998, a team drove up in a truck. They unloaded an apparatus: a mobile electrocution device, a pig-size box that forced the animals to walk over a wet metal plate while zapping an electric current through their heads. On the day the team arrived, Oosterlaken had 2,500 pigs on his farm, and his wife's parents had almost 2,000 more. He gathered the animals in his farmyard, and herded them through the apparatus, one by one. When he talks about it, even now, he weeps.

After the scouring of his farm, keeping out disease and keeping his pigs as healthy as possible became Oosterlaken's top priorities. He reorganized his operations on what he calls "the three-week system." Every three weeks, he would breed a different group of sows, guaranteeing a group of piglets—on his farm, at least 14 per sow—114 days later. The three-week gap followed the piglets through their lives, through weaning, raising, and removal from the farm to be fattened on other farms nearby. The gaps between birth cohorts let him group visits by outsiders—feed deliverers, veterinary technicians—into narrow windows of time, to reduce when contamination might be carried in unknowingly. When trucks drove onto the property, he insisted they park in one small area as far from the barns as possible.

Oosterlaken made other changes too. He allowed his sows to give birth in open pens instead of cages, and to keep their piglets with them. He fed the sows and the piglets together, on the floors of their pens instead of in troughs, so they would eat more slowly. Once the piglets were separated from their mothers, he kept the birth groups apart, not letting the herds mix in case one picked up an infection. He maintained the barns at a warmer tempera-ture than before, spending more to heat them, and he installed filters to scrub ammonia fumes from the air. He organized his barn into sectors and color-coded them—blue, yellow, brilliant red—and bought overalls and boots that matched so that he would know at a glance if anyone had worn outdoor clothes or shoes into the barn or moved from one group to another without changing gear.

Because he was committed to protecting his animals' health, Oosterlaken used antibiotics; it was a routine aspect of agricultural practice in the crowded pig-farming southeast. The Netherlands is too small to have land to spare for sprawling free-range operations, so making a virtue of necessity, it had become expert at producing livestock intensively. There were almost as many pigs in the Netherlands as there were people—14 million of them for 17 million residents—and in the mid-2000s, after it recovered from the shock of the swine fever epidemic, the country became the leading meat exporter in Europe.

That was accomplished without growth promoters. The shock of the discovery that avoparcin was creating VRE and undermining human medical care had finally prompted Europe to do what the Swann Commission had failed to accomplish decades before. In that time, a few countries had attempted within their own borders what Swann recommended: Sweden banned growth promoters in 1986 and preventive antibiotics in 1988, and Denmark followed in 1994. In 1997, the EU banned the use of avoparcin, and, in 1999, other antibiotics that were being used for growth promotion and that were similar or identical to human-use ones.

The Netherlands' farmers had not relinquished all antibiotics; as the European rules allowed, they used preventive doses, to defend their farms against outbreaks. They had lost so many pigs to disease that this seemed reasonable. But a fresh animal health emergency would demonstrate that those antibiotic controls were insufficient—and this danger would emerge not from across a national border but just down the road.

—‿‿—

THAT THE NETHERLANDS put such trust in farm antibiotics is surprising because it was cautious and proscriptive about human use. It used fewer antibiotics in medicine than any other European country—a quarter, for instance, of what people take in France. Government regulations defined which antibiotics doctors could prescribe, to keep resistant bacteria from occurring, and strict national standards for hospital hygiene aimed to keep them from spreading. The standards, known as "search and destroy," were created in 1988 and aimed at MRSA, the drug-resistant staph that had emerged in England in 1961 and swept the world. MRSA is a particular peril in hospitals because staph clings to skin. Any health care worker can pass it along if they are neglectful or distracted when it comes time to wash their hands, and many did. By the 1990s, MRSA was a deadly epidemic in hospitals worldwide.

But it was not in the Netherlands. The Dutch rules assumed that anyone might carry MRSA into a hospital without knowing it, and they required that anyone who worked in a hospital or was being admitted as a patient from certain high-risk situations had to be checked first to see whether they were carrying the bug. Someone who had been cared for in a hospital in another country had to go into an isolation room as soon as they entered a Dutch hospital, until tests proved they were not carrying the resistant germ. Anyone who was proved to be carrying resistant staph—whether it was a patient, or a physician, nurse, or low-level health worker—had to follow a mandatory regimen of showering with strong antiseptic soap and squirting antibiotic gel up their noses before being allowed into the hospital again.

The rules were harsh, and costly for hospitals to follow, but they worked. The check-in tests that hospitals performed showed that resistant staph, MRSA, was very rare, carried by fewer than one person out of 100. So it was an unpleasant surprise in October 2003 when Eric van den Heuvel, a pig farmer whose property is about 15 miles from Oosterlaken's, brought his daughter Eveline to a local hospital to be checked in advance of surgery. The next week, the hospital called him: She was carrying MRSA.

Eveline was two years old. She had been born with several serious defects in the structure of her heart; she had already had one emergency surgery, to repair a hole between the ventricles, and now she needed another. The hospital had checked her in accordance with the search-and-destroy rules, even though they did not expect to find anything, because she had not been in a hospital in a year and her family had not traveled to countries where MRSA was common. Yet the routine check showed she not only was carrying MRSA; she was colonized by a strain that had never before been recorded in the Netherlands. The hospital could not permit her to have surgery until she was clear. The staff there asked her parents to put her through the required routine of bacteria-chasing soap, nose gel, and antibiotics, while its epidemiologists pursued the resistant bacterium's origin.

They checked her family. Her father, Eric, and mother, Ine, were carrying the anomalous strain, and so was her 14-year-old brother, Gert; only her eight-year-old sister, Marieke, was not. The epidemiologists checked her parents' friends, a group of farmers who met once a month, rotating among each other's houses, to share tips on pig farming; six out of 23 of them were carrying it as well. Finally, lacking any other clues, they checked van den Heuvel's pigs, randomly choosing 30 of his 500 sows. One of the 30 carried the same novel strain.

There was something especially odd about the new MRSA, which, from its response to a widely used test, was called ST398. Scientists worldwide had kept close track of drug-resistant staph since it emerged in 1961 because it moved across the globe so rapidly and caused such explosive epidemics. This strain was resistant to tetracycline, which made no sense, because that drug was almost never used to treat it. Even in the United States, which in 2004 was sliding into an overwhelming MRSA epidemic, tetracycline was not the first drug doctors picked to treat a staph infection. But though the Netherlands was lowest on European charts for human use of antibiotics, it used more than any other EU country in its agriculture: Its farmers were giving more than 650,000 pounds of tetracycline per year to their pigs. MRSA, a human infection, had wandered into farm animals, picked up resistance from farm antibiotics, and then crossed back to humans again.

The Netherlands detected the new "pig MRSA" because it had so little of the usual form that the new strain stood out. But for that same reason—the absence of any competing MRSA strains already occupying the available living space on hands and noses and hospital counters—MRSA ST398 took off across the country like weeds springing up in a clear-cut forest.

Four months after Eveline van den Heuvel was diagnosed as carrying the germ—though she was never made sick by it—a new mother who lived in a town 60 miles away went to her local hospital with a vicious breast infection. It turned out to be pig MRSA, the first infection with the new strain that the Netherlands had yet seen, and her husband and baby daughter were carrying it too, though they were not sick. The husband was a pig farmer; three of his employees were also carrying the new MRSA, and so were eight out of 10 pigs that researchers picked at random from a herd of 8,000 swine. Next, a woman who lived more than 50 miles from either family—and had no contact with pigs or farmers in her everyday life—came back to a hospital where she had just received a kidney transplant. She was gravely ill with a pig MRSA infection in the lining of her heart. After that, there was a hospital outbreak—three patients with diabetes with deep foot ulcers caused by the pig strain and three other patients and five staff carrying it—and then an outbreak in a group home for the disabled and blind.

The Dutch suspected that MRSA ST398 was moving so quickly from town to town because it had already spread through the country's pigs—but with millions of pigs being raised in the Netherlands, that was difficult to prove. However, in such a small country, pigs were slaughtered in relatively few places, and nine abattoirs, which collectively processed about two-thirds of the pigs being killed nationwide, agreed to let the government check. Investigators found the novel MRSA strain in 40 percent of the hogs coming into slaughterhouses and on the premises of 80 percent of the farms sending pigs to them. By 2007, MRSA zoomed from one percent of staph infections in the Netherlands to 30 percent, and almost all of those infections were the pig strain.

The threat from hog farming was so obvious that the Dutch government felt compelled to act. It changed the search-and-destroy strategy. Now, anyone who checked into a hospital whose life included direct contact with pigs—a veterinarian, a farmer, a farmer's spouse or child—was immediately confined to an isolation room until tests proved he or she was clear. The strategy instantly stressed the health care system. The main hospital in Nijmegen, the big town close to the Oosterlakens and van den Heuvels and more than 7,000 other pig farms, ran out of isolation rooms to put its farmers in.

—⟋⟋⟍—

THE ST398 EPIDEMIC occurred after the European Union's 1999 partial ban on avoparcin and several growth promoters. It pushed Europe's nations to their next step: a complete ban on any antibiotic used as a growth promoter—veterinary, human, or shared—as of January 1, 2006.

This was a historic step, the first time that so many governments had agreed together to make antibiotic resistance a priority. But it was also insufficient, because the emergence of ST398 made clear that preventive use encouraged resistance just as much as growth promoters did. If there is an explanation, it is that a growth promoter ban was politically defensible, because growth promotion contributed nothing to animals' health. Preventive antibiotics did, even if their overuse increased the risk to human health at the same time. There must have also been an assumption that removing growth promoters from the market would cut farm antibiotic use. It did not.

After the 1999 ban, the Dutch government had created a surveillance project that monitored the remaining antibiotics given to livestock. It published yearly counts of antibiotic sales by feed manufacturers and antibiotic purchases made by farmers; it also included analyses of bacterial samples taken from a small number of meat animals after they were slaughtered. It created a more detailed picture of the interplay of animal health, antibiotic use, and resistance than any other government possessed, and it quickly flagged that the full ban in 2006 was not the triumph everyone assumed. Yes, growth promoters were no longer on the market. In the annual reports, the amounts sold in the Netherlands ticked down year by year—from more than 275 tons of growth promoters in 1999, the year of the partial ban, to no growth promoter sales at all in 2006. But the total amount of antibiotics sold for farm use did not change at all: From 1999 to 2006 and afterward, it stayed steady at more than 606 tons per year. Manufacturers had rebranded the farm drugs they were selling, changing the labels from growth promoters to preventive. In a similar manner to what had happened in the United Kingdom after the Swann Report, they had complied with the letter of the law but ignored the intent.

With such a large volume of antibiotics passing through farms, the Netherlands experienced more episodes of new forms of antibiotic resistance—this time, with a direct link to food. ESBL resistance, an acronym indicating enzymes that can disable penicillin and its relatives, and also several generations of the drug class cephalosporins, began appearing, in *E. coli* and *Salmonella* carried on chicken meat. (The Netherlands raises 100 million chickens per year.) Dutch researchers began hunting for a human connection, as Smith had done for fluoroquinolone resistance and Johnson had done for foodborne UTI bacteria. Two separate research groups bought chicken meat in various cities in the Netherlands and collected blood and fecal samples from hospital patients. They found the same things. Most of the chicken they tested was contaminated with ESBL-resistant bacteria—80 percent in one study, 94 percent in the other—and ESBL-resistant human infections were genetically identical to the chicken strains. In the same way that "pig MRSA" had demonstrated that the partial EU ban was insufficient, the ESBL epidemic showed that even the total growth-promoter ban did not go far enough to protect human health.

Spurred by that discovery, the Netherlands in 2009 created some of the most comprehensive controls ever enacted on antibiotic use in agriculture. The program clamped down on antibiotic sales, policed what farmers were doing on their properties, and minutely examined what veterinarians told and sold their farm clients. It could have been a political disaster. But the discoveries of widespread resistance had so shaken the country that farmers, veterinarians, and feed sellers—and the powerful national organizations representing them—did not resist. In fact, they helped design the program and spread the word about how necessary it was.

—⚭—

WHEN I MET GERBERT OOSTERLAKEN in November 2013, he, and all other Dutch farmers, had been living under some form of farm antibiotic restriction for almost seven years. I expected that someone so passionate about his work would chafe under the scrutiny and the tight controls on what he could give his pigs. But he took joy in it, viewing the rules as a guide back to a purer, more responsible way of farming.

"In 2006, I was on the board of the pig farmers' union," Oosterlaken said. We had just finished a tour of his barn and were sitting in a break room off the corridor that divided the color-coded sectors. To respect his strict biosecurity standards, which protect his herd against infection, I had had to shed every piece of my clothing, wash off my cosmetics, and shower and shampoo my hair; and change into a freshly laundered outfit of clothes the farm provided—a coverall, socks, and a sports bra and underwear, along with boots that had been wiped with disinfectant. He had poured me coffee that, judging by the smell, had been simmering for a while. With no hair dryer at hand, I had tied my hair back, and it was dripping down my neck.

"We made an agreement, us and the veterinarians, to try to use antibiotics on our farms better," he told me. "We wanted not to do some difficult things that would have taken years of investigation, but something that, when we talked about it today, we could introduce it to our farms tomorrow."

Oosterlaken had been very affected by the experience of the van den Heuvels, whom he knew from nearby farmers' groups. "When you hear that story, that his daughter had to go to the hospital and she couldn't be treated, that takes great effect on your own being," he said, turning his coffee cup around in his hands. "The thing you don't want to hear is that some of your own relatives, going into a hospital, get the message that they have a bacteria that is resistant to antibiotics. That would be terrible."

(Eveline van den Heuvel recovered; she had her delayed surgery in January 2004, after the decolonization treatment ended and she tested clear. She is 15 now and training to be a nursing home aide. The years after her MRSA scare were difficult for her father, Eric; he felt their experience had brought a stigma to the pig farming he had been so proud of. He stopped raising pigs in 2016 and now works for a Belgian company that makes cleaning solutions for barns, formulas that are loaded with good bacteria that appropriate the living space disease organisms would occupy.)

When the Dutch government's new policy created even tighter controls on antibiotics, Oosterlaken and the rest of the pig farmers' union welcomed the challenge—which was good, because there was no question that the new rules were challenging. Growth promoters had already been off the table, of course, but now preventive use also was banned. Antibiotics could be used only for treatment of disease after a veterinarian approved their use, and the drugs could not be ones that were used in human medicine as well. To prevent doctor shopping, each farmer was required to sign a contract with a single veterinarian and to register that relationship with the government, and all prescriptions were filed into a national database.

The antibiotics that farmers were still allowed to use were ranked into tiers. Some could be kept on farms in small amounts. Others could not be dispensed until a veterinarian had ordered a culture and a sensitivity test to prove their use was unavoidable. The government agencies and the major production groups, representing pig, cattle, and chicken growers, worked out a complicated algorithm that allowed them to define how often farmers were using antibiotics, and give them targets. Called the "animal defined daily dose," it effectively indicates for how many days in a year antibiotics are used on a farm. Nationwide, farmers who raise the same types of animals are judged on the basis of it, on a traffic light scale of green-yellow-red. To stay in the green zone, broiler farmers can use antibiotics for no more than 15 days in the year; pig breeders, 10. (Oosterlaken's score was 1.)

At the beginning of 2010, the government set its first targets: From the amounts used in 2009, it wanted to see farm antibiotic use cut by 20 percent in two years and 50 percent in three years, by 2013. Dutch farmers so embraced the system that they achieved the goal early. Across the country, they cut farm antibiotic use in half before the end of 2012. Across those same years, the antibiotic-resistant bacteria found in meat when animals were slaughtered declined as well.

Oosterlaken was quick to tell me that the commitment to relinquishing antibiotics had not harmed his bottom line. "You have better results, and everything works so much easier, when you have healthy pigs," he told me. "When they are healthy, they are giving me profit; sick pigs can't give you any profit."

He still felt he could do even better, and he was studying fresh ways to protect his herd from diseases: better pen design, maybe, or probiotics added to the feed or sprayed in the barn.

"If we want to be serious about having some medicine left over for our children, then we should do something about antibiotic use in our farms," he said. "That is why this is worth doing: for yourself, for the next generation, but also to get more profit out of your farm."

—◊◊◊—

WHILE ALL OF THE FARMERS in the Netherlands collaborated with the new system—though the government described the program as voluntary, it was not really possible to refuse—some struggled more than others. Even when farmers were skilled and experienced and followed the new rules in good faith, they could run into difficulties.

In the tiny town of Reek, just a few miles from Oosterlaken's property, brothers Rob and Egbert Wingens and their families run two broiler farms. They raise 250,000 chickens per flock, and eight flocks per year, of a fast-growing, broad-breasted hybrid sold by the international genetics company Aviagen, which long ago subsumed the Arbor Acres birds that came second in the Chicken of

Tomorrow. (Aviagen's birds, which originated with a Scottish company called Ross, are essentially the same as the Cornish cross raised by U.S. poultry farmers, which come from the company Cobb-Vantress and trace back to the Chicken of Tomorrow contest's first-place winners.)

Unusual for the pig-centric southeast, the Wingenses come from a chicken family. Their grandfather had a small flock and a small herd of pigs, and their father expanded both; he was raising 40,000 broilers per year and 700 pigs when he retired and sold his property to his sons. Egbert bought a second farm close by. Both have low, friendly buildings of warm red brick close to the narrow paved road out front; large, modern metal barns out back; and an abundance of whimsical chicken-related garden art made by their father to amuse himself in retirement.

Fast-growth hybrid birds—the Wingenses' will live 42 days before they are slaughtered—are completely legal under the new antibiotics rules, which say nothing about genetics. But they are culturally touchy in a society where supermarkets make buying decisions based on whether meat carries a *Beter Leven* (Better Life) animal welfare certification. Dutch welfare activists called the hybrid birds *plofkip*, "exploding chickens," for the speed at which they grow.

"I call them *bofkip*, 'lucky chicken,'" Rob countered. "They have been lucky for our family."

Judged by the new Dutch standards, the Wingenses are responsible farmers; their animal daily dose per year comes out to 13 days, well within the green zone. But it was a challenge to manage the health of their hybrid birds under the strict new regulations. "It is a good mind-set, to use less antibiotics," Egbert told me as we walked across the crunchy gravel separating the farmhouse from the chicken barns. "I don't want to use any antibiotics unless I have to."

Like almost all other broiler farmers, the Wingenses buy their birds as chicks, which are incubated at a Dutch hatchery and arrive when they are one day old. If the fluffball newborns arrive ill or weak, they cannot administer medication in the manner they formerly would have, even if they believe they know what is wrong. "From experience, we know that the first-choice drug that we are allowed to keep on hand will not work; we need the second choice or the third, but we cannot use those unless the veterinarian performs a test," Egbert said. "So we start with the first choice, and if that doesn't work we use the second, and if that doesn't work, by then we have the test result and we are allowed to use the third. But by that time you are seven, eight days further on, and you have a lot of dead chickens."

Chicks dying in their first days of life is a common problem—many hatcheries add an overage to a farmer's order, assuming some percentage won't make it—and giving birds a "good start" is an industry obsession. It obsesses the Wingenses as well, because with most antibiotic use prohibited, finding other ways to prop up the birds' health is critical to their bottom line. The brothers spend extra money to keep their barns warmer than standard, and they tinker with the content of the feed they use, developing a personal recipe that their feed mill assembles just for them. "There [are] a lot of problems you can solve with good feed," Rob assured me. But the advantage they gain from improving the feed never lasts, he added: "The breeding companies change the chickens genetically, every year, and when they change, they also need different food. It would be much easier if the companies were transparent about the changes they make."

When I visited, the Wingenses were in the middle of an experiment. They had no hope of breeding their own birds, because the genetics companies' tight hold on their intellectual property makes home-brewed fast-growth birds unfeasible. Instead, they had set out to improve their chickens' start in life, by hatching them on the farm. That allowed the brothers to prevent the stress and temperature variation of truck transport, and the possibility that some chicks in a batch were hatching early and going too long without food. They had installed a rig in one barn that would hold trays of eggs just inches above the floor, so the newborns could tumble out and eat right away, and they had filled them with eggs that were three days away from hatching.

They had done this a few times, and the first batch they had hatched themselves were now three weeks old. When Egbert pulled open the barn door, the half-grown birds, still just partially feathered, tumbled out inquisitively and pecked at the buckles on my boots.

"They were born in this barn and they will spend their whole lives there," he explained, gently dislodging a chicken that had hopped onto a piece of equipment. "You can see, they look good, and they are not stressed or frightened. And we were able to use less antibiotics in this barn."

CHAPTER 12

THE VIEW FROM THE BARN

THE WINGENSES AND OOSTERLAKENS had rejoiced in the innovation the new Dutch system afforded them, and in France, the Label Rouge farmers had chosen pieces of new technology with care, fitting them into a system that held old values at its core. U.S. poultry producers had fewer such options. Thanks to Jesse Jewell and the system of vertical integration he developed in the 1930s, companies make almost all the decisions that farmers once made for themselves. The 35 or so poultry companies scattered across the United States choose the bird varieties; receive the eggs and hatch them; purchase the feed ingredients, operate the feed-mixing mills, and deliver the feed to poultry growers; collect the birds, pay the farmers for them, and transport them to their own slaughter and processing plants; package the meat they become; and deliver or ship it wherever they have negotiated sales.

That apparent uniformity masks differences in companies' philosophies and procedures, including how and when they use antibiotics. Equally, it conceals how differently growers experience raising chickens: as a heritage or a vocation—or, for some, a task that no longer matches their values. For some farmers, antibiotic use has been like an extra knot in the string around a package, a final guarantee they had done what they could to produce a healthy animal. For others, it is like a fun-house mirror, reflecting back their actions and beliefs from a perspective they never considered before.

There are roughly 25,000 broiler producers in the United States, and no single farm can be a synecdoche for all of them. But among the many poultry producers I met, Leighton Cooley and his father, Larry, are representative of men who are comfortable with how they run their farm, even as their industry changes around them.

"We love what we do," Larry told me. We were sitting on aging couches in an outdoor pavilion next to Larry's white farmhouse in Roberta, Georgia, 85 miles south of Atlanta. It was a clear March morning in 2016, warm enough to forgo a jacket and breezy enough to keep early-arriving bugs at bay, and we had our backs to an old tractor as we watched the family's horses crop the grass. "We don't like people running it down."

Larry Cooley is 60 and has been raising chickens, or working construction to build chicken houses, for most of his adult life; his family has been in poultry for four generations, in Georgia and North Carolina. A wedding present of a piece of land launched him and his wife, Brenda, into poultry growing on their own; by April 1985, Larry had launched Cooley Farms, growing broilers for a succession of companies. Their son Leighton, 32, became part of the family firm after college, when his parents gave him a piece of land in turn; he used it as collateral to build his first four chicken houses. Now the family property holds 18 houses on three adjacent parcels of land, and in them the Cooleys grow approximately 500,000 chickens at a time, three million birds in a year.

Leighton is tall and broad and wears his hair short under a tractor cap. He once planned to become a high school teacher and coach, until the lure of being able to work outdoors all day won out over his love of football. Now he lives on one of the farm properties with his wife and three young sons and manages a small beef cattle operation along with the chickens. In conventional chicken production, he is a quiet celebrity. He serves as a young farmers' spokesperson for the American Farm Bureau, the largest conventional-farming organization in the country, and he appeared in a 2014 documentary, *Farmland,* that was financed by the U.S. Farmers and Ranchers Association and shown around the country.

The Cooleys raise chickens in the manner that is industry standard: 23,500 per barn, a space of 20,000 square feet. The birds live on litter, a six-inch layer of pine shavings combined with composted urine and manure that feels spongy and dense underfoot, like walking on thick rubber. They grow from fluff to full-sized in less than two months: 38 days to grow medium-sized birds for a fast-food chain or 50 days to make larger birds for supermarket chicken parts. They are raised in solid-walled barns, with no sunlight to disrupt a cycle of dark and electric light that dictates their sleeping, waking, and eating. Humming pipes overhead dispense feed into round red dishes that hang at ankle level. Water pipes run alongside the feed lines, studded with downward-facing nipples that the birds tap with their beaks. The breeze that coasts through from the fans is cooled by external piping in the blistering

Georgia summer and heated on the few days that start out really cold, keeping the barn temperature at 90 degrees when the birds are brand-new chicks and dropping into the 80s as they mature.

The Cooleys have let me into their barns several times over several years to help me understand what conventional poultry production looks like. A damp breeze wafts past, a steady eight miles an hour, strong enough to riffle the sleeves of my coverall and press the ties of heavy plastic booties against my shins. The garb keeps any outdoor contamination on my clothing from entering the barn's environment, protecting the chickens that are chuckling against my ankles. On this morning, the current crop is 42 days old; adult looking, red combed, white feathered on their heads and backs, with bare skin across their bellies. They do not have a lot of space to spare, but they can move freely through the barn; they move all at once, like a shoal of fish, and subside together in murmuring heaps. They are clean and upright and balanced and do not seem anxious or panicked or in pain.

Almost all the chickens eaten in the United States, and increasingly in the rest of the world, have been raised for decades in this manner: always indoors, always under artificial light, always eating only what the farmer supplies. It is what makes chickens a predictable crop and chicken meat a product that is consistent in texture, taste, and nutritional content across billions of birds grown in different locations and climates in every month of the year. It is the foundation of the chicken economy, and it is endorsed every day by anyone who ever bought a fast-food chicken sandwich, ordered wings in a bar, or picked up an extra tray of drumsticks because a supermarket put them on sale.

It is a means of doing business that has worked well for the Cooleys. They have turned their profits back into their farm and taken out loans to build new houses and add upgrades. They take pride in the results, and they trust their product. "I think people's picture of a conventional poultry farmer is a guy with a nasty, drabby house with lethargic, half-sick chickens," Leighton said. "And that's not us."

—⟋⟋⟋—

IF YOU DRIVE STRAIGHT EAST from Cooley Farms, you traverse the eastern half of the United States broiler belt. Meat chicken production may have been born on Delmarva, but the top chicken-producing states are all in the South now. Georgia is No. 1—if it were a freestanding country, it would be the fourth largest chicken economy in the world—followed by Alabama, Arkansas (home of Tyson Foods, the country's largest chicken company), North Carolina, and Mississippi. Not quite six hours from Roberta, a few hundred feet east of the North Carolina line, another chicken producer experiences the profession differently than the Cooleys do.

Craig Watts, who operates C&A Farm in the tiny town of Fairmont, is tall and lanky, with charcoal hair that he lets grow long and then periodically shears into a buzz cut; when the cut is fresh, his hair stands out from the corners of his head like a toddler's after a barbershop visit. He was 48 when I first met him in 2014, a first-generation broiler producer from a family that has been farming since before North Carolina was a state. He lives in a farmhouse built by his great-grandfather in 1901, on property that has been in his family for generations; his aunt, who lives a few fields over, preserves the family's original royal grant to the land, made in the 1700s when it was still within a British colony. His parents grew tobacco, but they both kept day jobs to bring in enough money to keep the household going. When Watts was in high school, his father had a heart attack; he survived, but stooping over row crops in the 100-degree heat was more effort than his weakened heart could stand. The family closed the metal-roofed witch's-hat shed they used for drying tobacco and rented out their fields.

Watts got a business degree in 1988 and went to work for a company that tested agricultural chemicals, visiting farmers to find out which insects were munching their soybeans and cotton and what pesticide residues were lingering in their soil. It was decent work, lucrative enough to get married on, but he disliked wearing a tie and performing the same tasks every day. To move up, he learned, he would have to move—probably to the firm's home office in the Midwest, a landscape and culture different from anything he knew. And then in 1992, a representative for a poultry integrator began chatting up local farmers, displaying charts of how much money they could make if they began raising broilers for a new Perdue Farms slaughter plant that was being built just over the South Carolina line.

Watts listened to the advance man's pitch, studied the spreadsheets, and agreed to sign the contract and build two houses that would hold 30,000 birds each. Three years later, he built two more; just afterward, the processing plant managers, who supervise the area for Perdue, asked him to upgrade the first two barns he built. Watts could bear the financing, because he owned his land and farmhouse free and clear, and he paid off the first two houses in 2001 and the second two in 2004. He began to feel he could get out from under the strict budget he had imposed on his wife and kids. But then his local slaughter complex asked him to upgrade his barns again.

"And you don't have to," he said. "Only they tell you, they won't bring you any more chicks unless you do."

He was fed up, even though he had paperwork to prove he was one of his complex's top local producers. He had spent $650,000 on his farm, and he had not had a raise in 10 years. He was caught in the grind of what is called the tournament system, which pits farmers in an area against each other. They are paid for their labor after the birds are collected, based on a formula that computes the weight the birds have gained against the feed it took to get them there. At the end of every flock cycle, the grower whose birds performed the best gets a bonus; the one whose birds scored lowest gets docked an equivalent amount.

Tournament is the economic basis of poultry production and originally was unique to the industry. (Hog and egg production are adopting the model now.) The industry insists that the system rewards farmers for good practices and reduces the amount of capital they would be forced to invest if they worked inde-pendently. Environmental advocates say it allows the companies to escape financial responsibility for pollution and the burden of accumulated waste. Economists who study contracts and game theory praise it for muting price risks while solving how to incen-tivize independent contractors. But other academics who have studied the practice use words like *plantation* and *sharecropping* to describe how it affects the farmers within it.

Growers appreciate the system's rewards as long as they rise above their local competitors. But as Watts kept farming, he felt the system punished him for things that were not under his con-trol: chicks that arrived puny, or never gained any weight, or responded badly to vaccines. In one ruinous cycle, chicks that were supposed to come from the hatchery to his farm in a several-hour trip were left packed in crates in a truck overnight. Three days after they were delivered, thousands of them died.

Watts began writing letters to local newspapers and testifying before Congress, objecting that farmers always seemed to be the ones to come up short. In 2010, he drove 500 miles across several states to speak at a hearing held by the Department of Justice and USDA, part of a listening tour put on by the first Obama administration to elicit livestock and poultry growers' frustrations. "We have no voice in an industry that we're so heavily invested in," he said that day. "Growers mortgage farms and homes based on an assumption that the relationship with the poultry company will be long-term and mutually beneficial, but what we get is an agreement with no security at all."

By 2014, Watts's frustration increased so greatly that he took a radical step. He invited a farm-animal rights organization, Compassion in World Farming, to make a video inside his barns. Welfare organizations who wanted to expose farm conditions had almost always depended on images filmed undercover—but here a farmer was walking them through the reality of poultry production, and appearing on camera to testify to what he experienced.

"If I am disagreeing with the system, I have to do what I can to change it," he told me, pulling his pickup out onto the two-lane road for the short trip to his four chicken sheds. "The public goes into the supermarket, and they see those packages with the white house and the red barn and the skipping cows and clucking chickens, and it's not like that. It's not like that at all."

—◊◊◊—

THERE WAS NO EXPLICIT CLAUSE in Watts's contract forbidding him from taking a film crew into his barns or discussing the conditions inside them, but it felt like breaching an industry norm;

Watts had counted on that to give the coming video impact. That sense of inhibition was part of a wider unwillingness to talk about what went on in poultry production, compounded of possessiveness over trade secrets and distrust of food-movement denunciations of conventional agriculture.

Antibiotic use fell under that code of silence too. For most of the life of the industry, from when the FDA licensed growth promoters in the 1950s into the 21st century, the details of antibiotic use in conventional agriculture were seldom disclosed. That was one reason the Union of Concerned Scientists' 2001 estimate of agricultural antibiotic use hit with the force that it did.

The first real disclosure would not come for a decade after that estimate, via a federal report mandated by a 2003 law called the Animal Drug User Fee Act (ADUFA). The law grew out of animal drug companies' impatience with the pace of new-drug approvals and created a mechanism for speeding up the process: The companies paid the FDA a user fee, and the FDA hired more staff to process paperwork. The arrangement was so popular—in five years, it earned the FDA an additional $43 million—that it was guaranteed reauthorization when the legislation came up for renewal in 2008. Members of Congress appalled by the long fight to force Baytril off the market spotted an opportunity. They shoehorned in a provision that any antibiotics manufacturer that wanted new drugs approved would have to give up some sales data in exchange.

The companies were willing to divulge very little information at first; the initial version of the "ADUFA Report" (technically, the "Summary Report on Antimicrobials Sold or Distributed for Use in Food-Producing Animals") was only four pages long. Still, its numbers were startling. It revealed that the Union of Concerned

Scientists' 2001 estimates had actually been conservative. In 2009, 28.8 million pounds of antibiotics were sold for use in U.S. livestock (and another 3.6 million were exported for animals raised elsewhere). Food animals were receiving antibiotics from almost every class on the market, either the exact drugs that humans also received or close chemical relatives: aminoglycosides such as streptomycin, cephalosporins, lincosamides such as clindamycin, macrolides such as azithromycin, penicillins, sulfa drugs, and tetracyclines (the most used category), as well as a category of drugs not used in humans, called ionophores.

Year after year since that first report was published, the amount of antibiotics used in animals has risen. In the 2015 numbers, published just before Christmas 2016, the total sold for animal use was 34.3 million pounds. The reports never make direct comparisons to human antibiotic use, but in 2011, the nonprofit Pew Charitable Trusts calculated equivalents using private records of prescription drug sales. In that year, American livestock received 29.9 million pounds, and human patients, one-fourth as much: 7.7 million pounds.

The ADUFA totals were not only the first glimpse the American public got of farm antibiotic use; they were the first crack in secrecy for many livestock growers too. There was no requirement that an integrator tell its contract farmers what it was giving their birds, and many did not.

"They wouldn't have told us," Larry Cooley had told me, looking back to the start of his growing career. "Maybe if we asked them direct. But we didn't know."

More scrupulous companies reported to their farmers via a "feed ticket," a receipt handed over when a load of feed is delivered. A feed ticket lists the percentages of protein, fat, and fiber in a feed recipe—broilers might be fed four different recipes over their short lives—and the amount of antibiotics added per ton of feed. Perdue Farms, the company both the Cooleys and Watts were growing for when I met them, gave its farmers feed tickets with those details, and Watts shared his with me, going back to the 1990s. They revealed a surprising trend: Slowly but perceptibly, the antibiotics given Watts's chickens had been changing.

At the start, the drugs were the historic problematic ones such as chlortetracycline, which Thomas Jukes used to launch growth promotion on the world. After that, the feed tickets listed bambermycin, permitted in the United States but banned in Europe since 2006. In Watts's records, bambermycin vanishes in the 2000s, replaced by the ionophores. That drug class dates back to the early days of the antibiotic era—but except for one version that cures fungal infections, called nystatin, ionophores have never really been used in humans. In poultry, they reduce the occurrence of a tough-to-eliminate parasitic disease called coccidiosis, common in crammed poultry houses, that inflames the intestines and opens the door to infections that can kill a bird. Poultry producers are so reliant on the drugs—after Europe instituted its growth promoter ban, ionophores for poultry were given an exemption—that they have been closely scrutinized for whether they stimulate resistance. Researchers have flagged only one potential problem: cross-resistance to bacitracin, which in the United States is used only in first-aid ointments. Unlike the other drugs that agriculture used so freely for decades, ionophores seem to pose no threat to human health.

The ionophores on Watts's field tickets were confirmation that at least one corner of poultry raising was heeding the message of decades of resistant outbreaks. Farmers could feel their industry shifting underneath them, in a manner that was good for public health but might prove difficult for them to absorb.

—ᘠᘠᘠ—

WATTS WAS ONE OF THOSE who struggled. He viewed the system of production that routine antibiotics made possible as fundamentally unsound. That was not, particularly, because it generated antibiotic resistance (though that concerned him). It was because conventional poultry raising felt cruel. With or without antibiotics, he realized, everything else in the birds' raising—their genetics, their short lives, the crowded conditions he was required to inflict—remained the same. Recognizing antibiotic use had led him into examining his birds' welfare, and when he confronted what he was required to do, he quailed.

In fall 2014, Watts walked me through one of his sheds, through 30,000 35-day-old chickens. It was hot in the house; a thermometer he laid on the litter read 88 degrees. It was an older barn compared to the Cooleys' buildings. The walls were not solid metal; originally, they had been open to the air but threaded with wire to keep wild birds out, and covered with roll-up tarps known as curtains. He had had to wall in the open sections in one of the rounds of upgrades, and now the only light came from yellow electric bulbs and a shaft of sunlight cutting through the casings of the fans. The barn smelled spicy with an edge of rot, but the air currents carried most of it away. Small feathers and litter scraps and dust hung and drifted, as though we were caught in a dung-scented snow globe.

Watts lifted one of the chickens and turned her upside down. Her belly was red and looked raw. Some of the birds in the barn lay with one leg twisted beneath them, stuck out at an awkward angle or stretched the wrong way behind their tails. He picked one up, displaying swollen, gumball-size joints. As we watched, another chicken with the same deformity stretched upward to reach one of the nipples on the water line strung across the barn— but the crippled leg made it impossible for the bird to balance, and it fell over without being able to drink.

"She's not going to be able to reach that, and she's never going to grow," he observed. "She'll suffer from this day forward. The kind thing to do would be to kill her."

Most of the birds we walked through avoided us, crowding against the walls. As they swirled out of our way, dead birds, stretched out and hollow, emerged from under their feet. One chicken wove unevenly across the empty space, stopped, and hunkered down and closed its eyes. Watts got down on his knees. He stroked its head, clucking soothingly, and dug his fingers into the feathers between the chicken's body and its wings.

"I wish I could raise them differently," he mused. "I'd open up those walls again, give them some natural ventilation. I'd give them some perches to sit on, something to do. A chicken wants to scratch and peck, walk around. Getting up, taking a drink of water, sitting right back down—that's not normal."

After 23 years, Watts knew poultry raising as thoroughly as a catechism—but he also understood that farming was an act of faith, and he was losing his faith. Before we went into the barn, he had shown me video from cameras he had rigged in the houses' rafters to alert him if anything went wrong overnight. The camera had captured company crews delivering his chicks—at night, when he was asleep, because it was cooler and the birds were calmer. In the delivery videos, the workers plucked crates of chicks from the back of a tractor, flipped them over at shoulder height, and shook them. The chicks hit hard, rolling like tiny tumbleweeds across the litter. Some of them bounced.

Watts must have watched the video a dozen times already, but he frowned as he saw it again. "I would never say I treat my birds badly, but the way the system is set up, I can't take good care of them," he said. He scratched the sick chicken's neck again; it opened its eyes, let out a faint squawk, and settled down against his legs. "The industry says that it cares about animal welfare; they don't."

He lifted the chicken with both hands, gently, so its legs extended, and set it back on its feet. It wobbled for a moment, balanced, and wandered away. He watched it go.

"I would love to see this done a different way, and if I could, I would do it different," he said. "But if I could get out of these contracts, I don't know that I would raise chicken again. The image and the reality, they're too far apart."

—⁂—

THE VIDEO THAT WATTS collaborated on, titled *Chicken Factory Farmer Speaks Out* and showing staggering chicks and distressed-looking birds, was released in December 2014 and went viral, with millions of views. It triggered an avalanche of news attention, led by coverage in the *New York Times*, and a 43-minute documentary by the bilingual network Fusion.

It also brought attention from the Perdue processing plant he reported to, which put him on a "performance improvement plan" and what he interpreted as harassment: multiple inspections when, he said, there were none before. (In February 2015,

Watts filed for legal protection as a whistle-blower.) A food industry nonprofit, the Center for Food Integrity, assembled a panel of animal health scientists to respond to the conditions the video recorded—and their responses captured the gulf between industry standards for chicken raising and what Watts aspired to do instead.

The experts agreed the birds appeared to be suffering, but said that Watts was to blame: He had allowed birds with deformities to live when he should have killed them to keep them from distress. They said up to 3 percent of birds in any flock—900 in a 30,000-bird barn—could be expected not to make it to slaughter age; some would die naturally, but if others appeared to suffer, it was the farmer's responsibility to kill them. To Watts's complaints that the birds did not walk around but squatted in place, the experts quoted industry expectations: "Broilers spend about 76 percent of their time sitting, 7 percent of their time standing idle on their feet, 3.5 percent standing preening, and 4.7 percent of their time standing eating." Responding to his desire to give his chickens fresh air and natural light, they sounded somewhat mystified: "Sunlight is not needed by the chickens to meet their vitamin D requirement, because the feed is supplemented with adequate levels of Vitamin D."

It came down, in the end, to what animal welfare meant. For Watts, deeply discontented with the underlying assumptions of poultry farming, it was honoring his thousands of chickens as individuals with intelligence and agency. He envisioned a different form of farming, in which he could offer them exercise and let them seek pleasure.

After visiting Watts, I went back to the Cooleys to understand what this looked like to people who were successful within the tournament system and at peace with the requirements of production. Larry Cooley had told me earlier that bad birds—presumably from bad genetics, but he did not have enough information to say—had happened to him too, before Leighton joined the business and before they signed up with Perdue. It was the luck of the draw. "Those birds got to 35 days old and they'd start dying," Larry had said. "We'd pick up thousands a day. There was nothing we could do."

Leighton agreed that some birds would die before their time. He collected dead birds daily and killed distressed ones; it was part of his morning chores, and each time I visited, he had done it before I arrived. He estimated that he would have to kill, or collect after death, 2 percent of a flock in any cycle. He did not enjoy it. "Culling is tough," he told me. We were standing in one of his barns, and birds had settled on the floor around us. Three of them were asleep on the toes of my boots. He indicated a runty chicken, two-thirds the size of nearby ones, that was stretching its full height but could not reach the water pipe, much like the chicken at Watts's farm. "That bird will die of thirst," he said. "You can't let them get like that. So that one's a cull."

On the Cooleys' farm, respecting animal welfare meant making sure the birds were dry, fed, calm, and not frightened or in distress. I asked him whether he ever considered adding what animal welfare activists call "enrichment"—room to flap around, something to climb or perch on, something to engage their small brains. He pushed back the brim of his cap. "Mercy," he said. "Why would a bird need intellectual stimulation? Or exercise? Most people exercise to lose weight. We don't want them doing that." He gestured to the investments they had made in the barn we were standing in: the automated waterers and feeders, the cooling systems, the alarms and fail-safes and cell phone relays that would alert him instantly if anything went wrong.

"Everything we do is to monitor the wellness of the bird," he said. "A bird deserves to be in a healthy house, in a good environment that is conducive to what they are here for. And that's to eat feed, and gain weight, and become a healthy, nutritious piece of meat."

In the ranks of American poultry farmers, most no doubt shared the Cooleys' views. A few probably felt like Watts did. The question was whether those two different concepts of quality of life, experienced by animals that were grown to be killed, could be brought into concordance and what role antibiotic use would play in highlighting those differences. Called out by activists and pushed by consumers, the American industry was just beginning on that journey. Surprisingly, some of its biggest companies would lead the way.

CHAPTER 13

THE MARKET
SPEAKS

THE DELMARVA PENINSULA, the lobster claw of land that attaches to the East Coast between Baltimore and Wilmington and reaches down to point at the shipyards of Newport News, is a deceptive place. Drive the deeply fringed coast, and you are in a world of salt air and small boats and mesh crab pots drying in driveways. But take U.S. Route 13, the road that threads the peninsula's spine through Delaware, the Eastern Shore of Maryland, and a scrap of Virginia, and the landscape might be Iowa: low golden fields, windbreaks of trees, and an utterly flat horizon. It is rural but so narrow—70 miles across at its widest point—that it is suited only for small farms. That includes the farm in Ocean View, Delaware, where Cecile Steele and her husband, Wilmer, launched the broiler industry in 1923. Thirty miles from there is the white clapboard, red-shuttered farmhouse in Salisbury, Maryland, where railroad agent Arthur W. Perdue started an egg business in 1920.

The sprawling complex that sits across the road from the preserved Perdue farm looks blandly like the headquarters of any agribusiness: tall metal sheds resembling warehouses, low brick buildings that might have come from an office park. But it too is not what it seems. The headquarters of Perdue Farms Inc.—the fourth largest chicken producer in the United States and one of the best known—marks the spot where industrial chicken production began to change its mind about antibiotic use.

Perdue's reassessment was an undisclosed project for more than a decade. It broke into the open in September 2014, when Jim Perdue, the company's chairman and Arthur's grandson, stood up at a press conference that he called in Washington, D.C., and said simply: "The message is that Perdue does not use growth promoting antibiotics, and we haven't since 2007."

Perdue's announcement was a stunning break from the chicken industry, which, like the rest of industrial agriculture, had been fighting antibiotic reform for decades. Everything he said next tore the rupture wide open: The company was giving no human-use antibiotics to 95 percent of the chickens it raised. Limited amounts of ionophores. No arsenic, an additive that was controversial but legal in 2014. And no antibiotics at all in its hatcheries. "It took 12 years and a lot of hard work," Perdue said that day. "But we found that we could raise healthy chickens with fewer antibiotics."

Six decades after the FDA's approval gave agriculture the gift of growth promoters, and almost four decades since the industry prevented the agency from taking them back, the enormous, fractious apparatus of farming was beginning to discover that the drugs might not have been worth the investment they demanded or the fights they provoked.

—◊—

IT MADE A CERTAIN SENSE that Perdue would be the first corporation to cross the industry line, since the company had always been a little quirky. That was best embodied in its second president, Frank Perdue, Arthur's son and Jim's father. Frank, who was born the year his father started the company and sacrificed college to join it when he was 19, had the inspiration to become its public face in the 1970s. He commissioned a New York agency to create an ad campaign that played his bantam-weight, big-nosed persona—he actually looked a little like a bird—against a pugnacious slogan: "It takes a tough man to make a tender chicken." The ads were funny, and countercultural enough to be a bit shocking. It was extraordinary at the time to use a CEO as a company's spokesperson instead of inventing a fictional character or hiring a celebrity, and it was unusual to ask consumers to care about the brand on a package of raw meat instead of its freshness or price. But the ads were instant cultural touchstones—Perdue's nerdy earnestness was endlessly referenced and parodied, a meme before the concept existed—and the campaign ran on television and in newspapers and magazines for 20 years.

His son, Jim, by contrast, came into the business after an attempt to have a career apart from the family. He earned a fisheries Ph.D., but changed course in 1983, taking a job in a Perdue processing plant and then in other divisions before rising to chairman in 1991. He doubled down on his father's connection with customers, studying the 3,000 comments that customers sent to the company every month and interrogating his top managers about them in a regular meeting and conference call.

"We started getting more and more questions about antibiotics," Bruce Stewart-Brown, Perdue's senior vice president for food safety,

quality, and live operations, told me. In the late 1990s, when consumer concerns about antibiotics began rising, he was moving out
of a job as a company field veterinarian and into overseeing the
health of the 600 million chickens and 10 million turkeys it raises
each year. "It occurred to us: People understand that if their child
gets sick and goes to the doctor, that they're going to get a prescription, and that they'll take it for a period of time, and the child will
get better. What they wouldn't understand is if you told them that
you're going to put an antibiotic in their child's breakfast cereal
every morning, and they're going to eat [that] for the rest of their
life. They wouldn't understand it for their child, and they probably
aren't going to understand it for their food. And we all kind of said,
You know what? That is something we could work on."

To determine whether it could give up antibiotics, Perdue had
to know for sure how much benefit they were getting out of the
drugs. The company set up a study, choosing 13 of its contract
farms in Delmarva, not far from headquarters, and six in eastern
North Carolina. On each farm, Perdue picked two houses that
were the same size and age and had the same structural features.
It devised a schedule that would allow it to bring new chicks to all
the houses at the same time and ordered the same feed for all of
them, with one difference. On each farm, one house of chickens
would receive the company's standard feeds, containing arsenic;
an ionophore; and the growth-promoter antibiotics bacitracin,
flavomycin, and virginiamycin. The chickens in the paired house
got the same feed formula without the growth promoting drugs.

Over three years, from October 1998 to September 2001,
Perdue evaluated seven million chickens in those paired flocks,
each of which was allowed to live 52 days. In 2002, it tabulated
the results. Between the birds that received antibiotics and those

that did not, the differences in feed conversion and weight at slaughter were hundredths of a decimal point. The number of chickens that died early differed by tenths of a point. There were no outbreaks of disease in the antibiotic-free flocks, and the number of carcasses rejected for internal signs of disease by USDA inspectors actually got better.

The results confirmed something that Perdue's veterinarians and growers had been suspecting for a while: At some point in the 50 years between Jukes's experiment and Perdue's test, growth promoters had lost the power the industry believed they had. A company that recognized that development and shaped its brand identity around it would gain a tremendous market advantage, because out in the world of food buying, big customers and small ones were rejecting meat raised with antibiotics. And after decades of delay, political change was moving against farm antibiotic use as well.

—✹—

THIRTEEN YEARS AFTER that discovery, Stewart-Brown walked me through the plants and farms that surround Perdue headquarters. It was June 2015, nine months since Jim Perdue's disclosure, and the company was about to announce a further milestone. In addition to forgoing antibiotics used in human medicine, it had also removed ionophores from the diets of 60 percent of its chickens. Perdue was calling the birds "no antibiotics ever," N.A.E.

Perdue does not own (though it once did) its own breeding company with its own great-grandparent and grandparent birds harboring a proprietary genetic mix. Instead, it commissions a chicken variety, almost like a recipe, from one of the international genetics companies. It buys parent birds from the big companies, which are installed on contract farms to lay the eggs that will become Perdue broilers. Protecting the health of the broilers so that they will not need antibiotics begins when they are still in their eggs. The last piece of the antibiotic program that the company relinquished—and, Stewart-Brown said, the hardest to do without—was the injection of gentamicin, a human-use antibiotic, into an embryonic chicken before it hatches. It is an industry standard practice to give the antibiotic at the same time as vaccines, to protect the chick from any infection coming through the tiny hole in the shell that the vaccination needle creates. On a hot Friday morning, a crew in coveralls and hair covers was going over a tall rack of pearly, ivory eggs with baby wipes, to clean them without removing their waxy protective coat.

"We had to get to where the eggs were clean enough that the antibiotic wasn't useful," Stewart-Brown said. "That meant they couldn't come into the hatchery with feces or debris or dirt—and that required empowering our hatchery manager to talk back to the managers of the breeder flocks, to tell them they were sending eggs that were too dirty. And that in turn forced the breeder managers to keep a closer eye on their flocks, because there are always some hens who would rather go lay their eggs on the floor over in the corner."

Vaccines are a more efficient protection against infection than antibiotics. If you administer them once, the immune system is forever primed to fight—but antibiotics must be given each time infections start, or continuously to prevent them beginning. Because they do not provoke resistance, vaccines are safer long term as well, but they are more expensive and require more precision to use.

Deep in the hatchery, Stewart-Brown showed me a laboratory-like clean room, devoted to compounding the vaccines the company now uses. The in-egg formulas, which protect against two viral diseases, were being mixed by people in sterile garb under laminar-flow hoods, safety cabinets that blow a smooth flow of air away from the containers to keep the contents sterile. In addition to the standard shot before hatching, delivered now without its antibiotic chaser, Perdue added other vaccines to the broilers' brief lives: two on their first day out of the shell and others further down the line. It also, unusually, began vaccinating the hens that laid the broilers.

"We believe that one of the best ways to reduce your need for antibiotics is to better vaccinate not just the chickens but the breeders," Stewart-Brown said. "If you vaccinate the hen, she passes the antibodies into the egg yolk, so the chickens absorb the protection she gives them." In 2002, according to internal data that he shared, Perdue was spending about one million dollars less than the industry average on vaccination. In 2013, it was spending four million dollars more.

Switching from antibiotics to vaccines provided a model for the next phase of the bird's lives. The company rewrote its feed recipes to remove antibiotics and include prebiotics and probiotics: organic acids, herbs, bacteria similar to the ones that turn milk into yogurt. It also took out proteins that the industry uses under the sanitized term *animal by-products:* rendered skins, hooves, and entrails left over from the slaughter of cows and hogs—and also, sometimes, left over from poultry. About one-third of a chicken's weight at slaughter is inedible, so the almost nine billion chickens killed each year in the United States leave behind billions of pounds of viscera, bones, and feathers that are recycled into a granulated feed additive. (In 2012, researchers found that "feather meal" originating with industrial poultry was carrying undetected antibiotic residues back into chickens' diets.) Perdue banned all that, along with "bakery meal," cheap fats and oils and expired baked goods sold by industrial bakeries.

A half-hour drive from the Perdue complex, the new feed clattered through chutes in the barns of the L. B. Collins farm in Gumboro, Delaware, down to the waiting beaks of 24-day-old chickens that were upright but reedy and not yet fully feathered. The birds had just passed a touchy milestone in their raising: At three weeks, they outgrew immunity passed on from their mothers. In the past, preventive antibiotics would have smoothed their way through the transition and sustained their systems through the metabolic load of digesting animal proteins. Now they were not necessary. "The quality of animal by-products is notoriously variable, and traceability, knowing where the components come from, is a nightmare," Stewart-Brown said, watching the birds rush the feeder. "And they can go rancid easily, which makes them irritating. We learned from our N.A.E. flocks that a more digestible diet makes it easier to reduce antibiotic use, and an all-vegetable diet is more digestible."

As we drove from farm to farm, owners explained other details to which reducing antibiotics forced them to pay attention: keeping chicken houses cleaner, composting litter between flocks and making sure it became hot enough to kill organisms, noticing if drinking water lines dribbled onto the floor and encouraged fungal growth. The ultimate benefit of removing antibiotics, Stewart-Brown said, was how it sharpened the company's awareness of what was happening on its farms, removing a filter that masked whether its chickens were healthy. "It became clear to me," he said, "that the more stuff you get out of the feed, the more you'll learn."

—⧟—

THE PLANS THAT PERDUE covertly executed as it moved the company away from routine antibiotics paralleled events in the wider world of food.

From the first generation of the food movement in the 1970s, small local companies—Bread & Circus in Massachusetts, for instance, and Mrs. Gooch's in California—had competed to buy and offer what at the time were limited supplies of organic meat and poultry. ("Organic" mostly overlaps with "antibiotic free," though the federal National Organic Standards created by the USDA in 2002 ban antibiotics only from the second day of a broiler's life.) On the supply side, sausage maker Applegate Farms was a pioneer; it began making antibiotic-free cured and processed meats in the 1990s. In food service, Panera Bread Co. began serving antibiotic-free chicken in 2004. But it was really the success of Whole Foods Market, founded in 1980, and Chipotle Mexican Grill, founded in 1993, that demonstrated how large the market for meat raised without antibiotics was likely to be.

Whole Foods pledged "no antibiotics" from the start, saying it would refuse animals that received growth promoters or preventive dosing, as well as animals that had received antibiotics to cure diseases. Chipotle founded its operations on a promise of "Food With Integrity": locally grown produce and meats from animals that lived in good welfare conditions and did not receive antibiotics. Both of those companies did so well that they were able to create their own supply chains of produce growers, processed-food makers, and livestock farmers; Whole Foods brought additional farmers into antibiotic-free growing by offering farm loans.

Their success did not immediately persuade very large food businesses to follow them; selling meat raised without antibiotics seemed as much a niche market as organic produce once had been. But out of public view, companies were perceiving the market growing, and laying groundwork to enter it. The first to go public, beating Perdue's announcement by months, was one that almost no one would have predicted. In February 2014, the southern sandwich chain Chick-fil-A declared that it would relinquish all antibiotics in its chicken within five years.

Chick-fil-A's headquarters lies just outside Atlanta's Hartsfield-Jackson International Airport, at the edge of an affluent suburbia that is the buckle of the Bible Belt. The company is privately held and openly Christian. It requires its restaurants to close on Sunday; its corporate motto, "To glorify God by being a faithful steward of all that is entrusted to us," is engraved on a plaque outside the headquarters' front door. Its chief executive officer generated tremendous publicity, most of it negative, by expressing biblically based opposition to same-sex marriage. But in a quiet, southern way, Chick-fil-A is a poultry powerhouse. By sales, it is both the eighth-largest fast food-chain in the United States and the largest with a menu based on chicken. Its sales are larger than the U.S. division of rival KFC, and its individual locations earn more per restaurant than McDonald's. And it has an almost obsessively loyal customer base, people who camp out overnight when new restaurants open and dress up as the chain's Holstein cow mascots, all in pursuit of its signature offering—a toasted, buttered bun cradling pickle slices and a fried boneless breast coated in salty spice. (The cows self-protectively beg customers to "Eat Mor Chikin"; apparently cows can't spell.)

Chicken is all that Chick-fil-A sells, barring beverages and salads and a few outlier breakfast items; its menu includes no burgers, no chili, no fried fish or shrimp. So the company keeps a close eye on where consumer preferences are headed. Thirty years after Whole Foods opened and almost 20 years after Chipotle debuted, those leanings were visibly shifting—not just in the choices of individual shoppers but also in the contracts written by big institutional buyers, which can create or change markets. In 2010, a coalition of 300 hospitals across the country announced they would no longer buy meat raised with routine antibiotic use. In 2011, the Chicago Public Schools, the third largest school district in the country, converted to antibiotic-free chicken. In 2013, the academic senate of the University of California, San Francisco (which, in addition to the university, operates the city's largest hospital) voted for its food procurement to go antibiotic free and urged the rest of the University of California system to follow its lead.

Chick-fil-A's announcement was the first signal that poultry production was breaking with the rest of the American meat industry; Perdue's, seven months later, was the second. One after another, major food service companies and poultry integrators fell in line behind them. McDonald's shook the market by announcing in March 2015 that it was going antibiotic free for chicken in all of its North American restaurants; Subway followed in October 2015. Costco put its buying power behind chicken raised without routine antibiotics in March 2015, and Walmart in May 2015. Poultry producer Pilgrim's Pride said in April 2015 that it would take 25 percent of its birds antibiotic free. Foster Farms, the company that had been dogged by foodborne illness outbreaks, committed in June 2015. And Tyson Foods, the largest chicken company in North America, announced in April 2015 that it had already eliminated human-use antibiotics in 80 percent of its broiler production (including its hatcheries), with the plan of being antibiotic free within two years.

Though they all phrased their moves as relinquishing routine antibiotics, the companies were not all committing to the same actions. Tyson continued to use ionophores, the drug family that the European Union had allowed in chickens when it banned growth promoters; McDonald's said it would accept use of those drugs by its suppliers. But Perdue committed to doing away with ionophores. Chick-fil-A, which buys chicken from Perdue and four other major companies, set a strict standard. It told its suppliers that it would not be acceptable to use antibiotics at any point in a bird's life—not even ionophores and not even for treatment of illness. Suppliers would have to undergo an annual audit to prove compliance.

Implementing that was not simple, and what Chick-fil-A went through to secure antibiotic-free chicken illustrates how the turn away from conventional raising will challenge the poultry industry. But the company presented its decision as a move the market demanded; in proprietary research, 70 percent of their customers said they were concerned about antibiotic use on farms. It could lead the way, or be left behind.

Chick-fil-A estimates that it buys roughly 250 million pounds of chicken per year. Before making its announcement, the chain's leadership met with all five of its poultry suppliers to see if they could meet the demand.

"In a perfect world, we'd be able to flip a switch," David Farmer, the chain's vice president for menu strategy and development, told me at its Atlanta headquarters a few months after the company announced its no-antibiotics pledge. "But that's not the reality. The goal is we're going to get there within five years, 20 percent of our supply per year."

To start, Farmer admitted, they would have to spend more money. Birds raised without routine antibiotics command a higher price, and it was a challenge to figure out whether any of that expense could be passed on to customers. Then the company would have to simplify complexities built into its sourcing, backstage decisions that would never be visible to a customer. For instance, Chick-fil-A had never embraced the industry trend of larger and larger chickens, since if a breast was too big, it would overwhelm the signature sandwich. But it also sells fried chicken tenders, and because tenders are actually breast muscles—the pectoralis minor, tucked up against the breastbone—ones that came from birds with the right size of breast for a sandwich were too small to make a good mouthful. A few years earlier, it had begun ordering only tenders from integrators whose farmers raised bigger birds. Now, confronting the relative scarcity of poultry raised without antibiotics, Chick-fil-A would have to embrace a beak-to-tail philosophy, with the goal of buying and using whole birds. Finding places to use the meat they had not previously bought might require debuting new dishes or changing the cooking procedures for items they already offered: adding antibiotic-free chicken to a soup in a restaurant, for example, instead of buying the soup from a contractor completely premade.

If, at the beginning of 2014, you had asked food movement leaders which company would lead the business away from antibiotics, it is a safe bet that no one would have nominated Chick-fil-A. It is axiomatic, if seldom spoken, that food activism arises in liberal coastal enclaves and seeps slowly into the red states. Chick-fil-A's executives and its core customers are churchgoing, big-box-shopping, college-football-watching conservatives, a constituency not necessarily attuned to animal welfare or antibiotic resistance. (It is likely that some of those churches preach suspicion of evolution, even though antibiotic resistance is evolution in real time.) But for just that reason, the company's conversion was thrilling to see. It demonstrated that concern about farm antibiotic use—and the changes in farm operations that reducing antibiotics would cause—could cross cultural fractures and party lines.

When I asked Farmer whether Chick-fil-A's move away from antibiotics meant it was endorsing the connection between farm use and resistance, he elided the question. "It was not our intent to enter into the scientific debate: Does this cause antibiotic resistance or not?" he told me. He framed the action instead as a task in line with the company's Christian focus, honoring the responsibility that the biblical book of Genesis gave humanity: over "the fish of the sea, and over the fowl of the air, and over the cattle, and over all the earth."

"It's not about shareholder value," Farmer told me. "It's about faithful stewardship. We're compelled to try to do the right thing for the right reasons."

Customer demand persuaded Chick-fil-A, Perdue, and the companies that followed after them to reassess antibiotic use. But cultural pressure might not have been enough to accomplish that without the influence of other events happening in parallel. After decades of stalemate, the politics of antibiotic use had changed as well.

ALMOST FROM THE TIME that FDA Commissioner Kennedy lost his fight to withdraw licenses for growth promoters, the political climate had been unfriendly to trying again. That was not only because Representative Whitten, the godfather of the FDA's budget, refused to allow any attempts—and kept refusing for his entire tenure in Congress, until he was gently muscled out in 1995 at the age of 85. (When he died a few months after leaving the Hill, he was the longest-serving member of the House of Representatives, having put in 53 years.) Any of the administrations that passed through the White House after Kennedy's 1977 disappointment could have resurrected the issue, but each of them had more urgent priorities.

Double-digit inflation, gasoline shortages, and the Iranian hostage crisis undermined President Jimmy Carter, Kennedy's boss, for the rest of his single term. His successors, Ronald Reagan and then George H. W. Bush, were pro-business, small-government Republicans, who would never have agreed to curbing the sales of major pharmaceutical companies. Bill Clinton, a Democrat, might have tried it—his attempt to create a national health care plan demonstrated he was willing to tackle complex issues—but he was in office only two years before the 1994 mid-term election gave Republicans their first congressional majority in 40 years. The power of that political bloc plus a persistent series of scandals prevented his implementing any other major reforms. George W. Bush, who followed Clinton, was another small-government conservative—and in his last year in office presided over a deep recession.

It took the arrival of President Barack Obama in January 2009 for the economy, the national mood, and the White House to align in a way that enabled policymakers to take on the pharma sector. Leading the charge was Representative Louise Slaughter, a

Democrat from upstate New York, who held a master's degree in public health; she was often described as "the only microbiologist in Congress." Her sister had died of pneumonia in childhood, so she took the threat of infectious diseases personally. Slaughter was an unrelenting champion of a bill called PAMTA, the Preservation of Antibiotics for Medical Treatment Act, which aimed to keep antibiotics crucial to human medicine from being used on farms. It never gained any traction and expired at the end of every two-year congressional term, but Slaughter kept doggedly reintroducing it. In July 2009, she held yet another hearing to launch the bill.

The new deputy director of the FDA, Dr. Joshua Sharfstein, appeared for the administration. He told the startled committee, "FDA supports ending the use of antibiotics for growth promotion and feed efficiency in the United States."

Like the European Union before it, the U.S. government had decided to move cautiously, aiming only to eliminate the least defensible form of farm antibiotic use. Nevertheless, agriculture and veterinary pharma ignited in outrage, insisting that science did not support a growth promoter ban.

"FDA has already made policy decisions on these critical issues absent any scientific basis or further dialogue between the agency and animal agriculture," the Animal Agriculture Coalition, an umbrella group of more than 40 powerful production organizations and the Animal Health Institute, protested in a quickly drafted letter to the FDA. "No conclusive scientific studies have been offered demonstrating the use of antibiotics on farms contributes significantly to an increase in human resistance," 20 of those organizations said in a statement sent directly to the White House. Given the decades of research since Levy's seminal experiment, these assertions seemed implausible, but public health insiders recognized the tactics. They were straight out of what was known as the "tobacco playbook," the lobbying script that had allowed cigarette companies to delay regulation for decades after smoking was linked to cancer, by insisting that not enough evidence existed and more research was needed.

Invoking the tobacco playbook signaled that agriculture and veterinary pharmaceutical industries would not give up antibiotics without a fight. The FDA's lawyers informed their leadership what the scope of that battle might be. When Kennedy had tried to ban growth promoters, there had been 62 antibiotic products at stake, from 16 manufacturers. Now, there were 287 antibiotic products, produced by 27 manufacturers, and the companies were prepared to take the agency to court over each individual drug. The battle to remove Baytril from the market just a few years earlier had consumed five years of court time. Another approach would be needed.

A year later, the FDA revealed what it would be. Rather than forcing manufacturers to withdraw their antibiotics, it would ask them to cooperate in a voluntary program of changing their labels so their drugs could not legally be used for growth promotion. The strategy was set forth in a wonkily titled agency document, Draft Guidance for Industry 209: "The Judicious Use of Medically Important Antimicrobial Drugs in Food-Producing Animals." Guidance 209 was 26 pages long, but it boiled down to the goals Sharfstein had stated in Slaughter's hearing: removing growth promoters from the American market, and putting any remaining antibiotics under veterinarians' control.

The term *Guidance* in the title was important, a regulatory term of art that signaled the document did not have the force of law; in fact, it carried the disclaimer "Contains non-binding recommendations" across the top of its title page. Advocates were unhappy; they wanted the kind of legal measure that Kennedy had tried to create. "Allowing this guidance to remain a voluntary policy will result in marginal, if any, reduction in injudicious use of antimicrobials," the Trust for America's Health, a public health nonprofit, said in one of more than 1,000 comments the FDA received. "The FDA needs to go beyond guidance to take enforceable action," agreed the Pew Charitable Trusts, which had been pushing the agency for years to assert stricter control. Meanwhile, the industry stuck to the playbook: "There are no peer-reviewed scientific studies that establish that judicious use of antibiotics in livestock increases antibiotic resistance in human infections," the Michigan Farm Bureau asserted in one of many near-identical comments. "Actions proposed in Guidance 209 are not based on demonstrated safety risk."

The FDA issued the final version of Guidance 209 in April 2012 and a companion document, Guidance 213, in December 2013. A third document setting out veterinarians' responsibilities, the Veterinary Feed Directive, was finalized in June 2015. The agency gave manufacturers three years to make the necessary changes, setting January 1, 2017, as the date when growth promoters would no longer be legal in the United States.

Despite how hard they had fought against the measures, animal antibiotics manufacturers fell in line with surprising speed. By April 2014, one drugmaker took its company and the three drugs it made out of the U.S. market; by June, 31 other drugs were removed from sale. With that, all 26 remaining manufacturers agreed to comply with the FDA's plan. After so many decades of opposing the smallest restrictions, pharma companies' sudden change of heart seemed like a collapse. But it was possible the industry had recognized what Perdue already perceived: growth promoters no longer worked.

—ɱ—

BACK IN 1948, Thomas Jukes's experimental feeding had more than doubled the weight of his chickens. That was in his lab, with chicks he had bred and fed for the purpose; no one expected the process to work quite that well in the real world. In the 1950s, though, penicillin and tetracycline gained farmers 10 percent more weight per bird in farm conditions, without any more feed. The newer antibiotics achieved in the 1970s and 1980s boosted the gains as high as 12 percent. By one 1970 estimate, using growth promoters saved the poultry industry $20 million in feed costs per year.

But by the 1990s, the benefit had dropped to a 4 percent average daily gain for pigs and 3 percent for broilers. The difference was so small that according to the USDA, some livestock producers already had noticed and were forgoing growth promoters. By 2011, 25 percent of feedlot cattle in the United States and 48 percent of the broilers (including much of Perdue, though the company had not yet disclosed its changes) were being raised without them. In 2015, USDA economists calculated that withdrawing growth promoters from U.S. farms would make a productivity difference of as little as one percent.

Scientists could only hypothesize about what was happening, since most of the research being done was within the proprietary bounds of meat companies. It was possible that the gut bacteria affected by the microdoses had finally become resistant to the drugs' effect and were no longer responding; at the start, Jukes had predicted that might happen. It was also possible that livestock had reached their maximum genetic potential, each species' limit of how big and how fast they could grow.

The most likely explanation was that growth promoters had never really conferred a benefit; instead, they had compensated for deficiencies in how farms were run, and those deficiencies no longer existed. In fact, in the 1950s, researchers had noted that when animals were raised in very clean conditions, in an experiment or on a farm, they gained less weight from growth promoters than average farm animals did. Seven decades later, better farm hygiene and monitoring, and industrially precise nutrition, had improved farms so much that growth promoters no longer made a difference.

But that did not mean that most of meat production was willing to relinquish all antibiotics. If growth promoters had lost their power, preventive antibiotics were still valuable, because they allowed producers to grow animals without worrying about disease risks. And the FDA guidances, like the European regulations before them, did not restrict preventive antibiotic use. As the Dutch government had found in 2008, unless other curbs were put in place, banning growth promoters would simply shift the market to use of preventive drugs. The "veterinary feed directive" the FDA had enacted was supposed to prevent that sort of shell game. But the Pew Trusts, which had insisted the guidances were not strict enough, uncovered that many new labels were so vague that they would allow one-third of the antibiotics still allowed to be used indefinitely, with no limitation on duration. That seemed like a contravention of the spirit of the new rules, an extension of growth promotion—and an invitation for antibiotic use, and resistance, to rise.

Meat production did not behave like an industry that was preparing to wean itself off antibiotics. From the time the first ADUFA report was published in 2009, antibiotic use in U.S. meat animals rose steadily. The most recent report, released in the last days of 2016, showed that sales for use in livestock rose 24 percent since the FDA began counting. (The report always lags by a year, so those data covered 2015, the middle year of the preparation period the FDA allowed industry before the guidances became final.) It was not possible to say which animals were receiving the drugs—that is, whether poultry was actually using fewer drugs and, if so, whether hogs and cattle were concomitantly using more—because the FDA data have never been broken out by species. The agency will add that detail for the first time in the report that will be published in December 2017. The edition of the report that will show whether the guidances actually worked as planned—by eliminating growth promoter use, and also reducing all antibiotic use on farms—will not arrive until the end of 2018.

—⁂—

PERDUE NEVER TALKED about growth promoters versus prevention. The company had decided not to make the distinction between types of antibiotic use, but just to do away with as much antibiotic use as it could. Stewart-Brown envisioned a customer reading the content of a feed ticket and spotting the name of a preventive antibiotic, and his having to explain the difference. It didn't seem worth the struggle. "We're trying to help people trust us," he told me.

In June 2016, the company announced a comprehensive animal welfare plan, the first among the largest poultry producers.

The plan created change in almost all of the aspects of broiler raising that animal welfare activists, and unhappy farmers such as Craig Watts, had found difficult to endure. The changes were so striking that they gained the endorsement of three major animal welfare organizations: Mercy for Animals, the Humane Society of the United States, and Compassion in World Farming, the group that had gone into Watts's barns with cameras.

The plan called for putting windows in barns to give birds natural light; installing perches and straw bales for them to climb on; increasing the lights-off hours in a 24-hour period to more resemble a normal day; even changing the system by which birds were killed, installing a gassing chamber that would lull them irreversibly to sleep. Perdue further committed to start reconsidering the fast-growth genetics that bring birds to slaughter weight so quickly. Though the company did not say so, the changes endorsed what European farms had been embodying since the European growth promoter ban became final a decade earlier. In October 2016, Perdue gave up even routine preventive use of the ionophore antibiotics that the European ban made room for and that other American integrators still allowed. The company said it would reserve ionophores only for barns and farms where the parasitic disease coccidiosis had been diagnosed. With that, the company took at least 95 percent of its birds—more, in some years—completely antibiotic free.

"We're talking about this as going back to the farm, to the way we used to do things," Jim Perdue told me. "Maybe they were a little smarter back then than we thought they were."

CHAPTER 14

THE PAST CREATES THE FUTURE

"I NEVER PLANNED ON CHICKENS," Will Harris III said.

We were sitting in a Jeep Wrangler, splashed halfway up the side panels with red mud. The Jeep was parked on the side of a deep green pasture, and in the pasture, there were several thousand broilers.

The birds were rusty feathered and glossy, with red combs and yellow legs. They were scratching and pecking in the wet grasses and lounging under rectangular awnings attached to cream-colored coops that looked like tiny garages. There were groupings of coops in the field, scattered as though they had popped up like mushrooms: six here, four there, another batch by the distant fence line. On the far side of the fence there were cattle, black with a glint of red where the sun caught their hides. Beyond the cattle was the heart of Harris's farm, White Oak Pastures: offices and corrals and USDA-approved abattoirs. Beyond those, out of sight from where we were parked, were almost 3,000 acres holding rabbits and sheep and pigs and goats, turkeys and ducks and geese, guinea hens and laying hens, vegetables, fruits and bees, and more broilers, all on lush grass.

Harris is the fourth generation to operate White Oak, which lies tucked into the thinly populated western edge of Georgia, an hour south of the military traffic to Fort Benning and 40 minutes north of the Florida state line. His family has been on the property since his great-grandfather, James Edward Harris, fled the collapse of the Confederacy and founded a subsistence farm in 1866 outside the town of Bluffton. Over the decades, it grew into a substantial cattle ranch, helped along by every 20th-century development that fueled the growth of American agriculture: chemical fertilizers and pesticides to maintain a monoculture of grass, and hormone implants, artificial insemination, and antibiotics to maintain a monoculture of cattle. Harris inherited that operation and expanded it, following every precept he had learned studying animal science at the University of Georgia. He had worked on the farm for almost 20 years when he began to hear a whisper of conscience: Maybe he should go a different way.

Over the course of the next 20 years, Harris and his wife and daughters and their employees—there are about 135 of them now—remade their single-species conventional farm into the largest certified organic property in the southeastern United States: a multispecies, pasture-based, zero-waste laboratory for sustainability and innovation. Including chickens in that process was crucial to the farm's success. It has proved that poultry can be shaken free not only from antibiotic use, but also from fast-growth genetics and industrialized production.

White Oak Pastures, and a few other businesses that have emerged without notice beyond the edges of the conventional industry, embody what poultry production can look like as it moves away from antibiotics. The new models are humane, personal, and ambitious. But they are not perfect. To different degrees, they demonstrate the limits of businesses that are not part of Big Chicken, and they pose questions—not yet answered—about how the market will respond to them.

"I never owned a creature with feathers before January 2010—not so much as a parakeet," Harris told me. "Then we bought a batch of 500, and now we have 60,000 on the ground at a time. But I can't say if this will get bigger. I think it will be slow."

—⟁—

HARRIS, WHO HAS PASSED 60, looks like a cattleman. He is sturdy and calm, wears a goatee and shaves his head, and is never seen in public without boots and a white Stetson, deeply creased from front to back. But he sounds like a forceful, slightly bawdy preacher—of sustainability, not religion. ("I was deep into the industrial model, but I am like a reformed prostitute now," he told me once. "I have the zeal of the converted." In his southwest Georgia accent, so different from the stereotypical southern drawl, the word hissed through his teeth: *zeeeeel.*) His change of heart was no lightning bolt on the road to Damascus. It arrived slowly over years, as he imagined the life of his farm from his animals' point of view.

"I had been taught that good animal welfare meant keeping them fed and watered and not intentionally inflicting pain or discomfort," he said when I first met him in mid-2012. "That's like saying good parenting would be taking your child and locking them in a closet. You give them plenty of food and leave the lights on and keep the heat at 72 degrees. They're not going to get bit by animals or stung by wasps or get their leg broke playing ball. So that's good parenting, right? But it's not. And good animal welfare is not just keeping them from suffering. It's creating an environment in which animals can express their instinctive behavior."

The first step in Harris's evolution away from his family's farming tradition was opening his corrals and putting his Angus-based herd on grass, forgoing the grain he had been feeding them and letting them get their nutrition naturally. Next to go were the hormones and antibiotics. Then he withdrew the synthetic fertilizers that kept his pastures green year-round.

And then he ran into trouble. There were plants—he would have called them weeds a short time before, when he was spraying every day to maintain his Tifton 85 Bermuda, a sterile hybrid hay—that the cows did not care to eat. When the animals chomped down the tasty competition, the weeds threatened to overwhelm the fields. So Harris purchased a flock of sheep to eat the weeds. That was a bold step for a cow guy; in the 19th century, cattle owners had driven sheep herders out of western states with violence. But his new sheep, a meat breed that did not need shearing, gave him a second animal to harvest and fit well into the farm. Too well, maybe: Both the weeds and the grasses were being eaten down now, and the pastures were covered with sheep dung and cow pies.

Enter the chickens. If they lived in the pastures as chickens evolved to do, they would forage for seeds and insects, breaking up the dung piles for tasty worms and fly larvae and contributing their own high-nitrogen droppings to encourage fresh vegetation.

Harris found a hatchery in Alabama that dealt in heritage birds and slow-growing hybrids and asked them to send some chicks. He got 508. He picked a pasture, set up a mobile coop, plopped in the birds, and waited to see what would happen. By the time they reached a marketable weight—12 weeks, twice as long as an industrial chicken—506 had survived, and the area within the fence was transformed, lush and green with no visible cow pies. Harris slaughtered the first batch and ordered more. After trying several varieties—including a fast-growing industrial bird, which he describes mostly by swearing—he settled on an energetic bird called the Red Ranger, a cross of several heritage breeds.

By adding the sheep and then the chickens, Harris was embarking on rotational grazing, a historic practice, lost after industrialization, that uses each species on a farm to augment or remedy the effect of whichever animal came through the fields before. He also suddenly had many more animals: not only the hundreds of calves he formerly would have sent to feedlots but thousands of chickens as well.

On a day when he was loading some of the calves into a semitrailer to send off for slaughter, he noticed, as if for the first time, that the ones on the lower level would have urine and dung cascading down on them throughout the ride. That did not accord with his emerging sense of animal welfare. To fix it, he spent millions, taking the extraordinary step of building his own USDA-inspected abattoirs in the center of the property, one for the cattle and a second one next to it for the birds. To make sure the slaughterhouses were humane, he hired animal welfare expert Temple Grandin to consult on their design. The abattoirs guaranteed that his animals would live their entire lives on the farm and would spend all of them on grass, except for their last few moments.

There are 10 species raised at White Oak now—five with four feet, five with two—and each earns its keep not just as a product but as a contributor to the farm's economic cycle. The goats might be the only animal that can eat faster than kudzu can grow; Harris uses them to clear overgrown fields and orchards before he moves the pigs in. The pigs in turn root up crusted-over fields so they can be replanted in the mix of grass species that replaced the Bermuda grass. Bones are dried in the fields and ground for fertilizer. Cow hides are tanned for rugs and made into purses; fat becomes soap and candles; tracheae, chicken feet, and other gristly bits are dehydrated to sell as pet chews. The rinse water from the abattoirs is sprayed on the fields. Viscera are dumped into tubs to make breeding grounds for high-protein fly larvae that are fed to the birds.

The chickens play an essential role. White Oak raises 260,000 broilers in a year and keeps a flock of 12,000 layer hens. They and the other birds—ducks, geese, turkeys, and guinea fowl, Harris's favorite—arrive as day-old chicks, spend three to four weeks in a brooder barn behind the main offices, and then live in the fields. Each new batch of broilers is deposited in a cluster of coops, far enough from the other clusters that the birds will not wander and mix. They are locked in for one night, and then left free to roam, though they naturally return each night to their housing for safety. Ranch hands bring them water and supplemental feed, and every two weeks, the coops are dragged 40 feet by a tractor so the birds will refresh a different piece of land.

While the birds benefit the farm, they themselves benefit from Harris's idea of better welfare. The slow-growth hybrids seem to have stronger immune systems; once out of the brooder, they do not randomly collapse and die as conventional birds do. And because they grow more slowly, they do not develop leg problems, and their hearts and circulation are not overstressed. Their main cause of death, before slaughter, is predation. Guardian dogs, Great Pyrenees and Akbash and Anatolian shepherds, protect the birds from coyotes and foxes—though in each flock, some are lost to owls and to bald eagles that roost in the farm's woods. When the chickens are slaughtered, their rates of foodborne organisms come in below federal standards—and with no antibiotics used on the farm, there is no antibiotic resistance.

There is only one challenge remaining: how to make them profitable.

—◊◊◊—

WHITE OAK SLAUGHTERS 5,000 birds weekly. Once each week, a USDA inspector assigned to the plant reaches into the bins of water and ice where the just-killed birds are cooling, pulls one out at random, drops it into a plastic bag filled with a liquid that is optimal for growing bacteria, massages the bird through the bag, drains off the growth medium, and sends it to a lab to be cultured. A White Oak worker does the same thing at the same time. The tests look for *Salmonella*, *Campylobacter*, and *E. coli*. White Oak is allowed to have up to five positive weeks, out of any 52, in which pathogens are found. In a year, they usually have one.

Brian Sapp, the farm's tall director of operations—he has a master's degree in meat science and grew up on a flower bulb farm in Florida—suggested it might be because they keep the birds moving during the day as they roam and month to month as they transfer pastures. "We can't control the environment the way you can in a closed house—the bedding, the air flow, and temperature control," he said. "But in a closed house, those birds are sitting on their excrement and on dead chickens, whereas if there are any pathogens our [animals] build up, in three weeks we've moved them off."

As in the Label Rouge program, fewer organisms come into the slaughterhouse with the chickens. But also, once in the White Oak processing plant, conditions make it less likely that bacteria will spread between birds. The number that White Oak kills and guts in a week, individually and by hand, equals what an industrial plant might handle in an hour, almost entirely by automation.

"We're looking at every bird, handling every bird two or three times," Sapp told me. "If there's something contaminated, we can immediately stop what we're doing, clean up, and start over. If you've got an evisceration line in a large plant, by the time somebody walks by and sees that a machine's not working right, you may have 300 birds that have been contaminated and you don't know where that 300 starts and when it stops."

The downside of all that handling is that labor is expensive. Harris estimates that White Oak's labor costs per bird are three times as high as conventional birds, which passes through to the retail price. "My grass-fed beef costs 30 percent more than the grain-fed beef at Whole Foods, but my chicken is 300 percent more—and the reason is, chicken lent itself to industrialization so much more," he said. "When we industrialized, we were able to take out labor costs, feed costs, land costs. When we put chicken back on pasture, we accept those costs back as well."

Whole Foods on the Atlantic coast is one of Harris's main retail channels, along with middlemen distributors, restaurants, and Internet sales. But Whole Foods shoppers are not what economists call "price sensitive"; they buy for ideology or identity or flavor as much as for cost. And White Oak's chickens are delicious, with lean meat and deep flavor similar to Label Rouge birds. But like those birds, they can be challenging to cook and to eat. An early collaboration with a chef who wanted to shape a fast-casual chain around them broke down when customers complained the flesh was chewy and fretted that the interior of the legs remained pink. (The color indicated the chicken got plenty of blood-pumping exercise when it was alive, but eaters worried it was raw.)

Working with chefs is a crucial part of the chicken project—not just for the birds they buy, but for the awareness they spread to their customers as well. Their needs are something that White Oak has had to learn. "We try to find chefs who are willing to celebrate inconsistency, because with pastured poultry, that is unavoidable," Harris's daughter Jenni told me. She is the middle child of his three daughters and the farm's director of marketing; everyone accepts that she will run the farm after him. "But I get it. They order a case of chicken from us, 12 birds, and in that box there are birds that are 3.1 pounds and 3.9 pounds. An industrial producer could control that better. Our birds are out burning calories, escaping from predators, hiding from the sun, taking dust baths, eating bugs and grubs of different types in different portions. From an animal welfare perspective that is excellent. But from a perspective of predictability is hard."

The result is that, after seven years, White Oak's chickens are still not paying for themselves. Harris says it is hard to know how much money the birds are losing him, because he does not break out his balance sheets by species. But he suspects that the farm's grass-fed beef, its signature product, pays for all the rest. He is okay with that. "I bet you we got two-point-something million dollars, maybe three million, invested in being in the chicken business, and I don't regret it," he said. "I believe the time will come that it will be profitable."

For other businesses pursuing a different vision of chicken raising, it already is.

—⁂—

"SMELL THAT," SCOTT SECHLER SAID. He pushed a round jar of opaque plastic, about four inches high, across his desk in my direction.

I unscrewed the top, looked down at the grainy contents, and took a cautious whiff. The smell was familiar, but it didn't fit with the surroundings, a second-floor office in a small town in the Mennonite part of Pennsylvania. I thought for a minute. "Pizza?"

"Oregano oil," he confirmed, nodding. "And fennel. And a little cinnamon." Several more jars stood on the desk; he nudged one with a broad finger. "This one has garlic. This one has yucca. This one is special for the organic market; it doesn't have fennel, so it's not as strong."

Sechler is a ruddy-faced, broad-shouldered man who looks like he likes to eat, but the spice combinations he was shoving my way had nothing to do with dinner. They were feed additives he had formulated for his chickens at Bell & Evans, one of the largest privately held poultry companies in the United States. They serve as natural equivalents to antibiotics: compounds that do not build up resistance yet allow Sechler to forgo growth promoters and preventive drugs.

Sechler had arrived at the spice mixes by poring over old poultry archives, interrogating old-timers, and experimenting. "I can't even tell you all the things we tried over the years," he told me.

"Fermented soybeans. Apple cider vinegar. We bought a trailer of garlic one time. People laughed at us. But we have eight formulas now—four for our organic birds and four for the nonorgranic—and we rotate them so that nothing in the houses or the environment can become immune."

Bell & Evans is not a niche producer, though it is small in comparison to the giants of poultry. The family-owned company, which Sechler runs with help from his children, Margo and Scott Jr. (known as Buddy), processes about 60 million birds a year. But it possesses what may be the longest record in the United States of raising broilers with no antibiotics and no disease outbreaks. It charges more for its chickens than any conventional producer, yet it has had more than 30 years of consistent growth. It is distributed through Whole Foods, the cult northeastern supermarket chain Wegmans, and small independent stores; Sechler turns away potential new customers every week. The way the business operates, and the attention to detail it requires, offer clues to what the remainder of the poultry world—and, under the new FDA rules, producers of hogs and cattle—will confront as they go antibiotic free.

Sechler grew up on a farm a half-hour's drive from Bell & Evans's headquarters in Fredericksburg, Pennsylvania. His father kept about 20 dairy cows and was a small-time middleman dealing in local chickens: buying them from farmers, butchering them, and selling them around. As a kid Sechler had his own coop of birds; one of his clearest memories is pulling open a bag of feed that his father had brought home and getting a blast of fish stench—fishmeal, his father explained. He asked why people gave fish to chickens, and his father said, "Because it's cheap." When Sechler was 14, his father fell ill. As he puts it now, "It was up to me to take things over, and so that's what I did." He had planned on college and law school. Instead, at age 16 he bought a tractor-trailer and started driving truckloads of chickens overnight across the Canadian border. When he was 24, one of the wholesalers he bought birds from asked if he would be interested in buying the business. At the time, it had 100 employees. Now it has about 1,700, which will rise to 3,000 when two new plants are completed.

He came in with an attitude: no junk. No fishmeal and no cheap thrown-away foods—no animal by-products, the organs, hides, and feathers thrown away by slaughterhouses and rendered, and no stale or spoiled goods discarded by industrial bakeries. In Pennsylvania, that includes pretzels. "If you have a chicken drinking more because there's too much salt in the diet, the droppings are wet and the chicken house gets ammoniated," Sechler said. "But people said, 'You've got to feed that to keep the price down.' Well, we took that out."

His next target was the grains they used in the feed. Some of their chickens were sickly, and he suspected the soybeans and corn they were fed were being stored carelessly and getting moldy. He stopped by the feed mill unannounced one morning and asked to inspect a shipment, expecting to smell mustiness. Instead, he inhaled something like superglue—the remains of hexane, a petroleum-based solvent used to extract soybean oil from the soybeans. He canceled the contracts and went in search of mechanically pressed soybeans instead, insisting they be raised in the United States. He threw out antibiotics next, including ionophores, and then he ruled out distillers' grains, the carbohydrate mash left over from ethanol manufacturing, which can contain trace antibiotics.

That gave him a bird with a diet he found acceptable. Then he went after the birds' living conditions. He set out rules for the company's more than 100 contract farmers, who all live within an hour's drive. Barns would have to be rebuilt, with concrete floors and windows cut into the sides. After each flock of chickens was slaughtered, every barn would be cleared of litter, scrubbed, and left empty for several weeks—a huge change from most of the industry, in which litter is composted within the barn but not replaced. When chicks went back in, they would be given the kind of enrichment that Craig Watts wanted and Leighton Cooley had questioned the need for: ramps and bales of straw to hop on and cardboard tubes to peck and play with.

The biggest innovations were at the end of the chickens' lives. In most big plants, chickens travel from farms in cages stacked in layers. They are pulled from transport crates and flipped upside-down while alive and alert, then put into ankle shackles attached to an always-moving chain. The chain drags them through an electrified bath to zap them into unconsciousness, then past a rotating knife that slits their throats, and finally into the scalding tank to loosen their feathers. Sechler paid millions to install a gas-stunning system. At his farms, chickens are inserted into containers like drawers that slot into racks like a massive office organizer. At the plant, the drawers are loaded into a tunnel filled with carbon dioxide that renders the birds permanently unconscious. On the far side, the chickens are fitted by their feet into the horizontal chain, and once dead and gutted, pulled slowly on a four-mile path through refrigerated chambers, "air chilling" (as the Label Rouge producers do) instead of cooling in a bath of chlorinated ice that could spread disease organisms. Meanwhile, the drawers keep going along a conveyor belt, traversing an arrangement like a giant dishwasher that sanitizes them before they are returned to the racks and the company-owned trucks.

Sechler drove me out to the parking lot behind the plant to watch the trucks be washed as well. "Have you ever been behind a chicken truck on a highway?" he asked. (I had followed plenty at this point, in several states.) "Those chickens look like they've been chocolate-dipped. That's how you get *E. coli* and *Salmonella* coming into the plant from the farm, and going back to the farm."

Sechler collects old chicken memorabilia; his office is covered in framed prints and plaques, and the cabinets hold four deep drawers of old poultry journals published as far back as the 1800s. He pulled out handfuls of them, spread them across his desk, and opened one that was marked 1947, one year before Thomas Jukes's experiment and the Chicken of Tomorrow contest transformed American poultry. "There is no sanitary measure more important than disinfection," he read. He flipped it shut again.

"Every magazine here, for a hundred years, tells you to clean out, disinfect, and give a chicken a good environment," he said. "People ask me, 'Where do you get your ideas?' You just have to go back and look at the history. A hundred years ago, they did a better job growing chickens than we do today."

Sechler's approach is validated by research. In January 2010, *Consumer Reports,* which has an ongoing project examining food safety, revealed that it had run tests on multiple brands of super-market chickens, looking for *Salmonella* and *Campylobacter.* Out of all the brands it tested, the magazine said, only the Bell & Evans chickens—whole broilers from the company's organic line—were completely free of both organisms. (But, the company added, it tested only eight Bell & Evans chickens, too few to guarantee that every bird the company produced would be equally clean.)

The approach seems to be endorsed by the marketplace as well. After a few wobbly years in the 1980s, when Sechler wondered whether his ideas had put him too far ahead of the market, the company has experienced at least 10 percent growth year over year. He is now building a new larger plant to process more birds and expand the company's cooked products, such as breaded breasts and tenders; a new rendering plant to create organic dog food from slaughterhouse discards; and a new hatchery furnished with Dutch systems that let chicks obtain food and water instantly, similar to what the Wingenses use. When he showed me the fittings—layered trays that cradle the eggs, trap the broken eggshells, and let the wet fluffy chicks drop gently onto a clean surface where they can eat and drink and stretch—he was as giddy as a kid with a new toy.

"I wouldn't say we're going back in time, exactly," he told me. "But in some ways we are, because we're going back to where there's more care for the animal, where we're not choosing the least-cost way of doing things. And at the end of the day we'll have better results."

———✧———

FOR ALL ITS ATTENTION to cleanliness and comfort, Bell & Evans buys mainstream genetics and raises its birds mostly indoors because Sechler thinks that is more sanitary. (Its organic chickens, about a third of its production, are given access to the outdoors.) That makes them a perfected version of conventional chicken: clean, high welfare, antibiotic free. Sechler hopes to break free of the transnational genetics suppliers; find his own sources; and buy longer-lived, sturdier birds. But even when he does, his chickens probably will not be the fundamentally different animal that White Oak's pastured birds are.

That was to Sechler's benefit, because White Oak's birds struggled to find a market. But in 2014, a White Oak supporter took awareness of their chickens to, literally, a higher level—about 30,000 feet higher.

"This rack is chicken stock," Linton Hopkins said, flicking broad fingers at industrial shelving stacked with bulging, vacuum-sealed bags. "This one is vegetables. This shelf is gravy. And over here we have the chicken breasts."

Hopkins, a bullet-headed, bulky man wearing a chef's jacket with all the buttons done up, was standing inside a walk-in refrigerator, enumerating the ingredients for chicken pot pies. There were a lot of ingredients, destined to make several hundred pies, and a good portion of them—the diced meat, the gravy, and the stock the vegetables had been poached in—all originated at White Oak. The walk-in was in a 2,000-square-foot prep kitchen, a few storefronts down from Restaurant Eugene, the flagship of Hopkins's small restaurant empire and one of Atlanta's most important restaurants.

But the hundreds of pies weren't headed down the street to the tiny restaurant's back door. Once they were assembled, in sturdy square ramekins with splash-proof slanted sides, they would be loaded into a truck and driven 15 miles south, to the industrial catering kitchen that serves Delta Air Lines. There they would be stacked into budge-proof trays and loaded into the business-class galleys of Delta flights headed across the Atlantic. Hopkins had won a cook-off to be Delta's next executive chef. He was building a menu that showcased small southern food producers, and White Oak was a cornerstone.

Hopkins is restaurant royalty in Atlanta. The son of a well-liked Emory University neurologist who continued to see patients long past his retirement age, he attended the Culinary Institute of

America and cooked in New Orleans and Washington, D.C. In 2004 he returned home to build Restaurant Eugene, a fine-dining jewel box set on the edge of Buckhead, in-town Atlanta's oldest-money neighborhood. The restaurant was small and perfect, with sincere, elegant service. But critics carped that it felt as old-fashioned as the graying neighborhood residents who flocked to it. When it had been open not quite a year, a burst pipe flooded the space—almost ruining it but giving Hopkins a chance to rethink. He refreshed the decor, reexamined the menu, and spun the restaurant's orientation toward a farm-focused version of southern fine dining, producing foie gras with pot likker and duck with beet caramel. He achieved a hit—and shortly afterward, the James Beard Foundation and *Food and Wine* Best New Chef awards. He was on his way.

It troubled him, though, that a high-end restaurant might be a small-scale producer's best channel to the public. As much as he wanted to reward his regulars with novelty and exclusivity, he wanted to democratize authentic food as well. He founded a farmers' market, which grew into Atlanta's largest. A grass-fed burger served in tiny quantities late at night at his second restaurant, a gastropub, became a cult object among city foodies; he spun it into a set of burger kiosks at the city's largest sports stadium. Hopkins is expansive and cheerful, but he brooded over how best to support artisans. Bad weather and busy weekends could deter farmers' market customers. He had seen growers sit an entire morning without a sale, and because people thought of outdoor markets as places for bargains, prices had to be kept low.

"Food is more important than restaurants," he told me. "How do we grow a sustainable system that doesn't depend on them? Can we make artisanal scale?"

In 2013, he got a chance to try. Delta, based in Atlanta, staged a "Cabin Pressure Cook Off," a reality TV-style web series, as an audition for the airline's next celebrity chef. (It already had two, Michelle Bernstein of Miami and Michael Chiarello of San Francisco, along with a master sommelier.) The goal was to find someone who would develop meals for what it calls "high-value customers," the front-of-the-plane big spenders who expect pampering as part of their $5,000 or $10,000 tickets. Hopkins won and decided to treat the airplane menus like restaurant ones, making them novel, seasonal, and dependent on small producers. (The airline, once it got over its shock, liked the concept so much that it dubbed the idea "farm to tray-table"—and trademarked the phrase.) Tuning a menu to the vagaries of weather and harvest might be normal in restaurants, but it would be radical in the skies. Airlines require food to be rigorously consistent, because even insignificant differences—an extra ounce on the chicken breast, another cookie on the coffee tray—can irritate customers and also add up, over a year, to carrying extra weight and burning extra fuel.

Hopkins was responsible for dinner and breakfast in the first-class cabins of the planes headed from Atlanta to Europe: about 3,920 meals each week, the equivalent of a medium-size restaurant. He elected to recreate a role that already exists in food service—the middleman aggregator and wholesaler, exemplified by the giant international corporation Sysco—but take on the logistics himself. That meant ordering cheeses to be made six months in advance to allow them time to age and recruiting a team of peach orchard owners because each variety would be at its best for only a few weeks. "People think that 'farm to table' means heirloom tomatoes in the summer," he told me. "But what it really means is raising standards throughout the entire process of sourcing."

White Oak was a crucial piece. Hopkins had known Will Harris since Eugene's earliest days, when Harris was raising only beef. Once White Oak ventured into growing chickens, Hopkins bought them, but Eugene's volume was small. This deal held the promise of bigger sales, and it reinforced the sustainable values that both men espoused because it gave the Harrises an opportunity to unload something of little use to them. "Chicken stock is one of those fundamental things you have to have in a kitchen," Hopkins said. "But large-scale food service uses chicken base, which has a lot of starches and chemicals and colorings. I'm a purist; I don't want to see that. The best chicken stock uses laying hens, and Will said he had all these hens that weren't laying any more that he really didn't know what to do with. So I offered to buy them."

On top of the hens, Hopkins bought White Oak's broilers. He added the legs and backs to the stock his sous chefs brewed in a freestanding tilt-kettle, a bathtub-sized stockpot heated with calibrated steam, and poached the breasts and carved them into chunks that were slipped beneath a puff-pastry lid. On board, it made a beautiful presentation, bubbling and crisp. The narrow, elegant menu that the flight attendants presented before dinner prominently showcased White Oak's name.

There was just one problem. The pot pies were one of four entrees on an evening meal that was served to between 30 and 40 business-class customers on 77 flights per week. Doing the math, that was probably fewer than 800 pies, each using just a few ounces of meat. It didn't add up to much chicken. "We're bringing in about 100 a week," Hopkins admitted.

It was a disappointment for Will and Jenni Harris. They valued their relationship with such an influential chef, and when he talked to them about the Delta plan, they couldn't help but envision big sales to follow. Once executed, though, it was not a major source of income. But it was a way to bring White Oak's products and values to the attention and the palates of connected, influential consumers—people who might remember the name attached to the chicken they ate, seek it out at home, and tell their friends.

—⟋𝔪⟍—

THE POPULARITY OF Bell & Evans chicken, and the persuasiveness of White Oak's pastoral image, open a space in the market for medium-scale businesses raising poultry without antibiotics. The endorsement of antibiotic-free raising by major poultry production companies such as Perdue and Tyson, and the embrace of antibiotic-free chicken by food-service giants such as McDonald's and Chick-fil-A, guarantee market share—and the consumer and activist pressure that persuaded the companies to make those changes makes it likely there will be buyers.

But some poultry companies rejected the trend, or followed it only reluctantly. In May 2015, Joe Sanderson, Jr., CEO of Sanderson Farms, one of the largest U.S. poultry companies, said he would stand fast in rejecting antibiotic control, telling the *Wall Street Journal*, "We have a duty to take care of the animals." In August 2016, the company released an ad campaign calling antibiotic-free raising a "gimmick." Even Donnie Smith, CEO of Tyson, told the *Guardian* in April 2016, "I don't see a problem" with antibiotic use and "I'm not sure any of the science I've reviewed points to a straight connection" to antibiotic resistance—despite the company's 2015 pledge to eliminate human-use antibiotics in its broilers. (Smith's successor, Tom Hayes, announced in February 2017 that the company had taken its broilers "N.A.E.," no antibiotics ever, as Perdue had done.) At the 2016 International Production and Processing Expo in Atlanta, the largest poultry industry meeting in the world, a Nebraska rancher and radio host named Trent Loos declared in a speech: "The consumer is leading us down a very dangerous path. We can either cave in, such as the United Kingdom has done, or put our feet down."

Barely a week goes by without one of the trade magazines for meat production running an article that explores, soothingly or alarmingly, what lies ahead. *PoultryUSA* headlined its April 2016 issue, "Secrets to Antibiotic-Free Poultry Production," only to begin its six-page article by admitting: "There is no single secret—success is mainly from superior execution of poultry production basics."

Nevertheless, it seems likely that in the United States, poultry will be the first sector of livestock production to abandon routine antibiotic use and will challenge hogs and cattle to follow. And they may: In February 2016, Tyson announced that it would convert 5 percent of its hog production, which totals more than 20 million hogs a year, to a vegetarian diet and antibiotic-free raising. However, that was just a few months after the National Pork Producers Council ran a full-page ad in the *Wall Street Journal* assailing Subway for announcing a no-antibiotics pledge for the chicken, pork, and beef it serves. "This policy decision could put our food supply in jeopardy," the group argued.

If the FDA rules remain in force and the companies uphold their commitments, then removing antibiotics from meat production will reduce the threat that antibiotic use poses to farm workers and farm neighbors. It will reduce the persistent pumping of resistant bacteria into the environment and the flow of unmetabolized antibiotics into watersheds, the intact compounds that force the further evolution of bacteria and may be affecting the microbiomes of us all. Most of all, it will greatly reduce the amount of resistant foodborne illness—and, just as much, the silent threat posed by resistance genes that migrate via plasmids away from the foodborne organisms that are carrying them, creating outbreaks of drug-resistant infections far from the farms that gave them life.

The big question that the reversal of routine antibiotic use poses is: Will it trigger further change in the lives of birds themselves? Can the industry that was created by Thomas Jukes's experiment—the development of the Chicken of Tomorrow, the rise of intensive production, the bird welfare issues that weighed so hard on Craig Watts—be reformed?

It is common to hear people say that to take away antibiotics is to dismantle what we think of as factory farming, since the loss of growth promoters and especially preventive antibiotics makes intensive indoor production unsustainable. And this is true—but only to a point. Chickens cannot be raised in the classic conditions of industrial confinement, with distorted genetics and neglected hygiene, without the crutch of routine antibiotic use. But the experiences of Perdue and Bell & Evans in the United States and farmers such as the Wingenses in the Netherlands demonstrate that chickens can be raised indoors in an intensive manner, with minimal or no antibiotics, if other compensatory moves are made: if the birds are given vaccine and other supplements, and space and a chance to exercise, and light. Those changes are intended to sustain chickens' immune systems in the absence of antibiotics, but they also improve birds' welfare and their experience—possibly even their enjoyment—of their lives. To remove antibiotics is not to ruin intensive agriculture; it is to bring intensive agriculture to a place where animal welfare becomes part of the business model. Once introduced, it breaks the old assumptions from the inside.

Most recently, it is fracturing the genetic hegemony of the short-lived, distorted conventional chicken. In March 2016, the Global Animal Partnership, an advocacy group that sets welfare standards for Whole Foods and other stores, announced that it had persuaded major retailers to turn away from the plofkip, the exploding chicken, and accept a slower-growing chicken breed. Instead of 38 to 42 days, the new birds will live 56 to 62. That is an improvement, though not close to the 84 days of a pastured bird or the 100 that Label Rouge lets some of its roasters live to. The new campaign affects only about 277 million of the almost 9 billion broilers raised in the United States each year. But a decade earlier, the same group engineered an introduction of cage-free eggs to United States food sales, also with Whole Foods. At the time, cage free seemed an incredibly niche concern, something that would be purchased by only a small subset of consumers. A decade later, 10 states have mandated cage-free raising and 35 major food companies have committed to it. It has become mainstream—and that success models a possible future for chickens raised without antibiotics and with better welfare as well.

In a sign of how easily the improvement will be to achieve, the slower-growing birds will be supplied by the same transnational genetics companies that produce short-lived industrial chickens. Better birds were always available. But no one felt it was possible to ask.

What the abandonment of antibiotics should do, if it succeeds, is to return diversity to chicken production—not just in genetics but in farm size, sustainability, price, and taste. That sounds like a wonky and qualified gain, the sort of thing that would be written in a consultant's report. But chicken is the most popular meat in the industrialized world, soon to become the most-eaten meat worldwide. To change the production of chicken is to change the meat economy of the planet and everything that it affects: land use, water use, waste disposal, resource consumption, the role of labor, concepts of animals' rights, and the diets of billions of people.

—⟋⟋⟍—

ON ANY DAY THAT HE IS HOME in Georgia, Will Harris likes to spend the last hours of daylight driving to each of his pastures and observing each of his herds and flocks, with a big glass of cheap Merlot tucked between his seat and the parking brake and a short-barreled rifle on the dashboard, handy for predators and snakes. On one of my visits, in 2015, we pulled up in his Wrangler to a field of broilers and parked to watch them as they wandered back to their nighttime shelters. The low sunset light glowed on their red feathers, and as they hopped into the trailers to cluster together, the breeze carried their soft, querulous conversations. I asked him to tell me again why they were so different from a supermarket bird.

He started to talk about industrial chicken and ran himself into a tangle of adjectives. He started again, and swore. He stopped, sighed, resettled his white Stetson up and down, and tried a third time.

"Here's the thing," he said. "They just bred all the chicken out of the chicken."

EPILOGUE

IT WAS AN UNSEASONABLY WARM Wednesday in September 2016, and steamy in the canyons between the skyscrapers that line the East Side of Manhattan, but the air was cool and crisp inside the modernist buildings that house the United Nations. Men in European-cut suits and women in practical flats moved quickly between the conference rooms where the heads and ambassadors of 193 governments were meeting in the annual General Assembly. The Assembly can be an anodyne gathering; the debates are hushed and abstract, full of minutiae on weapons treaties and border disputes. But this morning there was a humming energy in the building, and a surging crowd of visitors who would otherwise never have set foot inside. In an unexpected development, the UN was about to tackle the worldwide problem of antibiotic resistance, staging a "high-level meeting" to explore the threat.

This was unprecedented. The General Assembly almost never considers health problems. It had done so only three times since the UN was created in 1945, to assess the global burden of chronic diseases such as cancer, and to respond to the emergencies of Ebola and AIDS. There were plenty of people across the world who still had no idea that resistant bacteria were a problem and many more who did not know how widespread or urgent the problem was. Yet the UN had not waited for awareness to build. It had stepped out in front.

In the sleek, tall Trusteeship Chamber on the third floor of the UN tower, a wide auditorium paneled in pale wood and ringed near the ceiling with booths housing simultaneous translators, Secretary-General Ban Ki-moon, the UN's top official, leaned in to a microphone.

"Honorable ministers, excellencies, ladies and gentlemen," he said. "Antimicrobial resistance poses a fundamental, long-term threat to human health, sustainable food production and development. It is a very present reality—in all parts of the world; in developing and developed countries; in rural and urban areas; in hospitals; on farms and in communities. We are losing our ability to protect both people and animals from life-threatening infections."

By the end of the day—after panels of experts laid out the complexities and representatives of 70 governments, rich and poor, lined up to express concerns—the members of the General Assembly voted to act immediately. They committed to creating better surveillance and monitoring of new resistant infections and to supporting research and development of new drugs. They agreed that each government would immediately create a national plan to control antibiotic use and report back to the UN in 2018 on how far they had gotten in creating change. And they commissioned the UN to create an international coordinating body, similar to what it had done for AIDS decades before, that would monitor what happened next.

The declaration the governments voted in called antibiotic resistance "the greatest and most urgent global risk."

For scientists and strategists who had been laboring to make people understand the threat, the day felt like a triumph. It also fell frustratingly short. The declaration did not create any funding or set any usage limits. But the UN session enshrined antibiotic resistance as a grave global peril—and in every speech and statement, the session underlined that antibiotic overuse in agriculture was as crucial an aspect of the problem as misuse in medicine is. The warnings of decades were being listened to at last. Farm antibiotic use, and the need to rein it in, were finally on the global agenda.

—⁂—

THE UN SESSION SEEMED to come out of nowhere, but behind the scenes, the two years that preceded it were a time of extraordinary international momentum on resistance, unlike anything since the Swann Report prompted Kennedy to act in the 1970s. The FDA's sudden new guidances—arriving at the end of 2013, 36 years after Kennedy's defeat—were probably the spark. In September 2014, President Barack Obama issued an executive order making resistance a national priority and installed a new permanent body of experts within the government, the Presidential Advisory Council on Combating Antibiotic-Resistant Bacteria. Simultaneously, in Britain, Prime Minister David Cameron asked Lord Jim O'Neill, the former chief economist of Goldman Sachs, to make a hard economic—not just medical—case for taking action. The group O'Neill created, dubbed the Review on Antimicrobial Resistance, achieved an estimate of the global toll of resistance that became instant headline news: 700,000 deaths worldwide per year, due to rise to 10 million if nothing is done.

Early 2015 brought a second set of jaw-dropping numbers. A group of researchers looked at how earnings are rising in the developing world—in what O'Neill in his former job dubbed "the BRICs," for Brazil, Russia, India and China—and tried to calculate how extra income would affect demand for meat and consumption of farm antibiotics. They predicted that if nothing were done to change agricultural practices, factory farms would multiply so rapidly that in 15 years, they would use two-thirds more antibiotics than they do now: globally, 105,596 tons. By 2030, China would be feeding its meat animals 30 percent of all the antibiotics produced in the world.

Those dire estimates and the new actions by the United States and United Kingdom triggered action internationally. In January 2016, the World Economic Forum in Davos called for the development of new antibiotics and diagnostics and "more judicious use of antibiotics in livestock." In May 2016, the 194 governments in the governing body of the World Health Organization committed their countries to combating resistance. The same month, the G7 group of industrialized nations, meeting in Japan, affirmed that resistance has to become an international priority. Two weeks before the General Assembly met, the summit of the G20 nations, comprising Western countries and the developing world—and headed that year by China, the largest producer and consumer of antibiotics on the planet—said antibiotic resistance "poses a serious threat to public health, growth and global economic stability." The stage had been set for the United Nations to act.

—⁂—

IN THE MIDST OF THE GENERAL ASSEMBLY, I went in search of chicken.

New York magazine had noted that the city was in the midst of a roast chicken obsession. This was counterintuitive, because for a chef—or a diner—chicken is the safest choice; "I'll just have the chicken," in response to a complicated menu, is practically a punch line. Yet everything that had been happening out on poultry farms—forgoing antibiotics, letting birds live longer, giving them windows and light and access to the outdoors—were investments in the idea that chicken is not just the world's easiest, cheapest protein but something intrinsically worthwhile. If intensely competitive restaurants in one of the world's best restaurant cities were taking chicken seriously as a dish, perhaps the concept that chicken is worth the investment—of money, of attention, even of sympathy—was percolating down to plates.

As it turned out, there was plenty of extraordinary chicken in New York City: expensive birds stuffed with whole eggs and forcemeat or hoisted in front of hand-forged fire-rigs or presented on burning hay. What I hoped for, though, was not luxury but integrity: a bird that had experienced a longer, better life than a commodity chicken—one that had been allowed outdoors, permitted to exercise, and encouraged to feed itself. I wanted a bird, like my French market chicken, that I could imagine eating every day. I found it at Marlow & Sons, on the western edge of Brooklyn under the rising ramp of the Williamsburg Bridge. The tiny place, lit by dim bulbs and paneled in found wood, has been open since 2004. In that time its "brick chicken" has never been off the menu: half of an outdoor bird from a small property upstate, Snowdance Farm, pan fried under a weight for maximum contact between bird and pan. The skin was crunchy and dense, and the flesh had the savor and chew of muscles that had been used. It was salty and intense, brightened with lemon, leaking with juice. If it was not poulet crapaudine, it was close. It was delicious.

And it was a justification on a plate for what abandoning antibiotic-fueled farming has permitted: production that unwinds itself from the worst aspects of industrialization; that protects the health of the livestock, the farmer, the eater, and those with no connection to farms; that allows an animal's life, and the sacrifice of that life, to matter.

The political commitment expressed at the United Nations is the product of the evidence of decades: Raising meat animals with routine antibiotic use creates resistant bacteria that threaten to swamp the world. But the commitment of chefs and supermarkets and poultry producers to offering chicken raised without antibiotics is an expression of something else: the power of consumers to make a market change its mind. Without pressure from buyers—hospitals, school systems, parents—the behemoth of Big Chicken would never have changed course, no matter how many outbreaks were investigated or how many welfare scandals were exposed. Consumers who chose meat raised without routine antibiotics forced the United States to take the path that other countries had already followed. They forged ahead of both policymakers and producers, and pulled them along behind.

The challenge is to keep that momentum going. The antibiotic problem is not solved. It is not solved in the United States, where the new FDA rules still permit preventive antibiotic use with few limitations. It is not solved in Europe; just before the General Assembly met, a study found that one in four British supermarket chickens is carrying multidrug-resistant bacteria, and on the day of the UN meeting, researchers disclosed a new type of MRSA, causing infections in residents of Denmark, which might have entered that country on imported poultry. The problem is certainly not solved in South America and South Asia and China, where planners have difficulty imagining how to grow enough protein without the crutch of antibiotic use. Global events—the 2016 election results in the United States, the Brexit vote in the United Kingdom, the turn toward nationalism being expressed all over Europe—could raise political barriers that would stop this momentum dead. It is too soon right now to know.

Antibiotic resistance is like climate change: It is an overwhelming threat, created over decades by millions of individual decisions and reinforced by the actions of industries. It is also like climate change in that the industrialized West and the emerging economies of the global South are at odds. One quadrant of the globe already enjoyed the cheap protein of factory farming and now regrets it; the other would like not to forgo its chance. And it is additionally like climate change because individual action feels inadequate, like buying a fluorescent lightbulb while watching a polar bear drown.

Changing farm antibiotic use seems dauntingly difficult. But that does not mean it is not possible. The willingness to relinquish antibiotics of farmers in the Netherlands, as well as Perdue Farms and other companies in the United States, proves that industrial-scale production can be achieved without growth promoters or preventive antibiotics. The stability of Maïsadour and Loué and White Oak Pastures shows that medium-size and small farms can secure a place in a remixed meat economy. Whole Foods' pivot to slower-growing chicken—birds that share some of the genetics preserved by Frank Reese—illustrates that removing antibiotics and choosing birds that do not need them returns biodiversity to poultry production.

All of those achievements are signposts, pointing to where chicken, and cattle and hogs and farmed fish after them, need to go: to a mode of production where antibiotics are used as infrequently as possible—to care for sick animals, but not to fatten or protect them. That is the way antibiotics are now used in human medicine, and it is the only way that the utility of antibiotics and the risk of resistance can be adequately balanced.

This will not be easy to achieve. It will take time to persuade buyers in industrialized nations that the cheapest chicken is not the best chicken, and to encourage developing countries toward models of raising livestock that do not reproduce the worst excesses of factory farming. Those new models might be as high tech as in the Netherlands—or they might be founded on an indigenous low-intensity system, the South American or Asian equivalent of a Label Rouge farm. Yet it cannot take too much time, because as the relentless pace of bacterial evolution illustrates, there is not a lot of time to spare.

The goal is not only to change farming, and not only to enlarge the market for safe, sustainably raised meat. It is finally to demonstrate that Thomas Jukes's original choice—to feed the world with cheap protein at the risk of sickening the world with resistant bacteria—was a false choice. It is possible to have poultry production without antibiotic resistance and intensive farming without environmental destruction, and to use those lessons to transform the raising of other meat animals as well. It requires only the will. And a willingness to finally turn chicken back into chickens, again.

ACKNOWLEDGMENTS

A WRITER WHO SPENDS YEARS on a project cannot finish a book without the help of dozens, if not hundreds, of people. But this book exists particularly thanks to the generous storytelling of several individuals. The first to thank is my husband, Loren D. Bolstridge III. When we married a decade ago, I began to hear his family's stories of being small farmers in Maine going back into the 1800s. I was always a city kid, excluding the far suburbs of London, and this was a revelation. Next, Tara Smith, previously of the University of Iowa and now of Kent State University, introduced me to the mystery of antibiotic use in meat animals and the antibiotic resistance that resulted from that use. Dr. Robert Tauxe of the Centers for Disease Control and Prevention reminded me, each time we met, that I ought to be taking a closer look at *Salmonella*. And Holly Tucker, professor at Vanderbilt University and author of *Bloodwork* and *City of Light, City of Poison*, told me the sad but fascinating story of how her grandparents were driven inside their small home in southern Indiana when a giant turkey farm moved in next door, blanketing the hills with the smell of manure and, with its spotlights, chasing the stars away.

All of those stories—of the realities of farming, the effect of antibiotics, the dangers of foodborne illness, and the overwhelming impact of industrial poultry production—combined for me to produce this exploration of the entwined histories of agriculture and antibiotic use. I hope I have done justice to their accounts.

Those ideas became a book thanks to the unfailing advocacy of my agent, Susan Raihofer, and the astute stewardship of my editors, Hilary Black and Anne Smyth, at National Geographic Books. Susan Banta, fact checker extraordinaire, was diligent in keeping me from mistakes; whatever remain are my responsibility alone. I thank Pat Singer for introducing us.

I was very lucky to benefit from the support of two fellowship programs in researching this book. First, the Knight Science Journalism program at MIT, under its former director, Phil Hilts, gave me a year-long grant to explore antibiotic use. After that, the Schuster Institute for Investigative Journalism at Brandeis University, headed by Florence Graves with the indefatigable assistance of Claire Pavlik Purgus and Lisa Button, has generously supported me with amazing student assistants, especially Jay Feinstein, Aliza Heeren, and Madeline Rosenberg. I was very lucky also to have the support of these librarians: Alex Willett at Brandeis, Michelle Baildon at MIT, Fred Burchsted at Harvard University's Widener Library, and Daniel Hartwig and Tim Noakes at Stanford University's Department of Special Collections, who granted me access to the papers of Donald S. Kennedy. I also thank Cornell University's Albert R. Mann Library for allowing me access to the papers of Robert Baker.

One of the biggest challenges of writing a book is figuring out how to earn a living at the same time. A number of editors kindly helped. Thus, portions of this text had their origin, though they since have been rewritten, in stories for Sara Austin, formerly of *SELF* and now executive editor of *Real Simple;* Reyhan Harmanci, now at First Look Media, previously of Modern Farmer; Laura Helmuth, formerly of Slate, now of the *Washington Post;* Corby Kummer of the *Atlantic;* and Nancy Stedman of *More* magazine. Some of those stories were supported and placed by Sam Fromartz, Tom Laskawy, and the Food and Environment Reporting Network, who are ferocious partisans for reporting on food, agriculture, and the environment and a joy to work for. I also thank Bill Wasik and Claire Gutierrez of the *New York Times Magazine* for assigning me to investigate the 2015 avian influenza epidemic; an episode based on that story was cut for space, but the experience of reporting it was irreplaceable. (Thanks also to Eric Nelson, now of HarperCollins Publishers, for telling me years ago to look into chicken nuggets.)

This project was possible only because many people trusted me not only with their stories but also with their personal networks. I first must thank Dr. Stuart B. Levy of Tufts, the dean of antibiotic resistance studies in the United States; Michael T. Osterholm, of the University of Minnesota, who rang the alarm early on the hidden connections between resistant outbreaks; and Dr. Robert Tauxe, already mentioned, who has been a generous mentor as I sought to understand the impact of foodborne illness. There is no way to extend enough thanks to Lance B. Price, founder of the Antibiotic Resistance Action Center at the George Washington University, and Laura Rogers and Nicole Tidwell there.

Following them, in no order but alphabetical, because there is no way to rank their extraordinary assistance to me: at the ASPCA,

Suzanne Mcmillan; at Bell & Evans, Audrey King and Scott Sechler. At the CDC, Dr. Thomas R. Frieden, the director during the years I was working on this book; and Dr. Tom Chiller, Nicole Coffin, Elizabeth Lee Greene, Dana Pitts, Matthew Wise, Laura Gieraltowski, Dr. Jolene Nakao, and many others. At Compassion in World Farming, Leah Garces. At Cornell, Robert Gravani; thank you also to the family of Robert Baker, especially Dale and Michael Baker. At Delta Air Lines, Kate Modolo. At Farm Forward, Ben Goldsmith. At the U.S. Food and Drug Administration, Michael Taylor, Dr. William Flynn, and Megan Bensette; and also Thomas Grumbly, who was Donald Kennedy's right-hand man in the 1970s and now heads the Supporters of Agricultural Research Foundation. In France, Virginia Dae, Sabine Edelli, Marie Guyot, Maxime Quentin, Pascal Vaugarny, and the farmers of Maïsadour and Loué—and also, especially, my fellow fellow at MIT, Yves Sciama and his wife, Elise. At the Humane Society of the United States, Anna West. At National Geographic, Erika Engelhaupt, April Fulton, and Jamie Shreeve. At the Natural Resources Defense Council, Avinash Kar and Dr. David Wallinga. In the Netherlands, Dr. Jan Kluytmans, Dr. Dik Mevius, Dr. Andreas Voss, Gerbert Oosterlaken, Eric van den Heuvel, and Kor Mast. At the Pew Charitable Trusts, Allan Coukell, Dr. Karin Hoelzer, and Katherine Portnoy, and formerly Dr. Gail Hansen, Shelley Hearne, Alicia LaPorte, and Joshua Wenderoff. At the Review on Antimicrobial Resistance, Lord Jim O'Neill and Hala Audi, Will Hall, and Jeremy Knox, and also Dr. Jeremy Farrar of the Wellcome Trust and Tamar Ghosh of the Longitude Prize. At Small-r Films, Michael Graziano. At the Stone Barns Center for Food and Agriculture, Craig Haney, Martha Hodgkins, Fred Kirschenmann, and Laura Neil. Among the Waterkeeper Alliance, Kathy Phillips in Maryland and Larry

Baldwin and Rick Dove in North Carolina. In Washington State, Bill Marler and his staff and the indefatigable Dr. Reimert Ravenholt, who has now appeared in two of my books, and has been a delight to interview each time.

While reporting this project, I visited more than a dozen poultry farmers in a half-dozen states who asked to be anonymous in exchange for allowing me to witness large-scale poultry production. You know who you are, and I hope you know how grateful I am for your trust. Among the people in the industry whom I can name, I thank J. Craig Watts, Frank Reese, and Larry and Leighton Cooley; John and Brad Moline of West Liberty Foods in Iowa; David Pitman of Mary's Chicken in California; Paul Helgeson of Just Bare Chicken in Minnesota; and Stuart Joyce of Joyce Farms in North Carolina. Also, Dr. John Glisson of the U.S. Poultry and Egg Association, Tom Super of the National Chicken Council, Dr. David Swayne of the USDA Southeast Poultry Research Laboratory, Susan Sanchez of the University of Georgia College of Veterinary Medicine, and Jennifer Reinhard of the U.S. Farmers and Ranchers Association. I am especially grateful to the Harris family of White Oak Pastures in Georgia for their lessons in both animal husbandry and hospitality: Will Harris III and his wife, Yvonne; Jenni Harris and Amber Reece; Jodi Harris Benoit and John Benoit; and Brian Sapp and Frankie Darsey, along with Gretchen Howard and Melissa Libby. (Also, thank you, Helen Rosner, for telling me to go to Marlow & Sons.)

Writing is a solitary and difficult endeavor and no one gets through it without friends. Huge thanks to Richard Eldredge, Krista Reese and William Houston, Susan Percy, Dean Boswell, Mark Scott, Diane Lore, Frances Katz, Carol Grizzle, and Mike and Nancy Reynolds; the members of my writers' accountability groups, which confidentiality rules forbid me to name; and my siblings, Robert, Matthew, and Elizabeth McKenna, and our uncle, professor and author Fr. Robert Lauder, to whom this book is dedicated. (And thanks also to Walsh Whiskey Distillery of Carlow, Ireland, for Writers' Tears, which functions something like a vaccine.)

I began these acknowledgments with my husband, Loren, and it is appropriate to end with him also, since he is always my beginning and my end. I could not have undertaken this project without his love and support, and I would not have wanted to.

NOTES

Epigraphs

5 "Albert Camus": Camus, *La peste* (Paris: Éditions Gallimard, 1947). My translation.

5 "Henry Saglio": *Problems in the Poultry Industry. Part III. Hearing Before Subcommittee No. 6 of the Select Committee on Small Business*, p. 59.

Prologue

12 "63,151 tons": Van Boeckel et al., "Global Trends in Antimicrobial Use in Food Animals."

12 "wrote in 1971": Sawyer, *The Agribusiness Poultry Industry*, p. 225.

Chapter 1

17 "had never felt so sick": The reconstruction of Rick Schiller's experience is based on interviews with him; with his attorney, Bill Marler; with personnel from the Centers for Disease Control and Prevention, including Thomas Chiller, Jolene Nakao, Robert Tauxe, Matthew Wise, and Laura Gieraltowski; on CDC investigative documentation, along with documents from the California, Oregon, and Washington state health departments; on legal documents, including Marler, "Final Demand Letter to Ron Foster, President, Foster Farms Inc., in re: 2013 Foster Farms Chicken Salmonella Outbreak, Client: Rick Schiller"; and on contemporaneous news stories.

20 "that every year sickens": Majowicz et al., "The Global Burden of Nontyphoidal *Salmonella* Gastroenteritis."

22 "634 known victims": U.S. Centers for Disease Control and Prevention, "Multistate Outbreak of Multidrug-Resistant *Salmonella* Heidelberg Infections Linked to Foster Farms Brand Chicken (Final Update)."

22 "possibly thousands more": Voetsch et al., "FoodNet Estimate of the Burden of Illness Caused by Nontyphoidal Salmonella Infections in the United States." The CDC estimates that for every salmonella case confirmed by a laboratory, 38 others go unrecorded.

22 "the anomaly, PulseNet": U.S. Centers for Disease Control and Prevention, "PulseNet: 20 Years of Making Food Safer to Eat."

24 "'the greatest and most urgent global risk'": President of the General Assembly, "Draft Political Declaration of the High-Level Meeting of the General Assembly on Antimicrobial Resistance."

24 "at least 700,000": Review on Antimicrobial Resistance, "Antimicrobial Resistance: Tackling a Crisis for the Health and Wealth of Nations."

24 "23,000": U.S. Centers for Disease Control and Prevention, "Antibiotic Resistance Threats in the United States, 2013."

24 "more than 63,000": Titus and Center for Disease Dynamics, Economics and Policy, "The Burden of Antibiotic Resistance in Indian Neonates."

24 "two million annually": U.S. Centers for Disease Control and Prevention, "Antibiotic Resistance Threats in the United States, 2013."

25 "will cost the world": Review on Antimicrobial Resistance, "Antimicrobial Resistance: Tackling a Crisis for the Health and Wealth of Nations."

25 "for as long as antibiotics have existed": Time line reconstructed from U.S. Centers for Disease Control and Prevention, "Antibiotic Resistance Threats in the United States, 2013"; Marston et al., "Antimicrobial Resistance."

26 "Eighty percent": U.S. Centers for Disease Control and Prevention, "Antibiotic Resistance Threats in the United States, 2013."

26 "more than half": World Health Organization, "The Evolving Threat of Antimicrobial Resistance: Options for Action."

26 "most of those drugs": Marston et al., "Antimicrobial Resistance."

26 "nearly two-thirds": Ibid.

27 "26 percent": Food and Drug Administration, "National Antimicrobial Resistance Monitoring System (NARMS) Integrated Report 2012-2013."

27 "another resistant pathogen": Liu et al., "Emergence of Plasmid-Mediated Colistin Resistance Mechanism *mcr-1* in Animals and Human Beings in China"; Paterson and Harris, "Colistin Resistance."

29 "more than 30": Xavier et al., "Identification of a Novel Plasmid-Mediated Colistin-Resistance Gene, *mcr-2*, in *Escherichia coli*, Belgium, June 2016."

29 "a woman in Pennsylvania": Kline et al., "Investigation of First Identified *mcr-1* Gene in an Isolate from a U.S. Patient—Pennsylvania, 2016."

29 "men in New York and New Jersey": New York: Castanheira et al., "Detection of *mcr-1* Among *Escherichia coli* Clinical Isolates Collected Worldwide as Part of the SENTRY Antimicrobial Surveillance Program in 2014 and 2015"; New Jersey; Mediavilla et al., "Colistin- and Carbapenem-Resistant *Escherichia coli* Harboring *mcr-1* and bla_{NDM-5} Causing a Complicated Urinary Tract Infection in a Patient From the United States."

29 "a Connecticut toddler": Vasquez et al., "Investigation of *Escherichia coli* Harboring the *mcr-1* Resistance Gene—Connecticut, 2016."

30 "32.6 million pounds": Center for Veterinary Medicine, "2013 Summary Report on Antimicrobials Sold or Distributed for Use in Food-Producing Animals." "four times": Food and Drug Administration, "Drug Use Review."

30 "no new drugs": Infectious Diseases Society of America, "Bad Bugs, No Drugs."

31 "twice what it was": Humane Society of the United States, "The Welfare of Animals in the Chicken Industry" and "Welfare Issues Wth Selective Breeding for Rapid Growth in Broiler Chickens and Turkeys."

32 *"Fortune* magazine reported": "Antibiotics in the Barnyard."

32 "boasted in 1975": Cook, Bumgardner, and Shaklee, "How Chicken on Sunday Became an Anyday Treat."

Chapter 2

33 "It was Christmas Day": The reconstruction of Jukes's life and his experiments is based on a number of memoir articles he wrote toward the end of his life, including: "Some Historical Notes on Chlortetracycline"; "Adventures with Vitamins"; and "Vitamins, Metabolic Antagonists, and Molecular Evolution." See also Larson, "Pioneers in Science and Technology Series." Obituaries include Sanders, "Outspoken UC Berkeley Biochemist and Nutritionist Thomas H. Jukes Has Died at Age 93"; Maddox, "Obituary"; Carpenter, "Thomas Hughes Jukes (1906–1999)"; Crow, "Thomas H. Jukes (1906–1999)."

35 "Fourteen years earlier": Lax, *The Mold in Dr. Florey's Coat*, pp. 16–20

35 "One year previously": Lax, *The Mold in Dr. Florey's Coat*, pp.154–6

36 "Three months after": Lax, *The Mold in Dr. Florey's Coat*, pp.169–72

36 "back from the edge of death": Saxon, "Anne Miller, 90, First Patient Who Was Saved by Penicillin."

36 "the Cocoanut Grove nightclub disaster": Bud, *Penicillin*, pp. 55–9

36 "isolated streptomycin": Pringle, *Experiment Eleven.*

36 "crystallized chloramphenicol": Greenwood, *Antimicrobial Drugs*, pp. 219–22

37 "desperate to find": Maeder, *Adverse Reactions*, pp. 74–77

37 "dug from a field": Brown, "Aureomycin, Plot 23 and the Smithsonian Institution."

37 "filed for a patent": Duggar, Aureomycin and preparation of same.

37 "more than one billion pounds": Bugos, "Intellectual Property Protection in the American Chicken–Breeding Industry."

38 "identified their new compound": Rickes et al., "Comparative Data on Vitamin B_{12} From Liver and From a New Source, *Streptomyces griseus.*"

38 "he had set up an experiment": This description of Jukes's experiment is reconstructed from Stokstad et al., "The Multiple Nature of the Animal Protein Factor"; Stokstad and Jukes, "Further Observations on the 'Animal Protein Factor'"; Jukes, "Some Historical Notes on Chlortetracycline"; Jukes, "Vitamins, Metabolic Antagonists, and Molecular Evolution"; Larson, "Pioneers in Science and Technology Series."

40 "he guessed": Stokstad et al., "The Multiple Nature of the Animal Protein Factor."

41 "he apparently suspected": Stokstad and Jukes, "Further Observations on the 'Animal Protein Factor'"; Jukes, "Some Historical Notes on Chlortetracycline."

41 "horses that had been infected": "Animal Magicians."

41 "burned down an outbuilding": Jukes, "Adventures With Vitamins."

42 "state agricultural colleges": Jukes, "Antibiotics in Nutrition."

42 "They reported back": Dyer, Terrill, and Krider, "The Effect of Adding APF Supplements and Concentrates Containing Supplementary Growth Factors to a Corn-Soybean Oil Meal Ration for Weanling Pigs"; Lepley, Catron, and Culbertson, "Dried Whole Aureomycin Mash and Meat and Bone Scraps for Growing-Fattening Swine"; Burnside and Cunha, "Effect of Animal Protein Factor Supplement on Pigs Fed Different Protein Supplements."

43 "on the front page": Laurence, "'Wonder Drug' Aureomycin Found to Spur Growth 50%."

44 "into infections so serious": Loudon, "Deaths in Childbed from the Eighteenth Century to 1935"; Neushul, "Science, Government, and the Mass Production of Penicillin"; President's

Council of Advisors on Science and Technology, "Report to the President on Combating Antibiotic Resistance"; Surgeon-General's Office, "Report of the Surgeon-General of the Army to the Secretary of War for the Fiscal Year Ending June 30, 1921."

45 "a joyous overreaction": Falk, "Will Penicillin Be Used Indiscriminately?"; Brown, "The History of Penicillin From Discovery to the Drive to Production."

45 "their effect on viruses": Kaempffert, "Effectiveness of New Antibiotic, Aureomycin, Demonstrated Against Virus Diseases."

45 "liberated samples": Walker, "Pioneer Leaders in Plant Pathology."

45 "two years before": Moore and Evenson, "Use of Sulfasuxidine, Streptothricin, and Streptomycin in Nutritional Studies with the Chick."

46 "counted the published research": Jukes, "Antibiotics in Nutrition."

46 "already were giving livestock": Boyd, "Making Meat."

46 "two of Fleming's collaborators": Abraham and Chain, "An Enzyme From Bacteria Able to Destroy Penicillin."

46 "warned an audience in New York": "Penicillin's Finder Assays Its Future."

47 "a hospital in London": Barber, "The Waning Power of Penicillin."

47 "an epidemic in Australia": Rountree and Freeman, "Infections Caused by a Particular Phage Type of *Staphylococcus aureus*."

47 "crossed to the United States": Laveck and Ravenholt, "Staphylococcal Disease: An Obstetric, Pediatric, and Community Problem."

48 "were minuscule": White-Stevens, Zeibel, and Walker, "The Use of Chlortetracycline-Aureomycin in Poultry Production."

48 "were overruled": Jukes, "Some Historical Notes on Chlortetracycline."

48 "assumed this was part of the process": Jukes, "Public Health Significance of Feeding Low Levels of Antibiotics to Animals."

49 "with no advance public notice": Food and Drug Administration, "Certification of Batches of Antibiotic and Antibiotic-Containing Drugs"; also discussed in the excellent *Vermont Law Review* article: Heinzerling, "Undue Process at the FDA."

49 "Some researchers hypothesized": Freerksen, "Fundamentals of Mode of Action of Antibiotics in Animals."

50 "As more scientists studied": In addition to hundreds of individual research papers, growth promoters were examined intensively at two international conferences in the early years: the First International Conference on the Use of Antibiotics in Agriculture, Washington, D.C., 1956, and the University of Nottingham's Easter School in Agricultural Science, 1962.

51 "experimenters fed": The trials are all detailed in Jukes's 1955 monograph, "Antibiotics in Nutrition."

51 "experimenters added": Tarr, Boyd, and Bissett, "Antibiotics in Food Processing, Experimental Preservation of Fish and Beef with Antibiotics"; Deatherage, "Antibiotics in the Preservation of Meat"; Durbin, "Antibiotics in Food Preservation"; Barnes, "The Use of Antibiotics for the Preservation of Poultry and Meat."

52 "began to tell farmers": White-Stevens, Zeibel, and Walker, "The Use of Chlortetracycline-Aureomycin in Poultry Production."

52 "blessed the practice": Food and Drug Administration, "Exemption From Certification of Antibiotic Drugs for Use in Animal Feed and of Animal Feed Containing Antibiotic Drugs."

53 "Jukes challenged": Jukes, "Megavitamin Therapy"

Notes

53 "he also derided": Jukes: "DDT"; "The Organic Food Myth"; "Food Additives"; "Carcinogens in Food and the Delaney Clause."

53 "directed special rage": Conis, "Debating the Health Effects of DDT."

53 "the federal government had caved": Wang, *In Sputnik's Shadow*, p. 215

53 "mocked its author": Jukes, "A Town in Harmony."

54 "almost 40 years after": Jukes, "Some Historical Notes on Chlortetracycline."

54 "he declared": Jukes, "Antibiotics in Animal Feeds."

54 "he asked": Jukes, "The Present Status and Background of Antibiotics in the Feeding of Domestic Animals."

54 "maintained in the *New York Times*": Jukes, "Antibiotics and Meat."

54 "his last writing": Jukes, "Today's Non-Orwellian Animal Farm."

Chapter 3

56 "a luxury dish": Haley, *Turning the Tables*.

56 "farmers lacked a way": Sunde, "Seventy-Five Years of Rising American Poultry Consumption."

56 "colorfully named diseases": McGowan and Emslie, "Rickets in Chickens, With Special Reference to Its Nature and Pathogenesis."

56 "raising 'broiler' chickens": *Broiler* is a complex term. Now the poultry industry uses it to indicate chickens bred to be raised for meat, separate from layers, the birds that produce eggs. Originally, though, "broiler" indicated age and size, a bird young and tender enough to cook quickly under direct high heat, unlike old hens, which required the low, wet simmer of stewing to become palatable.

56 "in the entire United States": U.S. Department of Agriculture Bureau of Agricultural Economics, and Bureau of the Census, "United States Census of Agriculture 1950: A Graphic Summary."

56 "According to industry lore": Horowitz, "Making the Chicken of Tomorrow"; Bugos, "Intellectual Property Protection in the American Chicken–Breeding Industry."

56 "looked like a reliable alternative": "Problems in the Poultry Industry. Part II," p. 84.

57 "poultry's niche market": Godley and Williams, "The Chicken, the Factory Farm, and the Supermarket"; Sawyer, *The Agribusiness Poultry Industry*, pp. 47–9.

57 "made on his behalf": The ad was actually titled "A chicken for every pot"; it can be viewed in the online catalog of the National Archives, at http://research.archives.gov/description/187095.

57 "catering to the city": In what might be a New York version of a law of the universe—if something exists, someone will make a knockoff of it—the chicken trade became such a source of backdoor trading that it was denounced as "corrupt and vile" on the floor of the Senate. It led to a federal lawsuit over selling "unfit chickens," *ALA Schechter Poultry Corp. v. United States* (1935), a prosecution of two Brooklyn butcher brothers for underpricing their chickens. It ended up, on appeal, overturning key portions of President Franklin Roosevelt's New Deal.

57 "almost 90 million": Sawyer, *The Agribusiness Poultry Industry*, pp. 82–4.

58 "For almost a century": Sawyer, *Northeast Georgia*; "The New Georgia Encyclopedia"; Gisolfi, "From Crop Lien to Contract Farming"; Gannon, "Georgia's Broiler Industry."

60 "Georgia broiler sales rose": Gannon, "Georgia's Broiler Industry," p. 308.

61 "Outside of Delmarva": Striffler, *Chicken*, p. 43.

62 "more Cadillacs than Texas": Sawyer, *Northeast Georgia.*

62 "one billion broilers": Hansen and Mighell, *Economic Choices in Broiler Production.*

62 "a Who's Who of the poultry industry": "Problems in the Poultry Industry," Parts I–III.

66 "would follow the same path": Pew Environment Group, "Big Chicken."

66 "Today there are": National Chicken Council, "Broiler Chicken Industry Key Facts 2016."

68 "*Harper's Magazine* described": Soule, "Chicken Explosion."

69 "moving into the workforce": Toossi, "A Century of Change."

69 "too big for one couple": Horowitz, "Making the Chicken of Tomorrow."

69 "the way that fish sticks did": Josephson, "The Ocean's Hot Dog."

69 "found its disrupter": This reconstruction of Robert Baker's life and work is based on interviews with his widow, Jacoba Baker, son Dale Baker, and grandson Michael Baker; on a piece about his grandfather that Michael Baker wrote for the summer 2012 issue of Cornell's *Ezra Magazine* ("How 'Barbecue Bob' Baker Transformed Chicken"); on the contents of Baker's papers at Cornell's Albert R. Mann Library; and on his obituary in the *Cornell Chronicle* (Friedlander, "Robert C. Baker, Creator of Chicken Nuggets and Cornell Chicken Barbecue Sauce, Dies at 84") and the *New York Times* (Martin, "Robert C. Baker, Who Reshaped Chicken Dinner, Dies at 84").

70 "Baker's first experiments": All of Baker's experiments, with their recipes and marketing results, were published in Cornell University's *Agricultural Economics Research Bulletin* (AER) and, later, its *Miscellaneous Bulletin.* They include packaging with sauce, AER 55, December 1960; "Kid's Pack," AER 81, December 1961; hash, AER 151, August 1964; franks, AER 57, January 1961; chickalona, No. 83, AER 1962; breakfast sausage and burgers, MB 110, 1980; spaghetti sauce, MB 121, November 1981; meatloaf, AER 86, February 1961.

71 "chicken sticks": Marshall and Baker, "New Marketable Poultry and Egg Products: 12. Chicken Sticks."

72 "told in a company biography": Love, *McDonald's: Behind the Arches.*

72 "an earthquake in American diets": *Dietary Goals for the United States. Prepared by the Staff of the Select Committee on Nutrition and Human Needs, United States Senate.*

73 "Every year after that": "Per Capita Consumption of Poultry and Livestock, 1965 to Estimated 2016, in Pounds."

Chapter 4

75 "the Ukiah, California, *Daily Journal*": "Town and Country Market [Advertisement]."

75 "the Syracuse, New York, *Post-Standard*": "The Art of Pickin' Chicken Advertisement]."

76 "the *Odessa American* in Texas": "Fresh Food Plan Found."

76 "the *Kossuth County Advance*": "Pass the 'Acronized' Chicken, Please!"

76 "Vermont's *Bennington Banner* explained": Harris, "Home Demonstration."

76 "an invention of American Cyanamid": Kohler et al., "Comprehensive Studies of the Use of a Food Grade of Chlortetracycline in Poultry Processing."

77 "Lederle named the process": Just to complicate matters, at this point Lederle was also marketing an additional tetracycline formula, trademarked Achromycin. But the newspaper stories describing "Acronizing" make clear that the preservation process was using Aureomycin, not Achromycin.

77 "American Cyanamid had hired": "Advertising: Logistics to Fore in Big Move."

77 "fed Acronized and roasted chicken": Flanary, "Five Firms Entertain Food Editors."

78 "yet another 30-word order": Food and Drug Administration, Tolerances and exemptions from tolerances for pesticide chemicals in or on raw agricultural commodities; tolerance for residues of chlortetracycline.

78 "*Business Week* predicted": "Miracle Drugs Get Down to Earth."

78 "poultry was responsible": "Mandatory Poultry Inspection," pp. 104–105.

78 "less chicken than": "Problems in the Poultry Industry. Part I," p. 13.

79 "a separate series of Senate hearings": "Mandatory Poultry Inspection.," pp. 104–105.

80 "just as the war ended": Bud, *Penicillin*, pp. 82–83; Collingham, *The Taste of War*; Stone, "Fumbling With Famine"; Gerhard, "Food as a Weapon"; Fox, "The Origins of UNRRA."

80 "a 'meat famine' ": Norman, "G.O.P. to Open Inquiry into Meat Famine"; "Army Reduces Meat Ration as Famine Grows."

80 "urgently warned": "Report of the Special Meeting on Urgent Food Problems, Washington, D.C., May 20–27, 1946."

80 "hundreds of scientists": Farber, "Antibiotics in Food Preservation."

80 "Researchers promised": "Antibiotics and Food."

81 "would let Australia": Mrak, "Food Preservation."

81 "Canadian fishermen said": "Around Capitol Square."

81 "were shooting whales": "Whale Steak for Dinner."

81 "called their local media": Associated Press, "Tyler Firm to Preserve Chickens by Antibiotics"; "Acronize Maintains Poultry Freshness"; "New Poultry Process Will Be Used at Chehalis Plant."

81 "more than half of the slaughterhouses": "With Its New Farm & Home Division, Cyanamid Is Placing Increasing Stress on Consumer Agricultural Chemicals."

81 "Fish wholesalers": Associated Press, "Drug May Change Fish Marketing."

82 "*Business Week* documented": "Miracle Drugs Get Down to Earth."

83, "Reimert Ravenholt, a physician": This reconstruction of Ravenholt's experience is based on my interview with him; on his extensive online archive, Epidemic Investigations, maintained at http://www.ravenholt.com; on his interview with the Population and Reproductive Health Oral History Project; and on Laveck and Ravenholt, "Staphylococcal Disease," and Ravenholt et al., "Staphylococcal Infection in Meat Animals and Meat Workers."

83 "equally taken for granted": The recognition that staph can cause outbreaks outside of hospitals, in what medicine calls "the community"—that is, families, schools, and sports teams—would not be made for another 40 years: Herold et al., "Community-Acquired Methicillin-Resistant *Staphylococcus aureus* in Children With No Identified Predisposing Risk."

84 "maintained by the CDC": Since its founding in 1946, the CDC has changed its name several times while keeping the same acronym. It is now called the Centers for Disease Control and Prevention, but at the time of Ravenholt's training was known as the Communicable Disease Center.

87 "might eventually be explained": So, for instance, a 1968 outbreak of flu-like illness in the Pontiac Motors plant in Detroit, Michigan, known as Pontiac fever, was found to have been caused by the same bacterium that causes Legionnaires' disease—but that discovery had to wait until Legionnaires' disease itself, and the bacterium that causes it, were discovered in 1976: Kaufmann et al., "Pontiac Fever."

88 "would not coagulate": Curtis, "Food and Drug Projects of Interest to State Health Officers"; Welch, "Problems of Antibiotics in Food as the Food and Drug Administration Sees Them."

89 "penicillin allergies": Welch, "Antibiotics in Food Preservation"; "Antibiotics in Milk"; Garrod, "Sources and Hazards to Man of Antibiotics in Foods."

89 "the kinds of rashes": Vickers, Bagratuni, and Alexander, "Dermatitis Caused by Penicillin in Milk."

89 "the FDA tested milk": Welch, "Problems of Antibiotics in Food as the Food and Drug Administration Sees Them."

89 "in a special report": World Health Organization, "The Public Health Aspects of the Use of Antibiotics in Food and Feedstuffs."

89 "to discuss an urgent trend": Communicable Disease Center, "Proceedings, National Conference on Salmonellosis, March 11–13, 1964."

90 "confirmed Ravenholt's suspicions": Ng et al., "Antibiotics in Poultry Meat Preservation"; Njoku-Obi et al., "A Study of the Fungal Flora of Spoiled Chlortetracycline Treated Chicken Meat"; Thatcher and Loit, "Comparative Microflora of Chlor-Tetracycline-Treated and Nontreated Poultry With Special Reference to Public Health Aspects."

91 "the Pottstown, Pennsylvania, *Mercury*": "Consumer," "Chicken Flavor."

91 "the *Montana Standard-Post*": Reed, "Our Readers Speak."

91 "began to appear": "Quality Market [Advertisement]"; "Safeway [Advertisement]"; "Co-Op Shopping Center [Advertisement]."

91 "Capuchino Foods promised": "Capuchino Foods [Advertisement]."

91 "banned Acronized birds": Atkinson, "Trends in Poultry Hygiene."

91 "the agency canceled": Harold and Baldwin, "Ecologic Effects of Antibiotics."

92 "she worried aloud": Coates, "The Value of Antibiotics for Growth of Poultry."

92 "seeking to dismiss it": Hansard, Gastro-Enteritis (Tees-side).

92 "In October 1967": "The Diary of a Tragedy"; "The Men Who Fought It."

93 "one drug after another": Anderson, "Middlesbrough Outbreak of Infantile Enteritis and Transferable Drug Resistance."

94 "Ephraim Saul Anderson": Anderson's brilliance, and irascibility, were later detailed in obituaries in most of the major U.K. newspapers: Tucker, "ES Anderson: Brilliant Bacteriologist Who Foresaw the Public Health Dangers of Genetic Resistance to Antibiotics"; "Obituaries: E. S. Anderson: Bacteriologist Who Predicted the Problems Associated with Human Resistance to Antibiotics"; "Obituaries: E. S. Anderson: Ingenious Microbiologist Who Investigated How Bacteria Become Resistant to Antibiotics."

94 "Anderson had untangled": Anderson et al., "An Outbreak of Human Infection Due to *Salmonella* Typhimurium Phage Type 20a Associated With Infection in Calves."

95 "allowed his team to draw": Anderson, "The Ecology of Transferable Drug Resistance in the Enterobacteria."

96 "every single one": Anderson and Lewis, "Drug Resistance and Its Transfer in *Salmonella* Typhimurium."

96 "The source was": Watanabe and Fukasawa, "Episome-Mediated Transfer of Drug Resistance in Enterobacteriaceae. I."; Watanabe, "Infective Heredity of Multiple Drug Resistance in Bacteria"; Datta, "Transmissible Drug Resistance in an Epidemic Strain of *Salmonella* Typhimurium."

97 "what his lab was discovering": Anderson and Lewis, "Drug Resistance and Its Transfer in *Salmonella* Typhimurium"; Anderson, "Origin of Transferable Drug-Resistance Factors in the Enterobacteriaceae."

97 "10 times more cattle deaths": Dixon, "Antibiotics on the Farm—Major Threat to Human Health."

97 "from the Middlesbrough outbreak": Anderson, "Middlesbrough Outbreak of Infantile Enteritis and Transferable Drug Resistance.

Chapter 5

101 "The gray clapboard house": The reconstruction of the experiment on the Downings' property is based on interviews with Richard and Joan Downing; their daughter, Mary O'Reilly; Stuart Levy; and Levy's own descriptions in Levy, FitzGerald, and Macone, "Changes in Intestinal Flora of Farm Personnel After Introduction of a Tetracycline-Supplemented Feed on a Farm"; Levy, FitzGerald, and Macone, "Spread of Antibiotic-Resistant Plasmids From Chicken to Chicken and From Chicken to Man"; and Levy, *The Antibiotic Paradox.*

103 "40 percent": Office of Technology Assessment, Congress of the United States, "Drugs in Livestock Feed."

104 "persuaded friendly journalists": Those articles included, among others: "The Dangers of Misusing Antibiotics"; "Germ Survival in Face of Antibiotics"; Fishlock, "Government Action Urged on Farm Drugs."

104 "there were only three": It seems incredible now, but in the 1960s, British television had only three channels: BBC1, BBC2, and ITV (which stood for "Independent Television"), the only commercial channel. Channel 4, the first to disrupt that system, which got its name because it was, literally, the fourth channel in existence, didn't begin broadcasting until 1982 and Sky Television not until 1989.

104 "a scathing article": Dixon, "Antibiotics on the Farm—Major Threat to Human Health."

104 "delivered in 1962": "Antibiotics on the Farm"; Braude, "Antibiotics in Animal Feeds in Great Britain."

105 "a searing exposé": Harrison, *Animal Machines.*

105 "left Britons questioning": Sayer, "Animal Machines."

105 "a horrific blow": Reynolds and Tansey, "Foot and Mouth Disease."

105 "Anderson wrote": Anderson, "Transferable Antibiotic Resistance."

106 "The Swann committee": The testimony the committee heard and the calculations they conducted are all contained in the final report, known afterward as the Swann Report: Swann and Joint Committee on the Use of Antibiotics in Animal Husbandry and Veterinary Medicine, *Report. Presented to Parliament by the Secretary of State for Social Services, the Secretary of State for Scotland, the Minister of Agriculture, Fisheries and Food and the Secretary of State for Wales by Command of Her Majesty.*

108 "convened a meeting": National Research Council, *Proceedings of the First International Conference on the Use of Antibiotics in Agriculture, 19–21 October 1955.*

108 "Yet another committee": Committee on Salmonella, National Research Council, "An Evaluation of the Salmonella Problem."

108 "delivered to the FDA": Food and Drug Administration, "Report to the Commissioner of the Food and Drug Administration by the FDA Task Force on the Use of Antibiotics in Animal Feeds"; Lehmann, "Implementation of the Recommendations Contained in the Report to the Commissioner Concerning the Use of Antibiotics on Animal Feed."

109 "wrote a minority report": Solomons, "Antibiotics in Animal Feeds—Human and Animal Safety Issues."

109 "announced a compromise": Subcommittee on Oversight and Investigations, *Antibiotics in Animal Feeds Hearings Before the Subcommittee on Oversight and Investigations of the Committee on Interstate and Foreign Commerce.*

109 "blasted the FDA": Jukes, "Public Health Significance of Feeding Low Levels of Antibiotics to Animals."

110 "the son of a family doctor": Biographical details are drawn from interviews with Levy; White et al., *Frontiers in Antimicrobial Resistance*; and Azvolinsky, "Resistance Fighter."

110 "the *New England Journal of Medicine* was warning": "Infectious Drug Resistance."

111 "had identified the genes": Levy and McMurry, "Detection of an Inducible Membrane Protein Associated With R-Factor-Mediated Tetracycline Resistance."

118 "the new commissioner of the FDA": Donald S. Kennedy is not well, and his wife, Robin Kennedy, declined interviews on his behalf. The reconstruction of Kennedy's experience is based on interviews with his FDA chief of staff, Thomas Grumbly, now president of the Supporters of Agricultural Research Foundation; on federal records; on articles that Kennedy wrote years afterward, including Kennedy, "The Threat From Antibiotic Use on the Farm"; and on the Donald S. Kennedy papers in the Stanford University Archives.

119 "In a short statement": Kennedy's speech is preserved in the testimony within Subcommittee on Dairy and Poultry of the Committee on Agriculture, *Impact of Chemical and Related Drug Products and Federal Regulatory Processes*.

119 "almost every food animal": Subcommittee on Oversight and Investigations, *Antibiotics in Animal Feeds*.

120 "was 'totally unworkable' ": The industry organizations' protests are all preserved in the hearing record: Subcommittee on Dairy and Poultry of the Committee on Agriculture, *Impact of Chemical and Related Drug Products and Federal Regulatory Processes*.

120 "their case against growth promoters": Food and Drug Administration, Diamond Shamrock Chemical Co., et al.: "Penicillin-Containing Premixes," and Food and Drug Administration, Pfizer, Inc., et al.: Tetracycline (Chlortetracycline and Oxytetracycline)-Containing Premixes."

122 "for as long as Kennedy had been alive": Whitten began his political career in 1931 with his election to the Mississippi statehouse. Kennedy was born August 18, 1931.

Chapter 6

125 "The media barely covered": See, for instance: Lyons, "F.D.A. Chief Heading for Less Trying Job"; "Two Hands for Donald Kennedy."

126 "an odd form of pneumonia": Centers for Disease Control, "Pneumocystis Pneumonia—Los Angeles."

126 "Scott Holmberg had a vivid sense": The reconstruction of Scott Holmberg's career and experience in Minnesota is based on interviews with him and with his investigative partner, Michael T. Osterholm; on the paper he wrote about the investigation: Holmberg et al., "Drug-Resistant Salmonella From Animals Fed Antimicrobials"; and on news coverage afterward, including Sun, "Antibiotics and Animal Feed," "In Search of Salmonella's Smoking Gun," and "Use of Antibiotics in Animal Feed Challenged."

134 "Holmberg could see": Holmberg, Wells, and Cohen, "Animal-to-Man Transmission of Antimicrobial-Resistant Salmonella."

135 "called the findings": Sun, "Antibiotics and Animal Feed."

135 "on the front page": Russell, "Research Links Human Illness, Livestock Drugs."

136 "formally petitioned": Ahmed, Chasis, and McBarnette, "Petition of the Natural Resources Defense Council to the Secretary of Health and Human Services Requesting the Immediate Suspension of Approval of the Subtherapeutic Use of Penicillin and Tetracyclines in Animal Feeds."

136 "two massive new studies": National Research Council, "Effects on Human Health of Subtherapeutic Use of Antimicrobials in Animal Feeds"; Communicable Disease Control Section, Seattle-King County Department of Public Health, "Surveillance of the Flow of Salmonella and Campylobacter in a Community."

136 "In a blistering report": House Committee on Government Operations, "Human Food Safety and the Regulation of Animal Drugs."

138 "went into textbooks": Hennessy et al., "A National Outbreak of *Salmonella enteritidis* Infections From Ice Cream"; Centers for Disease Control and Prevention, "Four Pediatric Deaths From Community-Acquired Methicillin-Resistant *Staphylococcus aureus*"; Osterholm et al., "An Outbreak of a Newly Recognized Chronic Diarrhea Syndrome Associated With Raw Milk Consumption."

138 "Smith came from a farming family": The reconstruction of Kirk Smith's background, career, and investigation is based on interviews with him; with his supervisor at the time, Michael T. Osterholm; and on the paper he wrote about the investigation, Smith et al., "Quinolone-Resistant *Campylobacter jejuni* Infections in Minnesota, 1992-1998."

139 "extremely common on chicken": At the time that Smith was examining the Minnesota database, one out of five chickens slaughtered in the United States carried *Salmonella* and four out of five carried *Campylobacter*. USDA Food Safety and Inspection Service, "Nationwide Broiler Chicken Microbiological Baseline Data Collection Program, July 1994–June 1995."

139 "a big advance": Andersson, "Development of the Quinolones"; Andriole, "The Quinolones."

141 "a nationwide version": Gupta et al., "Antimicrobial Resistance Among *Campylobacter* Strains, United States, 1997–2001."

142 "In the Netherlands": Endtz et al., "Quinolone Resistance in *Campylobacter* Isolated From Man and Poultry Following the Introduction of Fluoroquinolones in Veterinary Medicine."

142 "In Spain": Jiménez et al., "Prevalence of Fluoroquinolone Resistance in Clinical Strains of *Campylobacter jejuni* Isolated in Spain"; Velázquez et al., "Incidence and Transmission of Antibiotic Resistance in *Campylobacter jejuni* and *Campylobacter coli*."

142 "In England": Piddock, "Quinolone Resistance and *Campylobacter* spp."; Gaunt and Piddock, "Ciprofloxacin Resistant *Campylobacter* spp. in Humans."

142 "the World Health Organization determined": World Health Organization, "Use of Quinolones in Food Animals and Potential Impact on Human Health."

142 "everywhere at once": Nelson et al., "Prolonged Diarrhea Due to Ciprofloxacin-Resistant *Campylobacter* Infection."

143 "In the late 1990s": Angulo et al., "Origins and Consequences of Antimicrobial-Resistant Nontyphoidal Salmonella."

143 "a novel strain of *Salmonella*": Threlfall et al., "Increasing Spectrum of Resistance in Multiresistant *Salmonella* Typhimurium"; Threlfall, Ward, and Rowe, "Multiresistant *Salmonella* Typhimurium DT 104 and *Salmonella bacteraemia*."

144 "soon arrived stateside": Centers for Disease Control and Prevention, "Multidrug-Resistant *Salmonella* Serotype Typhimurium—United States, 1996"; Cody et al., "Two Outbreaks of Multidrug-Resistant *Salmonella* Serotype Typhimurium DT104 Infections Linked to Raw-Milk Cheese in Northern California."

144 "a small family dairy farm": Spake, "O Is for Outbreak."

144 "zoomed from nonexistent": Glynn et al., "Emergence c ʿMultidrug-Resistant *Salmonella enterica* Serotype Typhimurium DT104 Infections in the United States."

144 "an emergency assessment": Hogue et al., "*Salmonella* Typhimurium DT104 Situation Assessment, December 1997."

145 "a long-delayed step": Food and Drug Administration, "Enrofloxacin for Poultry: Opportunity for a Hearing."

145 "an international story": Grady, "Bacteria Concerns in Denmark Cause Antibiotics Concerns in U.S."; O'Sullivan, "Seven-Year-Old Ian Reddin's Food Poisoning Put Family Life on Hold."

146 "in a 253-page report": National Research Council, "The Use of Drugs in Food Animals."

146 "The FDA estimated": U.S. Food and Drug Administration Center for Veterinary Medicine, "Human Health Impact of Fluoroquinolone Resistant *Campylobacter* Attributed to the Consumption of Chicken."

146 "an administrative law process": Nelson et al., "Fluoroquinolone-Resistant *Campylobacter* Species and the Withdrawal of Fluoroquinolones From Use in Poultry."

147 "lost its license": Kaufman, "Ending Battle With FDA, Bayer Withdraws Poultry Antibiotic."

148 "a devastating analysis": Mellon, Benbrook, and Benbrook, "Hogging It."

Chapter 7

150 "on the road to innovation": Smith and Daniel, *The Chicken Book*, pp. 237–9; Sawyer, *The Agribusiness Poultry Industry*, p. 26.

151 "a November 1944 poultry meeting": Seeger, Tomhave, and Shrader, "The Results of the Chicken-of-Tomorrow 1948 National Contest"; Shrader, "The Chicken-of-Tomorrow Program."

151 "droolingly described": Nicholson, "More White Meat for You."

152 "better poultry varieties": Boyd, "Making Meat"; Warren, "A Half-Century of Advances in the Genetics and Breeding Improvement of Poultry."

153 "That evening": Horowitz, "Making the Chicken of Tomorrow."

153 "Vantress won again": Shrader, "The Chicken-of-Tomorrow Program"; Bugos, "Intellectual Property Protection in the American Chicken–Breeding Industry."

154 "like growing hybrid soybeans or corn": How growers' history with hybrid corn set the stage for a new understanding of chicken is persuasively explained in Boyd, "Making Meat."

154 "hundreds of thousands of birds": Leeson and Summers, *Broiler Breeder Production*.

154 "after 1960": Bugos, "Intellectual Property Protection in the American Chicken–Breeding Industry."

155 "just three companies": Penn State Extension, "Primary Breeder Companies—Poultry."

155 "began keeping statistics": "U.S. Broiler Performance."

155 "average weight at slaughter": U.S. Department of Agriculture National Agricultural Statistics Service, "Poultry Slaughter 2014 Annual Summary."

155 "look nothing like": Zuidhof et al., "Growth, Efficiency, and Yield of Commercial Broilers from 1957, 1978, and 2005."

156 "twice the size": Schmidt et al., "Comparison of a Modern Broiler Line and a Heritage Line Unselected Since the 1950s."

156 "unbalances birds' bodies": Paxton, Corr, and Hutchinson, "The Gait Dynamics of the Modern Broiler Chicken"; Bessei, "Welfare of Broilers."

156 "Broilers can develop": ASPCA, "A Growing Problem. Selective Breeding in the Chicken Industry."

156 "difficulty walking": Danbury et al., "Self-Selection of the Analgesic Drug Carprofen by Lame Broiler Chickens"; McGeown et al., "Effect of Carprofen on Lameness in Broiler Chickens."

157 "where no one ends up by accident": The portrait of Frank Reese is based on interviews with him at Good Shepherd Poultry Ranch; on personal papers he shared with me; on his monograph, "On Animal Husbandry for Poultry Production"; on interviews with Ben Goldsmith and Andrew DeCoriolis of Farm Forward and Leah Garces of Compassion in World Farming; and on stories written about him, including O'Neill, "Rare Breed."

163 "have been shutting": Cloud, "The Fight to Save Small-Scale Slaughterhouses"; Janzen, "Loss of Small Slaughterhouses Hurts Farmers, Butchers and Consumers."

164 "the average chicken barn": Pew Environment Group, "Big Chicken."

164 "most broiler farms produced": MacDonald and McBride, "The Transformation of U.S. Livestock Agriculture."

165 "150 tons": Ritz and Merka, "Maximizing Poultry Manure Use Through Nutrient Management Planning."

165 "Lisa Inzerillo said": The description of the Inzerillos' experience is based on interviews with them at their home in Maryland and with volunteer Gabby Cammerata and Assateague coastkeeper Kathy Phillips; and on news coverage of local chicken farm negotiations, including: Gates, "Somerset Homeowners Clash With Poultry Farmer"; Kobell, "Poultry Mega-Houses Forcing Shore Residents to Flee Stench, Traffic"; Schuessler, "Maryland Residents Fight Poultry Industry Expansion"; Cox, "Why Somerset Turned Up the Heat on Chicken Farms."

165 "'no-land' farms": Pew Environment Group, "Big Chicken."

166 "an incremental improvement": WBOC-16, "Somerset County Approves New Poultry House Regulations."

166 "1.5 billion": Chesapeake Bay Foundation, "Manure's Impact on Rivers, Streams and the Chesapeake Bay"; Public Broadcasting System, "Who's Responsible For That Manure? Poisoned Waters."

167 "ignored or defied": Bernhardt et al., "Manure Overload on Maryland's Eastern Shore"; Bernhardt, Burkhardt, and Schaeffer, "More Phosphorus, Less Monitoring."

168 "Researchers have found": You, Hilpert, and Ward, "Detection of a Common and Persistent tet(L)-Carrying Plasmid in Chicken-Waste-Impacted Farm Soil"; Koike et al., "Monitoring and Source Tracking of Tetracycline Resistance Genes in Lagoons and Groundwater Adjacent to Swine Production Facilities Over a 3-Year Period"; Gibbs et al., "Isolation of Antibiotic-Resistant Bacteria from the Air Plume Downwind of a Swine Confined or Concentrated Animal Feeding Operation."

168 "trucks that bear chickens": Rule, Evans, and Silbergeld, "Food Animal Transport."

168 "carried away by flies": Graham et al., "Antibiotic Resistant Enterococci and Staphylococci Isolated From Flies Collected Near Confined Poultry Feeding Operations"; Ahmad et al., "Insects in Confined Swine Operations Carry a Large Antibiotic Resistant and Potentially Virulent Enterococcal Community."

169 "of people": Smith et al., "Methicillin-Resistant *Staphylococcus aureus* in Pigs and Farm Workers on Conventional and Antibiotic-Free Swine Farms in the USA"; Frana et al., "Isolation and Characterization of Methicillin-Resistant *Staphylococcus aureus* from

Pork Farms and Visiting Veterinary Students"; Rinsky et al., "Livestock-Associated Methicillin and Multidrug Resistant *Staphylococcus aureus* Is Present Among Industrial, Not Antibiotic-Free Livestock Operation Workers in North Carolina"; Castillo Neyra et al., "Multidrug-Resistant and Methicillin-Resistant *Staphylococcus aureus* (MRSA) in Hog Slaughter and Processing Plant Workers and Their Community in North Carolina (USA)"; Nadimpalli et al., "Persistence of Livestock-Associated Antibiotic-Resistant *Staphylococcus aureus* Among Industrial Hog Operation Workers in North Carolina over 14 Days"; Wardyn et al., "Swine Farming Is a Risk Factor for Infection With and High Prevalence of Carriage of Multidrug-Resistant *Staphylococcus aureus.*"

169 "poultry farm workers in Delmarva": Price et al., "Elevated Risk of Carrying Gentamicin-Resistant *Escherichia coli* Among U.S. Poultry Workers."

169 "who have never entered the premises": Casey et al., "High-Density Livestock Operations, Crop Field Application of Manure, and Risk of Community-Associated Methicillin-Resistant *Staphylococcus aureus* Infection in Pennsylvania"; Carrel et al., "Residential Proximity to Large Numbers of Swine in Feeding Operations Is Associated With Increased Risk of Methicillin-Resistant *Staphylococcus aureus* Colonization at Time of Hospital Admission in Rural Iowa Veterans."

169 "are complex": Deo, "Pharmaceuticals in the Surface Water of the USA"; Radhouani et al., "Potential Impact of Antimicrobial Resistance in Wildlife, Environment and Human Health"; Singh et al., "Characterization of Enteropathogenic and Shiga Toxin-Producing *Escherichia coli* in Cattle and Deer in a Shared Agroecosystem"; Smaldone et al., "Occurrence of Antibiotic Resistance in Bacteria Isolated from Seawater Organisms Caught in Campania Region"; Ruzauskas and Vaskeviciute, "Detection of the *mcr-1* Gene in *Escherichia coli* Prevalent in the Migratory Bird Species *Larus argentatus*"; Liakopoulos et al., "The Colistin Resistance *mcr-1* Gene Is Going Wild"; Simões et al., "Seagulls and Beaches as Reservoirs for Multidrug-Resistant *Escherichia coli.*"

170 "Possibly one-quarter or less": Chee-Sanford et al., "Occurrence and Diversity of Tetracycline Resistance Genes in Lagoons and Groundwater Underlying Two Swine Production Facilities"; Kumar et al., "Antibiotic Use in Agriculture and Its Impact on the Terrestrial Environment"; Marshall and Levy, "Food Animals and Antimicrobials."

170 "Some researchers suspect": The fullest exploration of these observations is in Blaser, *Missing Microbes;* but also see Cox and Blaser, "Antibiotics in Early Life and Obesity"; Cox et al., "Altering the Intestinal Microbiota During a Critical Developmental Window Has Lasting Metabolic Consequences"; Schulfer and Blaser, "Risks of Antibiotic Exposures Early in Life on the Developing Microbiome"; Blaser, "Antibiotic Use and Its Consequences for the Normal Microbiome."

Chapter 8

171 "had no way of knowing": The reconstruction of Rick Schiller's experience, and of the outbreak of which he was part, is based on interviews with him; with his attorney, Bill Marler; with personnel from the Centers for Disease Control and Prevention, including Drs. Thomas Chiller, Jolene Nakao, Robert Tauxe, and Matthew Wise and Laura Gieraltowski; on CDC investigative documentation and documents from the California Department of Public Health, the Oregon Public Health Division, and the Washington State Department of Health; on Schiller's medical records; on legal documents, includ-

ing Marler, "Final Demand Letter to Ron Foster, President, Foster Farms Inc., in re: 2013 Foster Farms Chicken Salmonella Outbreak, Client: Rick Schiller"; and on contemporaneous news stories. For Foster Farms, Inc.'s response, see the note below.

172 "In November 1992": Centers for Disease Control and Prevention, "Update: Multistate Outbreak of *Escherichia coli* O157:H7 Infections From Hamburgers—Western United States, 1992–1993"; Benedict, *Poisoned.*

173 "had been dropping": Sobel et al., "Investigation of Multistate Foodborne Disease Outbreaks"; Ollinger et al., "Structural Change in the Meat, Poultry, Dairy, and Grain Processing Industries."

173 "Not long before": Economic Research Service, "Tracking Foodborne Pathogens From Farm to Table."

174 "It was an expensive undertaking": Hise, "History of PulseNet USA."

174 "is indicative, not comprehensive": Food and Drug Administration, "National Antimicrobial Resistance Monitoring System (NARMS) Integrated Report 2012–2013"; Center for Veterinary Medicine, Centers for Disease Control and Prevention, and U.S. Department of Agriculture, "On-Farm Antimicrobial Use and Resistance Data Collection: Transcript of a Public Meeting, September 30, 2015."

175 "a company called Foster Farms": Foster Farms declined to make personnel available for interviews. In a statement emailed on March 3, 2016, Ira Brill, Foster Farms' director of marketing services, said, in part:

> Thank you for your interest in Foster Farms. After thinking further about your request to discuss Foster Farms food safety efforts, I decided to decline your invitation. . . .
>
> Since April 2014, Foster Farms has continuously maintained a Salmonella parts prevalence level of 5 percent, albeit the USDA parts standard that goes into effect in 2016 is 15.7 percent. Between Oct. 2013 and April 2014, Foster Farms drove down Salmonella at the parts level from approximately 25 percent in line with the USDA 2010/2011 industry benchmark study to 5 percent—no other company to my knowledge achieved a reduction in a food borne pathogen as rapidly or as significantly.
>
> In essence this was the result of three factors:
>
> 1. A multi-hurdle approach which sought to impact Salmonella at all critical points—from the breeder stock, through the growout, through the plant to the package. Overall this represented an investment of $75 million.
>
> 2. Intensive data management. In some instances as many as 8,000 microbiological samples from a single ranch complex were analyzed. Overall Foster Farms today conducts more than 135,000 microbiological tests annually, an increase of approximately 40 percent since 2013. The volume of data that Foster Farms reviews, in turn required more sophisticated data management infrastructure.
>
> 3. The advice of a Food Safety Advisory comprised of experts in all areas of the discipline drawn from government, industry and academia. While the 2013 food safety issues are well behind Foster Farms, the company remains committed to continuous improvement in this area. As an example we are conducting pioneering research in gene

sequencing to better understand virulence as it relates to specific Salmonella strains. Nor does Foster Farms view advancements in Salmonella control as proprietary; we continue to share our learning with the poultry industry, USDA and CDC.

175 "its 74th year in business": Holland, "After 75 Years, Foster Farms Remembers Its Path to Success."

175 "from amusing commercials": "Foster Farms—Road Trip [Advertisement]."

175 "made its first announcement": The CDC has placed all of the public reports it made during the 2013–2014 Foster Farms investigation online, listed under the dates when they were published, at http://www.cdc.gov/salmonella, in the single document: "Multistate Outbreak of Multidrug-Resistant *Salmonella* Heidelberg Infections Linked to Foster Farms Brand Chicken (Final Update)." The reports were made on October 8, 11, 18, and 30, 2013; November 19, 2013; December 19, 2013; January 16, 2014; March 3, 2014; April 9, 2014; May 27, 2014; and July 4 and 31, 2014.

178 "*E. coli* O157:H7 would become": Andrews, "Jack in the Box and the Decline of *E. coli*."

178 "would give the same designation": "USDA Takes New Steps to Fight *E. coli*, Protect the Food Supply."

178 "no federal power": Pew Charitable Trusts, "Weaknesses in FSIS's *Salmonella* Regulation."

178 "for more than a year": Jalonick, "Still No Recall of Chicken Tied to Outbreak of Antibiotic-Resistant *Salmonella*"; Bonar, "Foster Farms Finally Recalls Chicken"; Kieler, "Foster Farms Recalls Chicken After USDA Inspectors Finally Link It to Salmonella Case."

179 "In June 2012": Centers for Disease Control and Prevention, "Multistate Outbreak of *Salmonella* Heidelberg Infections Linked to Chicken (Final Update) July 10, 2013"; Centers for Disease Control and Prevention, "Outbreak of *Salmonella* Heidelberg Infections Linked to a Single Poultry Producer—13 States, 2012–2013."

179 "even earlier outbreaks": Oregon Public Health Division, "Summary of *Salmonella* Heidelberg Outbreaks Involving PFGE Patterns SHEX-005 and 005a. Oregon, 2004–2012."

180 "Investigators found": U.S. Department of Agriculture, "California Firm Recalls Chicken Products Due to Possible *Salmonella* Heidelberg Contamination."

181 "in some instances to zero": Charles, "How Foster Farms Is Solving the Case of the Mystery Salmonella."

183 "far-away episode": Agersø et al., "Spread of Extended Spectrum Cephalosporinase-Producing *Escherichia coli* Clones and Plasmids from Parent Animals to Broilers and to Broiler Meat in a Production Without Use of Cephalosporins"; Levy, "Reduced Antibiotic Use in Livestock"; Nilsson et al., "Vertical Transmission of *Escherichia coli* Carrying Plasmid-Mediated AmpC (pAmpC) Through the Broiler Production Pyramid."

184 "investing $75 million": See Brill statement above.

184 "down to 5 percent": Parsons, "Foster Farms Official Shares Data Management Tips, Salmonella below 5%."

185 "researchers in Sweden": Ternhag et al., "Short- and Long-Term Effects of Bacterial Gastrointestinal Infections."

185 "in western Australia": Moorin et al., "Long-Term Health Risks for Children and Young Adults After Infective Gastroenteritis."

185 "in Spain": Arnedo-Pena et al., "Reactive Arthritis and Other Musculoskeletal Sequelae Following an Outbreak of *Salmonella* Hadar in Castellon, Spain."

186 "asked researchers to track": Clark et al., "Long Term Risk for Hypertension, Renal Impairment, and Cardiovascular Disease After Gastroenteritis From Drinking Water Contaminated with *Escherichia coli* O157:H7."

Chapter 9

189 "The first signals of that epidemic": The story of the slow recognition of foodborne urinary tract infections is reconstructed from multiple interviews over years with Amee Manges, James R. Johnson, and Lance B. Price.

190 "At the University of California, Berkeley": Manges et al., "Widespread Distribution of Urinary Tract Infections Caused by a Multidrug-Resistant *Escherichia coli* Clonal Group."

191 "They were an outbreak": Stamm, "An Epidemic of Urinary Tract Infections?"

192 "in December 1986": Eykyn and Phillips, "Community Outbreak of Multiresistant Invasive *Escherichia coli* Infection."

192 "Ten weeks later": Wright and Perinpanayagam, "Multiresistant Invasive *Escherichia coli* Infection in South London."

192 "in the era before": The *Lancet* did not begin publishing online until 1996.

192 "More than a year later": Phillips et al., "Epidemic Multiresistant *Escherichia coli* Infection in West Lambeth Health District."

194 "an entire third category": Russo and Johnson, "Proposal for a New Inclusive Designation for Extraintestinal Pathogenic Isolates of *Escherichia coli.*"

195 "They calculated in 2003": Russo and Johnson, "Medical and Economic Impact of Extraintestinal Infections Due to *Escherichia coli.*"

195 "compared the urine samples": Manges et al., "Widespread Distribution of Urinary Tract Infections Caused by a Multidrug-Resistant *Escherichia coli* Clonal Group."

196 "had been rising": Sanchez, Master, and Bordon, "Trimethoprim-Sulfamethoxazole May No Longer Be Acceptable for the Treatment of Acute Uncomplicated Cystitis in the United States."

197 "told its members in 2011": Gupta et al., "Managing Uncomplicated Urinary Tract Infection—Making Sense Out of Resistance Data."

197 "began to investigate": Jakobsen et al., "*Escherichia coli* Isolates From Broiler Chicken Meat, Broiler Chickens, Pork, and Pigs Share Phylogroups and Antimicrobial Resistance With Community-Dwelling Humans and Patients With Urinary Tract Infection"; Jakobsen et al., "Is *Escherichia coli* Urinary Tract Infection a Zoonosis?"

198 "decided to research": Johnson et al., "Isolation and Molecular Characterization of Nalidixic Acid-Resistant Extraintestinal Pathogenic *Escherichia coli* From Retail Chicken Products."

198 "Johnson and his collaborators": Johnson et al., "Contamination of Retail Foods, Particularly Turkey, From Community Markets (Minnesota, 1999–2000) With Antimicrobial-Resistant and Extraintestinal Pathogenic *Escherichia coli.*"

199 "In a second study": Johnson et al., "Antimicrobial-Resistant and Extraintestinal Pathogenic *Escherichia coli* in Retail Foods."

199 "the team recruited the employees": Johnson et al., "Antimicrobial Drug-Resistant *Escherichia coli* From Humans and Poultry Products, Minnesota and Wisconsin, 2002–2004."

200 "a research team in Jamaica": Miles et al., "Antimicrobial Resistance of *Escherichia coli* Isolates From Broiler Chickens and Humans."

200 "researchers in Spain": Johnson et al., "Similarity Between Human and Chicken *Escherichia coli* Isolates in Relation to Ciprofloxacin Resistance Status."

200 "Between 2009 and 2014": Hannah et al., "Molecular Analysis of Antimicrobial-Susceptible and -Resistant *Escherichia coli* From Retail Meats and Human Stool and Clinical Specimens in a Rural Community Setting"; Giufre et al., "*Escherichia coli* of Human and Avian Origin"; Kaesbohrer et al., "Emerging Antimicrobial Resistance in Commensal *Escherichia coli* With Public Health Relevance"; Literak et al., "Broilers as a Source of Quinolone-Resistant and Extraintestinal Pathogenic *Escherichia coli* in the Czech Republic"; Lyhs et al., "Extraintestinal Pathogenic *Escherichia coli* in Poultry Meat Products on the Finnish Retail Market"; Sheikh et al., "Antimicrobial Resistance and Resistance Genes in *Escherichia coli* Isolated From Retail Meat Purchased in Alberta, Canada"; Aslam et al., "Characterization of Extraintestinal Pathogenic *Escherichia coli* Isolated From Retail Poultry Meats From Alberta, Canada."

200 "a large study in Denmark": Jakobsen et al., "*Escherichia coli* Isolates From Broiler Chicken Meat, Broiler Chickens, Pork, and Pigs Share Phylogroups and Antimicrobial Resistance With Community-Dwelling Humans and Patients With Urinary Tract Infection"; Jakobsen et al., "Is *Escherichia coli* Urinary Tract Infection a Zoonosis?"

200 "data from 11 countries": Vieira et al., "Association Between Antimicrobial Resistance in *Escherichia coli* Isolates From Food Animals and Blood Stream Isolates From Humans in Europe."

201 "she showed that": Manges et al., "Retail Meat Consumption and the Acquisition of Antimicrobial Resistant *Escherichia coli* Causing Urinary Tract Infections"; Vincent et al., "Food Reservoir for *Escherichia coli* Causing Urinary Tract Infections"; Bergeron et al., "Chicken as Reservoir for Extraintestinal Pathogenic *Escherichia coli* in Humans, Canada"; Aslam et al., "Characterization of Extraintestinal Pathogenic *Escherichia coli* Isolated From Retail Poultry Meats From Alberta, Canada."

202 "In the summer of 2015": The reconstruction of the discovery of *mcr-1* is based on multiple interviews with Timothy Walsh, Lance B. Price, and Robert Skov.

203 "Walsh's analysis revealed": Yong et al., "Characterization of a New Metallo-β-Lactamase Gene, *bla*NDM-1, and a Novel Erythromycin Esterase Gene Carried on a Unique Genetic Structure in *Klebsiella pneumoniae* Sequence Type 14 from India."

204 "By the end of 2012": Berrazeg et al., "New Delhi Metallo-Beta-Lactamase Around the World."

204 "In March 2013": Department of Health, "Antimicrobial Resistance Poses 'Catastrophic Threat,' Says Chief Medical Officer."

204 "In September 2013": U.S. Centers for Disease Control and Prevention, "Press Briefing Transcript—CDC Telebriefing on Today's Drug-Resistant Health Threats."

205 "put the drug on its warning list": Paterson and Harris, "Colistin Resistance."

205 "was already emerging": Kempf et al., "What Do We Know About Resistance to Colistin in Enterobacteriaceae in Avian and Pig Production in Europe?"; Catry et al., "Use of Colistin-Containing Products Within the European Union and European Economic Area (EU/EEA)."

206 "released their findings": Liu et al., "Emergence of Plasmid-Mediated Colistin Resistance Mechanism *mcr-1* in Animals and Human Beings in China."

206 "more than 30 countries": Xavier et al., "Identification of a Novel Plasmid-Mediated Colistin-Resistance Gene, *mcr-2*, in *Escherichia coli*, Belgium, June 2016."

206 "in others": Skov and Monnet, "Plasmid-Mediated Colistin Resistance *(mcr-1* Gene)"; Rapoport et al., "First Description of *mcr-1*-Mediated Colistin Resistance in Human Infections Caused by *Escherichia coli* in Latin America."

Chapter 10

211 "It was lunchtime": Descriptions of the operations of the Landes and Loué cooperatives in France and of the Label Rouge structure are based on interviews in France with Maxime Quentin, Bernard Tauzia, and Jean-Marc Durroux; Pascal Vaugarny, Stéphane Brunet, Alain Allinant, and Christophe Chéreau; and Sabine Edelli at l'Institut national de l'origine et de la qualité and Marie Guyot at le Syndicat national des labels agricoles de France.

212 "a strict set of standards": Stevenson and Born, "The 'Red Label' Poultry System in France."

213 "program was born": The history of the Label Rouge system is based on interviews and on accounts in these books: *Les fermiers de Loué;* and Saberan and Deck, *Landes en toute liberté.*

213 "the physical and economic devastation": Hoffmann, "The Effects of World War II on French Society and Politics"; Kesternich et al., "The Effects of World War II on Economic and Health Outcomes Across Europe."

214 "The rate of *Salmonella*": Westgren, "Delivering Food Safety, Food Quality, and Sustainable Production Practices."

219 "warned in a veterinary journal": Linton, "Antibiotic Resistance."

219 "so many prescriptions were being written": Braude, "Antibiotics in Animal Feeds in Great Britain."

219 "furiously demanded in an editorial": "Why Has Swann Failed?"

220 "just one year after the drug's debut": Jevons, " 'Celbenin'-Resistant Staphylococci."

220 "increased 100-fold": Kirst, Thompson, and Nicas, "Historical Yearly Usage of Vancomycin."

220 "Something was, in agriculture": Witte, "Impact of Antibiotic Use in Animal Feeding on Resistance of Bacterial Pathogens in Humans"; Wegener et al., "Use of Antimicrobial Growth Promoters in Food Animals and *Enterococcus faecium* Resistance to Therapeutic Antimicrobial Drugs in Europe"; Witte, "Selective Pressure by Antibiotic Use in Livestock."

Chapter 11

227 "You can see a long way": Descriptions of intensive but antibiotic-free farming in the Netherlands are based on interviews with farmers Gerbert Oosterlaken, Eric van den Heuvel (with translation assistance from Kor Mast), and Rob and Egbert Wingens; and with Jan Kluytmans, Andreas Voss, Hetty van Beers, Joost van Herten, Dik Meevius, and Albert Meijering.

231 "In that time": Cogliani, Goossens, and Greko, "Restricting Antimicrobial Use in Food Animals."

231 "the EU banned": Bonten, Willems, and Weinstein, "Vancomycin-Resistant Enterococci"; Casewell, "The European Ban on Growth-Promoting Antibiotics and Emerging Consequences for Human and Animal Health."

232 "created in 1988": Souverein et al., "Costs and Benefits Associated With the MRSA Search and Destroy Policy in a Hospital in the Region Kennemerland, the Netherlands."

232 "The Dutch rules assumed": Wertheim et al., "Low Prevalence of Methicillin-Resistant *Staphylococcus aureus* (MRSA) at Hospital Admission in the Netherlands"; Vos and Verbrugh, "MRSA."

233 "an unpleasant surprise": Voss et al., "Methicillin-Resistant *Staphylococcus aureus* in Pig Farming."

234 "This strain was resistant": de Neeling et al., "High Prevalence of Methicillin Resistant *Staphylococcus aureus* in Pigs."

234 "it used more": van Geijlswijk, Mevius, and Puister-Jansen, "[Quantity of veterinary antibiotic use]"; Grave, Torren-Edo, and Mackay, "Comparison of the Sales of Veterinary Antibacterial Agents Between 10 European Countries"; Grave et al., "Sales of Veterinary Antibacterial Agents in Nine European Countries During 2005–09."

234 "wandered into farm animals": Price et al., "*Staphylococcus aureus* CC398."

234 "a new mother": Huijsdens et al., "Community-Acquired MRSA and Pig-Farming."

235 "a woman who lived more": Ekkelenkamp et al., "Endocarditis Due to Methicillin-Resistant *Staphylococcus aureus* Originating From Pigs."

235 "After that": Wulf and Voss, "MRSA in Livestock Animals: An Epidemic Waiting to Happen?"; Fanoy et al., "An Outbreak of Non-Typeable MRSA Within a Residential Care Facility."

235 "Investigators found": Neeling et al., "High Prevalence of Methicillin Resistant *Staphylococcus aureus* in Pigs."

235 "to 30 percent": Huijsdens et al., "Molecular Characterisation of PFGE Non-Typable Methicillin-Resistant *Staphylococcus aureus* in the Netherlands, 2007."

236 "a complete ban": European Parliament, and Council of the European Union, Regulation (EC) No. 1831/2003 of the European Parliament and of the Council of 22 September 2003 on additives for use in animal nutrition.

237 "not the triumph": Ministry of Economic Affairs, "Reduced and Responsible: Policy on the Use of Antibiotics in Food-Producing Animals in the Netherlands."

237 "began appearing, in *E. coli* and *Salmonella*": Dierikx et al., "Increased Detection of Extended Spectrum Beta-Lactamase Producing *Salmonella enterica* and *Escherichia coli* Isolates From Poultry."

237 "Dutch researchers began hunting": Overdevest, "Extended-Spectrum B-Lactamase Genes of *Escherichia coli* in Chicken Meat and Humans, the Netherlands"; Leverstein-van Hall et al., "Dutch Patients, Retail Chicken Meat and Poultry Share the Same ESBL Genes, Plasmids and Strains"; Kluytmans et al., "Extended-Spectrum-Lactamase-Producing *Escherichia coli* From Retail Chicken Meat and Humans."

240 "achieved the goal early": National Institute for Public Health and the Environment and Stichting Werkgroep Antibioticabeleid, "Nethmap/MARAN 2013."

Chapter 12
245 "almost all the decisions": Pew Charitable Trusts, "The Business of Broilers."

246 "among the many poultry producers I met": The portrait of conventional intensive poultry production in this chapter is drawn from my visits to more than a dozen growers of chickens and turkeys who are not named and from interviews with Larry and Leighton Cooley and J. Craig Watts on their farms.

249 "the top chicken-producing states": U.S. Poultry & Egg Association, "Industry Economic Data."

251 "the economic basis of poultry production": An excellent book-length examination of the direct and hidden costs of the tournament system is Leonard, *The Meat Racket*.

251 "adopting the model now": Martinez, "A Comparison of Vertical Coordination in the U.S. Poultry, Egg, and Pork Industries."

251 "Environmental advocates say": Pew Environment Group, "Big Chicken."

251 "Economists who study": Knowber, "A Real Game of Chicken"; Vukina and Foster, "Efficiency Gains in Broiler Production Through Contract Parameter Fine Tuning."

251 "other academics": Khan, "Obama's Game of Chicken."

252 "began writing letters": Watts, "Easing the Plight of Poultry Growers"; *Arbitration: Is It Fair When Forced?*

252 "drove 500 miles": Tobey, "Public Workshops."

253 "shoehorned in a provision": Center for a Livable Future, "Industrial Food Animal Production in America," p. 7.

253 "only four pages long": The first few ADUFA reports (all available at https://www.fda .gov/ForIndustry/UserFees/AnimalDrugUserFeeActADUFA/ucm042896.htm) were revised by the FDA in September 2014 as data became accessible. That original four-page report grew to 26 pages, and the most recent, 2015 iteration was 58 pages long.

254 "In the 2015 numbers": Center for Veterinary Medicine, "2015 Summary Report on Antimicrobials Sold or Distributed for Use in Food-Producing Animals."

254 "calculated equivalents": Pew Campaign on Human Health and Industrial Farming, "Record-High Antibiotic Sales for Meat and Poultry Production."

255 "banned in Europe": Singer and Hofacre, "Potential Impacts of Antibiotic Use in Poultry Production"; Marshall and Levy, "Food Animals and Antimicrobials."

255 "back to the early days": Butaye, Devriese, and Haesebrouck, "Antimicrobial Growth Promoters Used in Animal Feed."

255 "reduce the occurrence": Chapman, Jeffers, and Williams, "Forty Years of Monensin for the Control of Coccidiosis in Poultry."

255 "only one potential problem": Vitenskapkomiteen for mattrygghet (Norwegian Scientific Committee for Food Safety), "The Risk of Development of Antimicrobial Resistance With the Use of Coccidiostats in Poultry Diets."

258 "video that Watts collaborated on": Compassion in World Farming, *Chicken Factory Farmer Speaks Out.*

258 "avalanche of news attention": Kristof, "Abusing Chickens We Eat"; "Cock Fight: Meet the Farmer Blowing the Whistle on Big Chicken."

259 "filed for legal protection": Food Integrity Campaign, "Historic Filing."

259 "The experts agreed": Center for Food Integrity, "Expert Panel Examines Broiler Farm Video."

Chapter 13

264 "began to change its mind": The reconstruction of how Perdue Farms moved away from antibiotic use is based on multiple interviews, in Salisbury, Maryland, and elsewhere, with Jim Perdue, chairman, and Bruce Stewart-Brown, senior vice president for food safety, quality, and live operations.

264 "controversial but legal": The feed drug Roxarsone, which contained organic arsenic, was withdrawn from the market by its manufacturer, Pfizer, in April 2015.

265 "was best embodied": Rogers, "Broilers"; Sloane, "I Turned My Father's Tiny Egg Farm Into a Poultry Powerhouse and Became the Face of an Industry."

265 "came into the business": Strom, "Into the Family Business at Perdue."

266 "set up a study": Engster, Marvil, and Stewart-Brown, "The Effect of Withdrawing Growth Promoting Antibiotics From Broiler Chickens."

267 "about to announce": PR Newswire, "After Eliminating Human Antibiotics in Chicken Production in 2014, Perdue Continues Its Leadership."

270 "researchers found": Love et al., "Feather Meal."

272 "The company": Schmall, "The Cult of Chick-Fil-A."

272 "biblically based opposition": O'Connor, "Chick-Fil-A CEO Cathy."

272 "By sales": "The QSR 50."

273 "a coalition of 300 hospitals": Eng, "Meat With Antibiotics off the Menu at Some Hospitals."

273 "the Chicago Public Schools": "305,000 K-12 Students in Chicago Offered Chicken Raised Without Antibiotics."

273 "the University of California, San Francisco": Fleischer, "UCSF Academic Senate Approves Resolution to Phase Out Meat Raised With Non-Therapeutic Antibiotics."

273 "One after another": Natural Resources Defense Council, "Going Mainstream."

274 "what Chick-fil-A went through": The account of Chick-fil-A's conversion to antibiotic-free poultry is based on interviews with David Farmer and others at Chick-fil-A headquarters in Atlanta.

276 "gave humanity": The case for a religious and politically conservative approach to animal welfare is made well in Scully, *Dominion*, written by a former speechwriter for President George W. Bush.

277 "It took the arrival": This account of the creation of the FDA's long-delayed move against agricultural antibiotic use is based on interviews with former FDA staff Thomas Grumbly (now president of the Supporters of Agricultural Research Foundation), Michael Blackwell (now senior director for veterinary policy at the Humane Society of the United States), and Michael R. Taylor (now a senior fellow with Freedman Consulting); with William Flynn, deputy director for science policy in the FDA's Center for Veterinary Medicine; with Representative Louise Slaughter; and with Jonathan Kaplan and Avinash Kar of the Natural Resources Defense Council and Laura Rogers, formerly of the Pew Charitable Trusts and now deputy director of the Antibiotic Resistance Action Center at the George Washington University.

278 "in a quickly drafted letter": Animal Agriculture Coalition, "AAC Followup Letter to Margaret A. Hamburg, MD, Commissioner, Joshua M. Sharfstein, MD, Deputy Commissioner, Food and Drug Administration."

278 "in a statement sent directly": American Association of Avian Pathologists et al., "Letter to Melody Barnes, Assistant to the President, the White House."

279 "the 'tobacco playbook'": Brownell and Warner, "The Perils of Ignoring History"; Malik, "Catch Me if You Can: Big Food Using Big Tobacco's Playbook?"

279 "Now, there were 287": Center for Veterinary Medicine, "FDA Update on Animal Pharmaceutical Industry Response to Guidance #213."

279 "a wonkily titled agency document": Center for Veterinary Medicine, "Guidance for Industry #209."

280 "Advocates were unhappy": Trust for America's Health, "Comment on the Judicious Use of Medically Important Antimicrobial Drugs in Food-Producing Animals—Draft

Guidance"; Pew Charitable Trusts, "Comment on the Judicious Use of Medically Important Antimicrobial Drugs in Food-Producing Animals—Draft Guidance."

280 "the Michigan Farm Bureau asserted": Michigan Farm Bureau, "Comment on the Judicious Use of Medically Important Antimicrobial Drugs in Food-Producing Animals—Draft Guidance."

280 "a companion document": Center for Veterinary Medicine, "Guidance for Industry #213: New Animal Drugs and New Animal Drug Combination Products Administered in or on Medicated Feed or Drinking Water of Food-Producing Animals: Recommendations for Drug Sponsors for Voluntarily Aligning Product Use Conditions with GFI #209."

280 "With that": Center for Veterinary Medicine, "FDA Secures Full Industry Engagement on Antimicrobial Resistance Strategy."

281 "10 percent more weight": Coates, "The Value of Antibiotics for Growth of Poultry."

281 "boosted the gains": Office of Technology Assessment, U.S. Congress, "Drugs in Livestock Feed"; Graham, Boland, and Silbergeld, "Growth Promoting Antibiotics in Food Animal Production."

281 "By one 1970 estimate": Boyd, "Making Meat."

281 "But by the 1990s": Laxminarayan, Teillant, and Van Boeckel, "The Economic Costs of Withdrawing Antimicrobial Growth Promoters From the Livestock Sector."

281 "In 2015": Sneeringer et al., "Economics of Antibiotic Use in U.S. Livestock Production."

282 "many new labels were so vague": Hoelzer, "Judicious Animal Antibiotic Use Requires Drug Label Refinements."

283 "rose 24 percent": Center for Veterinary Medicine, "2015 Summary Report on Antimicrobials Sold or Distributed for Use in Food-Producing Animals."

Chapter 14

285 "Will Harris III said": This account of the development of White Oak Pastures is based on multiple interviews in Bluffton, Georgia, and Atlanta with Will Harris, Jenni Harris, Brian Sapp, John Benoit, and Frankie Darsey.

294 "Scott Sechler said": This account of the development of Bell & Evans is based on multiple interviews in Fredericksburg, Pennsylvania, and Atlanta with Scott Sechler, Scott Sechler, Jr., and Margo Sechler.

298 "only the Bell & Evans chickens": "Chicken Safety."

300 "Linton Hopkins said": This description of how Chef Linton Hopkins brought White Oak Pastures into Delta Air Lines is based on interviews with Linton Hopkins; his executive chef, Jason Paolini; Delta personnel at Restaurant Eugene and at Delta; and the Harrises.

302 "a reality-TV style web series": Levere, "A Cook-Off Among Chefs to Join Delta's Kitchen."

304 "said he would stand fast": Bunge, "Sanderson Farms CEO Resists Poultry-Industry Move to Curb Antibiotics."

304 "a 'gimmick' ": Alonzo, "Sanderson Calls Antibiotic-Free Chicken a 'Gimmick' "; Sanderson Farms, *The Truth About Chicken—Supermarket.*

304 "told the *Guardian*": Levitt, " 'I Don't See a Problem.' "

305 "Smith's successor, Tom Hayes": "Tyson Foods New Leaders Position Company for Future Growth."

305 "declared in a speech": Plantz, "Consumer Misconceptions Dangerous for American Agriculture."

305 "Tyson announced": "Tyson Fresh Meats Launches Open Prairie Natural Pork"; Shanker, "Just Months After Big Pork Said It Couldn't Be Done, Tyson Is Raising up to a Million Pigs Without Antibiotics."

305 "ran a full-page ad": National Pork Producers Council, "Dear Subway Management Team and Franchisee Owners [Advertisement]."

305 "the persistent pumping": You and Silbergeld, "Learning from Agriculture"; Davis et al., "An Ecological Perspective on U.S. Industrial Poultry Production"; You et al., "Detection of a Common and Persistent *tet*(L)-Carrying Plasmid in Chicken-Waste-Impacted Farm Soil."

307 "announced that it had persuaded": "Global Animal Partnership Commits to Requiring 100 Percent Slower-Growing Chicken Breeds by 2024."

307 "it has become mainstream": Roth, "What You Need to Know About the Corporate Shift to Cage-Free Eggs."

Epilogue

309 "a 'high-level meeting'": President of the General Assembly, "Programme of the High Level Meeting on Antibiotic Resistance."

310 "leaned in to a microphone": United Nations Secretary-General, "Secretary-General's Remarks to High-Level Meeting on Antimicrobial Resistance [as Delivered]."

311 "'the greatest and most urgent global risk'": President of the General Assembly, "Draft Political Declaration of the High-Level Meeting of the General Assembly on Antimicrobial Resistance."

311 "a new permanent body of experts": President Barack Obama, Executive Order 13676—Combating Antibiotic-Resistant Bacteria.

312 "instant headline news": Review on Antimicrobial Resistance, "Antimicrobial Resistance."

312 "if nothing were done": Van Boeckel et al., "Global Trends in Antimicrobial Use in Food Animals."

312 "the G7 group": "G7 Ise-Shima Leaders Declaration."

312 "the G20 nations": "G20 Leaders' Communiqué."

BIBLIOGRAPHY

"A Growing Problem. Selective Breeding in the Chicken Industry: The Case for Slower Growth." ASPCA, November 2015. https://www.aspca.org/sites/default/files/chix_white_paper_nov2015_lores.pdf.

Abbey, A., et al. "Effectiveness of Acronize Chlortetracycline in Poultry Preservation Following Long Term Commercial Use." *Food Technology* 14 (December 1960): 609–12.

Abdula, Nazira, et al. "National Action for Global Gains in Antimicrobial Resistance." *Lancet* 387, no. 10014 (January 2016): e3–5.

Abraham, E. P., and E. Chain. "An Enzyme From Bacteria Able to Destroy Penicillin." *Nature* 146 (December 5, 1940): 837.

"'A Chicken for Every Pot'" *New York Times*, October 30, 1928. Advertisement. https://research.archives.gov/id/187095.

"Acronize Maintains Poultry Freshness." *Florence (SC) Morning News*, June 22, 1956.

"Action Sought on Antibiotics After Babies' Deaths." *Times (London)*, April 14, 1969.

"Advertising: Logistics to Fore in Big Move." *New York Times*, January 24, 1957.

Agersø, Yvonne, et al. "Spread of Extended Spectrum Cephalosporinase-Producing *Escherichia coli* Clones and Plasmids From Parent Animals to Broilers and to Broiler Meat in a Production Without Use of Cephalosporins." *Foodborne Pathogens and Disease* 11, no. 9 (September 2014): 740–46.

Ahmad, Aqeel, et al. "Insects in Confined Swine Operations Carry a Large Antibiotic Resistant and Potentially Virulent Enterococcal Community." *BMC Microbiology* 11, no. 1 (2011): 23.

Ahmed, A. Karim, et al. "Petition of the Natural Resources Defense Council to the Secretary of Health and Human Services Requesting the Immediate Suspension of Approval of the Subtherapeutic Use of Penicillin and Tetracyclines in Animal Feeds." November 20, 1984.

Aldous, Chris. "Contesting Famine: Hunger and Nutrition in Occupied Japan, 1945-1952." *Journal of American-East Asian Relations* 17, no. 3 (September 1, 2010): 230–56.

Aleccia, JoNel. "Foster Farms Salmonella Outbreaks: Why Didn't USDA Do More?" NBCNews
.com, December 19, 2013. http://www.nbcnews.com/health/foster-farms-salmonella
-outbreaks-why-didnt-usda-do-more-2D11770690

Alonzo, Austin. "Sanderson Calls Antibiotic-Free Chicken a 'Gimmick.'" WATTAgNet.com,
August 1, 2016. http://www.wattagnet.com/articles/27744-sanderson-calls-antibiotic
-free-chicken-a-gimmick.

American Association of Avian Pathologists et al. "Letter to Melody Barnes, Assistant to the
President, the White House." August 14, 2009. http://www.nmpf.org/sites/default/files/
Industry-White-House-Antibiotic-Letter-081409.pdf.

Anderson, Alicia D., et al. "Public Health Consequences of Use of Antimicrobial Agents in
Food Animals in the United States." *Microbial Drug Resistance* 9, no. 4 (2003): 373–79.

Anderson, E. S. "Drug Resistance in *Salmonella* Typhimurium and Its Implications." *British
Medical Journal* 3, no. 5614 (August 10, 1968): 333–39.

———. "Middlesbrough Outbreak of Infantile Enteritis and Transferable Drug Resistance."
British Medical Journal 1 (February 3, 1968): 293.

———. "Origin of Transferable Drug-Resistance Factors in the Enterobacteriaceae." *British
Medical Journal* 2, no. 5473 (November 27, 1965): 1289–91.

———. "Salmonellosis in Livestock." *Lancet* 2, no. 7768 (July 15, 1972): 138.

———. "The Ecology of Transferable Drug Resistance in the Enterobacteria." *Annual Review
of Microbiology* 22 (1968): 131–80.

———. "Transferable Antibiotic Resistance." *British Medical Journal* 1, no. 5591 (March 2,
1968): 574–75.

Anderson, E. S., and N. Datta. "Resistance to Penicillins and Its Transfer in Enterobacteriaceae." *Lancet* 1, no. 7382 (February 20, 1965): 407–9.

Anderson, E. S., et al. "An Outbreak of Human Infection Due to *Salmonella* Typhimurium
Phage-Type 20a Associated with Infection in Calves." *Lancet* 1, no. 7182 (April 22,
1961): 854–58.

Anderson, E. S., and M. J. Lewis. "Characterization of a Transfer Factor Associated with Drug
Resistance in *Salmonella* Typhimurium." *Nature* 208, no. 5013 (November 27, 1965):
843–49.

———. "Drug Resistance and Its Transfer in *Salmonella* Typhimurium." *Nature* 206, no. 984
(May 8, 1965): 579–83.

Anderson, E. S. "One Fine Day." *New Scientist,* January 19, 1978.

Andersson, M. I. "Development of the Quinolones." *Journal of Antimicrobial Chemotherapy*
51, no. 90001 (May 1, 2003): 1–11.

Andrews, James. "Jack in the Box and the Decline of *E. coli.*" *Food Safety News,* February 11,
2013. http://www.foodsafetynews.com/2013/02/jack-in-the-box-and-the-decline-of
-e-coli/#.WKO_UfONtF8.

Andriole, Vincent T. "The Quinolones: Past, Present, and Future." *Clinical Infectious Diseases*
41, Suppl. 2 (2005): S113–S119.

Angulo, F. J., et al. "Origins and Consequences of Antimicrobial-Resistant Nontyphoidal
Salmonella: Implications for the Use of Fluoroquinolones in Food Animals." *Microbial
Drug Resistance* 6, no. 1 (2000): 77–83.

Angulo, F. J., et al. "Evidence of an Association Between Use of Anti-Microbial Agents in
Food Animals and Anti-Microbial Resistance Among Bacteria Isolated From Humans
and the Human Health Consequences of Such Resistance." *Journal of Veterinary*

Bibliography

Medicine. B, Infectious Diseases and Veterinary Public Health 51, no. 8–9 (November 2004): 374–79.

Animal Agriculture Coalition. "AAC Followup Letter to Margaret A. Hamburg, MD, Commissioner, Joshua M. Sharfstein, MD, Deputy Commissioner, Food and Drug Administration." July 16, 2009. http://www.aavld.org/assets/documents/AAC%20followup%20letter%20to%20FDA%20-%20FINAL%20-%20July%2016%202009.pdf.

"Animal Magicians: A $60,000 Horse, and Rabbits on a Modern Laboratory Farm Help Produce Precious Serum That Saves Many Human Lives." *Popular Science* (February 1942).

"Antibiotic Is Approved." *New York Times,* October 4, 1956.

"Antibiotic on Human Food." *Science News-Letter* 68, no. 24 (1955): 373.

"Antibiotics and Food." *Chemical and Engineering News,* December 12, 1955.

"Antibiotics as Food Preservatives." *British Medical Journal* 2, no. 4997 (1956): 870.

"Antibiotics in Milk." *British Medical Journal* 1, no. 5344 (June 8, 1963): 1491–92.

"Antibiotics in the Barnyard." *Fortune,* March 1952.

"Antibiotics on the Farm." *Nature* 219 (July 13, 1968): 106–7.

"Antibiotics Used to Preserve Food." *New York Times,* October 20, 1956.

Arcilla, Maris S., et al. "Dissemination of the *mcr-1* Colistin Resistance Gene." *Lancet Infectious Diseases* 16, no. 2 (February 2016): 147–49.

Årdal, Christine, et al. "International Cooperation to Improve Access to and Sustain Effectiveness of Antimicrobials." *Lancet* 387, no. 10015 (January 2016): 296–307.

"Army Reduces Meat Ration as Famine Grows." *Chicago Tribune,* September 29, 1946, sec. 1.

Arnedo-Pena, A., et al. "Reactive Arthritis and Other Musculoskeletal Sequelae Following an Outbreak of *Salmonella* Hadar in Castellon, Spain." *Journal of Rheumatology* 37, no. 8 (August 1, 2010): 1735–42.

"Around Capitol Square." *Burlington (NC) Daily Times,* November 21, 1956.

"The Art of Pickin' Chicken." *(Syracuse, NY) Post-Standard,* July 3, 1957. Advertisement.

Aslam, Mueen, et al. "Characterization of Extraintestinal Pathogenic *Escherichia coli* Isolated From Retail Poultry Meats From Alberta, Canada." *International Journal of Food Microbiology* 177 (May 2, 2014): 49–56.

Associated Press. "Antibiotics Will Be Used in Animal Feeds for '77." *New York Times,* September 24, 1977.

———. "DNA Evidence Links Drug-Resistant Infection to Dairy Farm," October 4, 1999.

———. "Drug May Change Fish Marketing." *Fairbanks (AK) Daily News-Miner.* December 1, 1955.

———. "F.D.A. to Order Big Cuts in Penicillin for Animals." *New York Times,* April 16, 1977.

———. "He's Used to Having the Feathers Fly : Chicken Farmer Frank Perdue Is No Stranger to Controversy." *Los Angeles Times,* November 30, 1991.

———. "Tyler Firm to Preserve Chickens by Antibiotics." *Corpus Christi (TX) Caller Times.* December 18, 1955.

Atkinson, Joe W. "Trends in Poultry Hygiene." *Public Health Reports (1896-1970)* 72, no. 11 (1957): 949–56.

"Aureomycin Keeps Poultry Fresh." *Denton (MD) Journal,* December 16, 1955.

Ayres, J. C. "Use of Antibiotics in the Preservation of Poultry." In *Antibiotics in Agriculture: Proceedings of the University of Nottingham Ninth Easter School in Agricultural Science,* edited by M. Woodbine, 244–71. London: Butterworths, 1962.

Azad, M. B., et al. "Infant Antibiotic Exposure and the Development of Childhood Over-weight and Central Adiposity." *International Journal of Obesity* 38, no. 10 (October 2014): 1290–98.

Azvolinsky, Anna. "Resistance Fighter." *Scientist,* June 1, 2015.

Bailar, John C., and Karin Travers. "Review of Assessments of the Human Health Risk Asso-ciated with the Use of Antimicrobial Agents in Agriculture." *Clinical Infectious Diseases* 34, Suppl. 3 (June 1, 2002): S135–43.

Baker, Michael. "How 'Barbecue Bob' Baker Transformed Chicken." *Ezra,* Summer 2012.

Baker, O. E., and United States. *A Graphic Summary of Farm Animals and Animal Products: (Based Largely on the Census of 1930 and 1935).* Miscellaneous Publication/U.S. Depart-ment of Agriculture, no. 269. Washington, D.C.: U.S. Department of Agriculture, 1939. https//catalog.hathitrust.org/Record/009791326.

Barber, Mary. "The Waning Power of Penicillin." *British Medical Journal* 2, no. 4538 (Decem-ber 2, 1947): 1053.

Barnes, E. M. "The Use of Antibiotics for the Preservation of Poultry and Meat." *Bibliotheca Nutritio Et Dieta* 10 (1968): 62–76.

Barnes, Kimberlee K., et al. "A National Reconnaissance of Pharmaceuticals and Other Organic Wastewater Contaminants in the United States—I) Groundwater." *Science of the Total Environment* 402, no. 2–3 (September 1, 2008): 192–200.

Barza, Michael. "Potential Mechanisms of Increased Disease in Humans From Antimicrobial Resistance in Food Animals." *Clinical Infectious Diseases* 34, Suppl. 3 (June 1, 2002): S123–25.

Barza, Michael, and Karin Travers. "Excess Infections Due to Antimicrobial Resistance: The 'Attributable Fraction.'" *Clinical Infectious Diseases* 34, Suppl. 3 (June 1, 2002): S126–30.

Bates, J. "Epidemiology of Vancomycin-Resistant Enterococci in the Community and the Relevance of Farm Animals to Human Infection." *Journal of Hospital Infection* 37 (October 1997): 89–101.

Batt, Angela L., et al. "Evaluating the Vulnerability of Surface Waters to Antibiotic Contam-ination From Varying Wastewater Treatment Plant Discharges." *Environmental Pollu-tion* 142, no. 2 (July 2006): 295–302.

Benarde, M. A., and R. A. Littleford. "Antibiotic Treatment of Crab and Oyster Meats." *Applied Microbiology* 5, no. 6 (November 1957): 368–72.

Benedict, Jeff. *Poisoned: The True Story of the Deadly E. coli Outbreak That Changed the Way Americans Eat.* Buena Vista, VA: Inspire Books, 2011.

Bergeron, Catherine Racicot, et al. "Chicken as Reservoir for Extraintestinal Pathogenic *Escherichia coli* in Humans, Canada." *Emerging Infectious Diseases* 18, no. 3 (March 2012): 415–21.

Bernhardt, Courtney, et al. "More Phosphorus, Less Monitoring." Environmental Integrity Project, September 8, 2015. http://www.environmentalintegrity.org/reports/more-phosphorus-less-monitoring/.

Bernhardt, Courtney, et al. "Manure Overload on Maryland's Eastern Shore." Environmental Integrity Project, December 8, 2014. http://www.environmentalintegrity.org/news/new-report-manure-overload-on-marylands-eastern-shore/.

Berrazeg, M., et al. "New Delhi Metallo-Beta-Lactamase Around the World: An eReview Using Google Maps." *Eurosurveillance* 19, no. 20 (May 22, 2014).

Bibliography

Bessei, W. "Welfare of Broilers: A Review." *World's Poultry Science Journal* 62, no. 3 (September 2006): 455–66.

Blaser, M. J. "Antibiotic Use and Its Consequences for the Normal Microbiome." *Science* 352, no. 6285 (April 29, 2016): 544–45.

Blaser, Martin J. "Who Are We? Indigenous Microbes and the Ecology of Human Diseases." *EMBO Reports* 7, no. 10 (October 2006): 956–60.

———. *Missing Microbes: How the Overuse of Antibiotics Is Fueling Our Modern Plagues.* New York: Holt and Co., 2014.

Blaser, Martin J., and Stanley Falkow. "What Are the Consequences of the Disappearing Human Microbiota?" *Nature Reviews Microbiology* 7, no. 12 (2009): 887–94.

Bogdanovich, T., et al. "Colistin-Resistant, *Klebsiella pneumoniae* Carbapenemase (KPC)-Producing *Klebsiella pneumoniae* Belonging to the International Epidemic Clone ST258." *Clinical Infectious Diseases* 53, no. 4 (August 15, 2011): 373–76.

Bonar, Samantha. "Foster Farms Finally Recalls Chicken." *LA (CA) Weekly*, July 7, 2014. http://www.laweekly.com/content/printView/4829157.

Bondt, Nico, et al. "Trends in Veterinary Antibiotic Use in the Netherlands 2004-2012." Wageningen, NL: LEI, Wageningen University, November 2012. http://www.wur.nl/upload_mm/8/7/f/e4deb048-6a0c-401e-9620-fab655287fbc_Trends%20in%20use%202004-2012.pdf.

Bonten, Marc J. M., et al. "Vancomycin-Resistant Enterococci: Why Are They Here, and Where Do They Come From?" *Lancet Infectious Diseases* 1, no. 5 (2001): 314–25.

Boyd, William. "Making Meat: Science, Technology, and American Poultry Production." *Technology and Culture* 42, no. 4 (2001): 631–64.

Braude, R. "Antibiotics in Animal Feeds in Great Britain." *Journal of Animal Science* 46, no. 5 (May 1978): 1425–36.

Braude, R., et al. "The Value of Antibiotics in the Nutrition of Swine: A Review." *Antibiotics and Chemotherapy* 3, no. 3 (March 1953): 271–91.

Brown, J. R. "'Aureomycin, Plot 23 and the Smithsonian Institution,' Excerpted From JR Brown, '100 Years—Sanborn Field: A Capsule of Scientific Agricultural History in Central Missouri,' " MU in Brick and Mortar, 2006. http://muarchives.missouri.edu/historic/buildings/Sanborn/files/aueromycin.pdf.

Brown, Kevin. "The History of Penicillin From Discovery to the Drive to Production." *Pharmaceutical Historian* 34, no. 3 (September 2004): 37–43.

Brownell, Kelly D., and Kenneth E. Warner. "The Perils of Ignoring History: Big Tobacco Played Dirty and Millions Died. How Similar Is Big Food?" *Milbank Quarterly* 87, no. 1 (March 2009): 259–94.

Bruinsma, Jelle. "World Agriculture: Towards 2015/2030: An FAO Perspective." Food and Agriculture Organisation of the United Nations. Accessed February 17, 2017. http://www.fao.org/docrep/005/y4252e/y4252e05b.htm.

Bud, Robert. *Penicillin: Triumph and Tragedy.* New York: Oxford University Press, 2006.

Bugos, Glenn E. "Intellectual Property Protection in the American Chicken–Breeding Industry." *Business History Review* 66, no. 1 (March 1992): 127–68.

Bunge, Jacob. "Sanderson Farms CEO Resists Poultry-Industry Move to Curb Antibiotics." *Wall Street Journal*, May 20, 2015. http://www.wsj.com/articles/sanderson-farms-ceo-resists-poultry-industry-move-to-curb-antibiotics-1432137667.

Burnside, J. E., and T. J. Cunha. "Effect of Animal Protein Factor Supplement on Pigs Fed Different Protein Supplements." *Archives of Biochemistry* 23, no. 2 (September 1949): 328–30.

Butaye, P., et al. "Antimicrobial Growth Promoters Used in Animal Feed: Effects of Less Well Known Antibiotics on Gram-Positive Bacteria." *Clinical Microbiology Reviews* 16, no. 2 (April 1, 2003): 175–88.

"Capuchino Foods Advertisement." San Mateo (CA). *The Post.* January 17, 1962.

Carpenter, Kenneth J. "Thomas Hughes Jukes (1906–1999)." *Journal of Nutrition* 130, no. 6 (2000): 1521–23.

Carrel, Margaret, et al. "Residential Proximity to Large Numbers of Swine in Feeding Operations Is Associated With Increased Risk of Methicillin-Resistant *Staphylococcus aureus* Colonization at Time of Hospital Admission in Rural Iowa Veterans." *Infection Control and Hospital Epidemiology* 35, no. 2 (February 2014): 190–92.

Casewell, M. "The European Ban on Growth-Promoting Antibiotics and Emerging Consequences for Human and Animal Health." *Journal of Antimicrobial Chemotherapy* 52, no. 2 (July 1, 2003): 159–61.

Casey, Joan A., et al. "High-Density Livestock Operations, Crop Field Application of Manure, and Risk of Community-Associated Methicillin-Resistant *Staphylococcus aureus* Infection in Pennsylvania." *Journal of the American Medical Association Internal Medicine* 173, no. 21 (November 25, 2013): 1980.

Castanheira, Mariana, et al. "Detection of *mcr-1* Among *Escherichia coli* Clinical Isolates Collected Worldwide as Part of the SENTRY Antimicrobial Surveillance Program in 2014 and 2015." *Antimicrobial Agents and Chemotherapy* 60, no. 9 (September 2016): 5623–24.

Castanon, J. I. R. "History of the Use of Antibiotic as Growth Promoters in European Poultry Feeds." *Poultry Science* 86, no. 11 (November 1, 2007): 2466–71.

Castillo Neyra, et al. "Multidrug-Resistant and Methicillin-Resistant *Staphylococcus aureus* (MRSA) in Hog Slaughter and Processing Plant Workers and Their Community in North Carolina (USA)." *Environmental Health Perspectives*, February 7, 2014.

Catry, Boudewijn, et al. "Use of Colistin-Containing Products within the European Union and European Economic Area (EU/EEA): Development of Resistance in Animals and Possible Impact on Human and Animal Health." *International Journal of Antimicrobial Agents* 46, no. 3 (September 2015): 297–306.

CDC Foodborne Diseases Active Surveillance Network. "FoodNet 2012 Surveillance Report (Final Report)." Atlanta, GA: U.S. Department of Health and Human Services, Centers for Disease Control and Prevention, 2014. https://www.cdc.gov/foodnet/PDFs/2012_annual_report_508c.pdf.

" 'Celbenin'-Resistant Staphylococci." *British Medical Journal,* January 14, 1961, 1113–14.

Center for Food Integrity. "Expert Panel Examines Broiler Farm Video," April 19, 2016. http://www.foodintegrity.org/wp-content/uploads/2016/04/ACRP-broiler-video-04-19-16-FINAL.pdf.

Center for a Livable Future. "Industrial Food Animal Production in America: Examining the Impact of the Pew Commission's Priority Recommendations." Baltimore, MD: Johns Hopkins Bloomberg School of Public Health, October 2013. http://www.jhsph.edu/research/centers-and-institutes/johns-hopkins-center-for-a-livable-future/_pdf/research/clf_reports/CLF-PEW-for%20Web.pdf.

Bibliography

Center for Veterinary Medicine, Food and Drug Administration, U.S. Department of Health and Human Services. "2009 Summary Report on Antimicrobials Sold or Distributed for Use in Food-Producing Animals (Revised September 2014)." Washington, D.C.: Food and Drug Administration, December 9, 2010. http://www.fda.gov/downloads/ForIndustry/UserFees/AnimalDrugUserFeeActADUFA/UCM231851.pdf.

———. "2010 Summary Report on Antimicrobials Sold or Distributed for Use in Food-Producing Animals (Revised September 2014)." Washington, D.C.: Food and Drug Administration, October 28, 2011. http://www.fda.gov/downloads/ForIndustry/UserFees/AnimalDrugUserFeeActADUFA/UCM277657.pdf.

———. "2011 Summary Report on Antimicrobials Sold or Distributed for Use in Food-Producing Animals (Revised September 2014)." Washington, D.C.: Food and Drug Administration, February 5, 2013. http://www.fda.gov/downloads/ForIndustry/UserFees/AnimalDrugUserFeeActADUFA/UCM338170.pdf.

———. "2012 Summary Report on Antimicrobials Sold or Distributed for Use in Food-Producing Animals." Washington, D.C.: Food and Drug Administration, September 2014. http://www.fda.gov/downloads/ForIndustry/UserFees/AnimalDrugUserFeeActADUFA/UCM416983.pdf.

———. "2013 Summary Report on Antimicrobials Sold or Distributed for Use in Food-Producing Animals." Washington, D.C.: Food and Drug Administration, April 2015. http://www.fda.gov/downloads/ForIndustry/UserFees/AnimalDrugUserFeeActADUFA/UCM440584.pdf.

———. "2014 Summary Report on Antimicrobials Sold or Distributed for Use in Food-Producing Animals." Washington, D.C.: Food and Drug Administration, December 2015. http://www.fda.gov/downloads/ForIndustry/UserFees/AnimalDrugUserFeeActADUFA/UCM476258.pdf.

———. "2015 Summary Report on Antimicrobials Sold or Distributed for Use in Food-Producing Animals." Washington, D.C.: Food and Drug Administration, December 2016. http://www.fda.gov/downloads/ForIndustry/UserFees/AnimalDrugUserFeeActADUFA/UCM534243.pdf.

———. "FDA Announces Final Decision About Veterinary Medicine." July 28, 2005. https://www.fda.gov/NewsEvents/Newsroom/PressAnnouncements/2005/ucm108467.htm.

———. "FDA Secures Full Industry Engagement on Antimicrobial Resistance Strategy," June 30, 2014. http://www.fda.gov/AnimalVeterinary/NewsEvents/CVMUpdates/ucm403285.htm.

———. "FDA Update on Animal Pharmaceutical Industry Response to Guidance #213." June 11, 2015. http://www.fda.gov/AnimalVeterinary/SafetyHealth/AntimicrobialResistance/JudiciousUseofAntimicrobials/ucm390738.htm.

———. "FDA Update on Animal Pharmaceutical Industry Response to Guidance #213," March 26, 2014. http://www.fda.gov/AnimalVeterinary/SafetyHealth/AntimicrobialResistance/JudiciousUseofAntimicrobials/ucm390738.htm.

———. "Guidance for Industry #152: Evaluating the Safety of Antimicrobial New Animal Drugs With Regard to Their Microbiological Effects on Bacteria of Human Health Concern." Washington, D.C.: Food and Drug Administration, October 23, 2003. http://www.fda.gov/downloads/AnimalVeterinary/GuidanceComplianceEnforcement/GuidanceforIndustry/UCM052519.pdf.

———. "Guidance for Industry #209 : The Judicious Use of Medically Important Antimicrobial Drugs in Food-Producing Animals." Washington, D.C.: Food and Drug Administration, April 30, 2012. http://www.fda.gov/downloads/AnimalVeterinary/GuidanceCompliance Enforcement/GuidanceforIndustry/UCM216936.pdf.

———. "Guidance for Industry #213: New Animal Drugs and New Animal Drug Combination Products Administered in or on Medicated Feed or Drinking Water of Food-Producing Animals: Recommendations for Drug Sponsors for Voluntarily Aligning Product Use Conditions With GFI #209." Washington, D.C: Food and Drug Administration, December 2013. http://www.fda.gov/downloads/AnimalVeterinary/GuidanceCompliance Enforcement/GuidanceforIndustry/UCM299624.pdf.

———. "Human Health Impact of Fluoroquinolone Resistant *Campylobacter* Attributed to the Consumption of Chicken." October 18, 2000. https://www.fda.gov/downloads/ AnimalVeterinary/SafetyHealth/RecallsWithdrawals/UCM152308.pdf

———. "Product Safety Information—Questions and Answers Regarding 3-Nitro (Roxarsone)." April 2015. http://www.fda.gov/AnimalVeterinary/SafetyHealth/ProductSafety Information/ucm258313.htm.

Center for Veterinary Medicine, Food and Drug Administration, Centers for Disease Control and Prevention, and U.S. Department of Agriculture. "On-Farm Antimicrobial Use and Resistance Data Collection: Transcript of a Public Meeting, September 30, 2015." Washington, D.C., September 30, 2015. https://www.regulations.gov/document? D=FDA-2015-N-2768-0011.

Centers for Disease Control and Prevention, U.S. Department of Health and Human Services. "Antibiotic Resistance Threats in the United States, 2013." Atlanta, GA: U.S. Department of Health and Human Services, Centers for Disease Control and Prevention, November 2013. https://www.cdc.gov/drugresistance/threat-report-2013/.

———."Four Pediatric Deaths From Community-Acquired Methicillin-Resistant *Staphylococcus aureus*—Minnesota and North Dakota, 1997-1999." *Morbidity and Mortality Weekly Report* 48, no. 32 (August 20, 1999): 707–10.

———. "Multidrug-Resistant *Salmonella* Serotype Typhimurium—United States, 1996." *Morbidity and Mortality Weekly Report* 46, no. 14 (April 11, 1997): 308–10.

———. "Multistate Outbreak of *Salmonella* Heidelberg Infections Linked to Chicken (Final Update) July 10, 2013," July 10, 2013. https://www.cdc.gov/salmonella/heidelberg-02-13/index.html.

———. "Multistate Outbreak of Multidrug-Resistant *Salmonella* Heidelberg Infections Linked to Foster Farms Brand Chicken (Final Update)." Atlanta, GA, July 31, 2014. https:// www.cdc.gov/salmonella/heidelberg-10-13/index.html.

———. "Nosocomial Enterococci Resistant to Vancomycin—United States, 1989-1993." *Morbidity and Mortality Weekly Report* 42, no. 30 (August 6, 1993): 597–99.

———. "Outbreak of *Salmonella* Heidelberg Infections Linked to a Single Poultry Producer—13 States, 2012–2013." *Morbidity and Mortality Weekly Report* 62, no. 27 (July 12, 2013): 553–56.

———."Pneumocystis Pneumonia—Los Angeles." *Morbidity and Mortality Weekly Report* 30, no. 21 (June 5, 1981): 250–52.

———. "Press Briefing Transcript—CDC Telebriefing on Today's Drug-Resistant Health Threats." Centers for Disease Control and Prevention, September 16, 2013. https:// www.cdc.gov/media/releases/2013/t0916_health-threats.html.

Bibliography

——. "PulseNet: 20 Years of Making Food Safer to Eat." Accessed April 3, 2016. https://www.cdc.gov/pulsenet/pdf/pulsenet-20-years_4_pg_final_508.pdf.

——. "Update: Multistate Outbreak of *Escherichia coli* O157:H7 Infections From Hamburgers—Western United States, 1992–1993." *Morbidity and Mortality Weekly Report* 42, no. 14 (April 16, 1993): 258–63.

Cetinkaya, Y., et al. "Vancomycin-Resistant Enterococci." *Clinical Microbiology Reviews* 13, no. 4 (October 2000): 686–707.

Chaney, Margaret S. "The Role of Science in Today's Food." *Marriage and Family Living* (May 1957): 142-149.

"Changes in Methods of Marketing Are on the Way." *Salisbury (MD) Times,* December 4, 1956.

Chapman, H. D., et al. "Forty Years of Monensin for the Control of Coccidiosis in Poultry." *Poultry Science* 89, no. 9 (September 2010): 1788–1801.

Charles, Dan. "How Foster Farms Is Solving the Case of the Mystery Salmonella." The Salt, National Public Radio, August 28, 2014. http://www.npr.org/sections/the salt/2014/ 08/28/342166299/how-foster-farms-is-solving-the-case-of-the-mystery -salmonella.

——. "Is Foster Farms a Food Safety Pioneer or a Persistent Offender?" The Salt, National Public Radio, July 9, 2014. http://www.npr.org/sections/thesalt/2014/07/09/330160016/ is-foster-farms-a-food-safety-pioneer-or-a-persistent-offender.

Chee-Sanford, J. C., et al. "Occurrence and Diversity of Tetracycline Resistance Genes in Lagoons and Groundwater Underlying Two Swine Production Facilities." *Applied and Environmental Microbiology* 67, no. 4 (April 1, 2001): 1494–1502.

Cherrington, John. "Why Antibiotics Face Their Swann Song." *Financial Times,* November 11, 1969.

Chesapeake Bay Foundation. "Manure's Impact on Rivers, Streams and the Chesapeake Bay," July 28, 2004. http://www.cbf.org/document.doc?id=137.

"Chlortetracycline as a Preservative." *Public Health Reports (1896–1970)* 71, no. 1 (1956): 66.

"Clamp down." *Economist,* November 22, 1969.

Clark, W. F., et al. "Long Term Risk for Hypertension, Renal Impairment, and Cardiovascular Disease After Gastroenteritis From Drinking Water Contaminated With *Escherichia coli* O157:H7: A Prospective Cohort Study." *British Medical Journal* 341, (November 17, 2010): c6020.

Cloud, Joe. "The Fight to Save Small-Scale Slaughterhouses." *Atlantic,* May 24, 2010. https://www.theatlantic.com/health/archive/2010/05/the-fight-to-save-small-scale-slaughter houses/ 57114/.

Coates, M. E. "The Value of Antibiotics for Growth of Poultry." In *Antibiotics in Agriculture: Proceedings of the University of Nottingham Ninth Easter School in Agricultural Science,* edited by M. Woodbine. London: Butterworths, 1962, 203–208.

Coates, M.E., et al. "A Mode of Action of Antibiotics in Chick Nutrition." *Journal of the Science of Food and Agriculture* (January 1952): 43–48.

"Cock Fight: Meet the Farmer Blowing the Whistle on Big Chicken." *Fusion Interactive,* February 2015. http://interactive.fusion.net/cock-fight/.

Cody, S. H., et al. "Two Outbreaks of Multidrug-Resistant *Salmonella* Serotype Typhimurium DT104 Infections Linked to Raw-Milk Cheese in Northern California." *Journal of the American Medical Association* 281, no. 19 (May 19, 1999): 1805–10.

Cogliani, Carol, et al. "Restricting Antimicrobial Use in Food Animals: Lessons From Europe." *Microbe* 6, no. 6 (2011): 274–79.

Cohen, M. L., and R. V. Tauxe. "Drug-Resistant Salmonella in the United States: An Epidemiologic Perspective." *Science* 234, no. 4779 (November 21, 1986): 964–69.

Collignon, Peter. "Fluoroquinolone Use in Food Animals." *Emerging Infectious Diseases* 11, no. 11 (November 2005): 1789–92.

Collignon, Peter, et al. "Human Deaths and Third-Generation Cephalosporin Use in Poultry, Europe." *Emerging Infectious Diseases* 19, no. 8 (August 2013): 1339–40.

Collingham, E. M. *The Taste of War: World War Two and the Battle for Food.* London: Allen Lane, 2011.

Combs, G. F. "Mode of Action of Antibiotics in Poultry." In *Proceedings of the First International Conference on the Use of Antibiotics in Agriculture, 19-21 October 1955,* pp. 107–26. Washington, D.C.: National Academy of Sciences, 1956.

Comery, R., et al. "Identification of Potentially Diarrheagenic Atypical Enteropathogenic *Escherichia coli* Strains Present in Canadian Food Animals at Slaughter and in Retail Meats." *Applied and Environmental Microbiology* 79, no. 12 (June 15, 2013): 3892–96.

Committee on Salmonella, National Research Council. *An Evaluation of the Salmonella Problem.* Washington, D.C.: National Academy of Sciences, 1969.

Committee on the Judiciary, U.S. Senate. *Arbitration: Is It Fair When Forced? 2011. Hearing Before the Committee on the Judiciary, U.S. Senate,* 112th Cong., 1st sess., Pub. L. No. J-112-47, 2011. https://www.gpo.gov/fdsys/pkg/CHRG-112shrg71582/html/CHRG-112shrg71582.htm.

Communicable Disease Center, U.S. Department of Health, Education and Welfare. *Proceedings, National Conference on Salmonellosis, March 11-13, 1964.* Atlanta, GA: U.S. Department of Health, Education and Welfare, March 1965.

Communicable Disease Control Section, Seattle-King County Department of Public Health. "Surveillance of the Flow of *Salmonella* and *Campylobacter* in a Community. Prepared for U.S. Department of Health and Human Services, Public Health Service, Food and Drug Administration Bureau of Veterinary Medicine." Seattle: Seattle-King County Department of Public Health, August 1984.

Compassion in World Farming. *Chicken Factory Farmer Speaks Out.* 2014. https://www.youtube.com/watch?v=YE9l94b3x9U.

Conis, Elena. "Debating the Health Effects of DDT: Thomas Jukes, Charles Wurster, and the Fate of an Environmental Pollutant." *Public Health Reports* 125, no. 2 (April 2010): 337–42.

"Consumer." "Chicken Flavor." *Mercury.* December 30, 1959.

Cook, Robert E., et al. "How Chicken on Sunday Became an Anyday Treat." In *That We May Eat: The Yearbook of Agriculture—1975.* Washington, D.C.: U.S. Department of Agriculture, 1975, 125–32.

"Co-Op Shopping Center." *(Eau Claire, WI) Daily Telegram,* June 17, 1964. Advertisement.

Cox, Jeremy. "Why Somerset Turned Up the Heat on Chicken Farms." Delmarva Media Group, June 13, 2016. http://www.delmarvanow.com/story/news/local/maryland/2016/06/10/why-somerset-turned-up-heat-chicken-farms/85608166/.

Cox, Laura M., and Martin J. Blaser. "Antibiotics in Early Life and Obesity." *Nature Reviews Endocrinology* 11, no. 3 (December 9, 2014): 182–90.

Bibliography

Cox, Laura M., et al. "Altering the Intestinal Microbiota During a Critical Developmental Window Has Lasting Metabolic Consequences." *Cell* 158, no. 4 (August 2014): 705–21.

Crow, James F. "Thomas H. Jukes (1906–1999)." *Genetics* 154, no. 3 (2000): 955–56.

Cunha, T. J., and J. E. Burnside. "Effect of Vitamin B$_{12}$, Animal Protein Factor and Soil for Pig Growth." *Archives of Biochemistry* 23, no. 2 (September 1949): 324–26.

Curtis, Jack M. "Food and Drug Projects of Interest to State Health Officers: Antibiotics and Food." *Public Health Reports (1896–1970)* 71, no. 1 (1956): 50–51.

Danbury, T. C., et al. "Self-Selection of the Analgesic Drug Carprofen by Lame Broiler Chickens." *Veterinary Record* 146 (March 11, 2000): 307–11.

"Dangerous Contaminated Chicken." *Consumer Reports* (January 2014). http://www.consumer reports.org/cro/magazine/2014/02/the-high-cost-of-cheap-chicken/index.htm

"The Dangers of Misusing Antibiotics." *Guardian (Manchester)*, February 3, 1968.

Dar, Osman A, et al. "Exploring the Evidence Base for National and Regional Policy Interventions to Combat Resistance." *Lancet* 387, no. 10015 (January 2016): 285–95.

Das, Pamela, and Richard Horton. "Antibiotics: Achieving the Balance Between Access and Excess." *Lancet* 387, no. 10014 (January 2016): 102–4.

Datta, N. "Transmissible Drug Resistance in an Epidemic Strain of *Salmonella* Typhimurium." *Journal of Hygiene* 60 (September 1962): 301–10.

Davis, Meghan F., et al. "An Ecological Perspective on U.S. Industrial Poultry Production: The Role of Anthropogenic Ecosystems on the Emergence of Drug-Resistant Bacteria From Agricultural Environments." *Current Opinion in Microbiology* 14, no. 3 (June 2011): 244–50.

Dawson, Sam. "Food Research." Massillon (OH). *Evening Independent*, June 1, 1956.

———. "New Methods to Keep Food Under Study." *Freeport (IL) Journal-Standard*, June 1, 1956.

Deatherage, F. E. "Antibiotics in the Preservation of Meat." In *Antibiotics in Agriculture: Proceedings of the University of Nottingham Ninth Easter School in Agricultural Science*, edited by M. Woodbine, 225–43. London: Butterworths, 1962.

———. Method of preserving meat. U.S. Patent 2786768 A, filed May 12, 1954, and issued March 26, 1957. http://www.google.com/patents/US2786768.

———. "Use of Antibiotics in the Preservation of Meats and Other Food Products." *American Journal of Public Health and the Nation's Health* 47, no. 5 (May 1957): 594–600.

Dechet, Amy M., et al. "Outbreak of Multidrug-Resistant *Salmonella enterica* Serotype Typhimurium Definitive Type 104 Infection Linked to Commercial Ground Beef, Northeastern United States, 2003-2004." *Clinical Infectious Diseases* 42, no. 6 (March 15, 2006): 747–52.

Deo, Randhir P. "Pharmaceuticals in the Surface Water of the USA: A Review." *Current Environmental Health Reports* 1, no. 2 (June 2014): 113–22.

Department of Health. "Antimicrobial Resistance Poses 'Catastrophic Threat,' Says Chief Medical Officer." Gov.uk, March 12, 2013. https://www.gov.uk/government/news/antimicrobial-resistance-poses-catastrophic-threat-says-chief-medical-officer--2.

"The Diary of a Tragedy." *Evening Gazette (Middlesbrough)*, March 7, 1968.

Dierikx, Cindy, et al. "Increased Detection of Extended Spectrum Beta-Lactamase Producing *Salmonella enterica* and *Escherichia coli* Isolates From Poultry." *Veterinary Microbiology* 145, no. 3-4 (October 2010): 273–78.

Dietary Goals for the United States. Prepared by the Staff of the Select Committee on Nutrition and Human Needs, United States Senate. 2nd ed. Washington, D.C.: Government Printing Office, 1977.

Dixon, Bernard. "Antibiotics on the Farm—Major Threat to Human Health." *New Scientist,* October 5, 1967.

Dow Jones. "U.S. Approves Antibiotic Drug to Preserve Uncooked Poultry." *Bridgeport (CT) Telegram,* December 1, 1955.

Du, Hong, et al. "Emergence of the *mcr-1* Colistin Resistance Gene in Carbapenem-Resistant Enterobacteriaceae." *Lancet Infectious Diseases* 16, no. 3 (March 2016): 287–88.

Duggar, Benjamin M. Aureomycin and preparation of same. Patent U.S. Patent 2,482,055 A, filed February 11, 1948, and issued September 13, 1949. http://www.google.com/patents/US2482055.

Durbin, C. G. "Antibiotics in Food Preservation." *American Journal of Public Health and the Nation's Health* 46, no. 10 (October 1956): 1306–8.

Dyer, I. A., et al. "The Effect of Adding APF Supplements and Concentrates Containing Supplementary Growth Factors to a Corn-Soybean Oil Meal Ration for Weanling Pigs." *Journal of Animal Science* 9, no. 3 (August 1950): 281–88.

Eckblad, Marshall. "Dark Meat Getting a Leg Up on Boring Boneless Breast." *Wall Street Journal,* April 16, 2012. http://www.wsj.com/articles/SB10001424052702304587704577333923937879132.

"Economic Report on Antibiotics Manufacture." U.S. Federal Trade Commission, June 1958. http://hdl.handle.net/2027/mdp.39015072106332.

Economic Research Service, U.S. Department of Agriculture. "Tracking Foodborne Pathogens From Farm to Table." Washington, D.C.: U.S. Department of Agriculture Economic Research Service, January 9, 1995. https://www.ers.usda.gov/webdocs/publications/mp1532/32452_mp1532_002.pdf.

Ekkelenkamp, M. B., et al. "Endocarditis Due to Methicillin-Resistant *Staphylococcus aureus* Originating From Pigs." *Nederlands Tijdschrift Voor Geneeskunde* 150, no. 44 (November 4, 2006): 2442–47.

Elfick, Dominic. "A Brief History of Broiler Selection: How Chicken Became a Global Food Phenomenon in 50 Years." Aviagen International, 2013. http://cn.aviagen.com/assets/Sustainability/50-Years-of-Selection-Article-final.pdf.

Endtz, Hubert P., et al. "Quinolone Resistance in *Campylobacter* Isolated from Man and Poultry Following the Introduction of Fluoroquinolones in Veterinary Medicine." *Journal of Antimicrobial Chemotherapy* 27, no. 2 (1991): 199–208.

Eng, Monica. "Meat With Antibiotics off the Menu at Some Hospitals." *Chicago Tribune,* July 20, 2010. http://articles.chicagotribune.com/2010-07-20/health/ct-met-hospital-meat-20100718_1_antibiotic-free-antibiotics-for-growth-promotion-food-producing-animals.

Engster, H. M., et al. "The Effect of Withdrawing Growth Promoting Antibiotics From Broiler Chickens: A Long-Term Commercial Industry Study." *Journal of Applied Poultry Research* 11, no. 4 (December 1, 2002): 431–36.

European Centre for Disease Prevention and Control, European Food Safety Authority, and European Medicines Agency. "ECDC/EFSA/EMA First Joint Report on the Integrated Analysis of the Consumption of Antimicrobial Agents and Occurrence of Antimicrobial Resistance in Bacteria from Humans and Food- Producing Animals." *EFSA Journal* 13, no. 1 (January 30, 2015): 4006–20.

European Medicines Agency. "Updated Advice on the Use of Colistin Products in Animals Within the European Union: Development of Resistance and Possible Impact on Human and Animal Health." London: European Medicines Agency, July 27, 2016. http://www.ema.europa.eu/docs/en_GB/document_library/Scientific_guideline/2016/07/WC500211080.pdf

———. "Use of Colistin Products in Animals Within the European Union: Development of Resistance and Possible Impact on Human and Animal Health." London: European Medicines Agency, July 19, 2013. http://www.ema.europa.eu/docs/en_GB/document_library/Report/2013/07/WC500146813.pdf.

European Parliament, and Council of the European Union. "Regulation (EC) No. 1831/2003 of the European Parliament and of the Council of 22 September 2003 on Additives for Use in Animal Nutrition." Regulation (EC) No. 1831/2003 § (2003). http://eur-lex.europa.eu/legal-content/EN/TXT/?uri=CELEX%3A32003R1831.

Eykyn, S. J., and I. Phillips. "Community Outbreak of Multiresistant Invasive *Escherichia coli* Infection." *Lancet* 2, no. 8521–22 (December 20, 1986): 1454.

Falk, Leslie A. "Will Penicillin Be Used Indiscriminately?" *Journal of the American Medical Association*, March 17, 1945.

Falkow, Stanley. "Running Around in Circles: Following the 'Jumping Genes' of Antibiotic Resistance." *Infectious Diseases in Clinical Practice* 9 (April 2000): 119–22.

"Famed Poultry Experts to Judge COT Finals at Hollidaysburg." *Altoona (PA) Tribune*, April 9, 1956.

Fanoy, E., et al. "An Outbreak of Non-Typeable MRSA within a Residential Care Facility." *European Communicable Disease Bulletin* 14, no. 1 (January 8, 2009).

Farber, L. "Antibiotics in Food Preservation." *Annual Review of Microbiology* 13, no. 1 (October 1959): 125–40.

"Farmyard Use of Drugs." *Times (London)*, November 21, 1969.

Ferguson, Dwight D., et al. "Detection of Airborne Methicillin-Resistant *Staphylococcus aureus* Inside and Downwind of a Swine Building, and in Animal Feed: Potential Occupational, Animal Health, and Environmental Implications." *Journal of Agromedicine* 21, no. 2 (April 2, 2016): 149–53.

Fish, N. A. "Health Hazards Associated with Production and Preparation of Foods." *Revue Canadienne de Santé Publique* 59, no. 12 (December 1968): 463–66.

Fishlock, David. "Government Action Urged on Farm Drugs." *Financial Times,* May 1, 1968.

———. "Closer Control of Farm Antibiotics." *Financial Times,* November 21, 1969.

Flanary, Mildred E. "Five Firms Entertain Food Editors." Long Beach (CA). *Independent*, October 3, 1957.

Fleischer, Deborah. "UCSF Academic Senate Approves Resolution to Phase Out Meat Raised With Non-Therapeutic Antibiotics." UCSF Office of Sustainability, May 2013. http://sustainability.ucsf.edu/1.353.

"Food Additives Bills." *Journal of Agricultural and Food Chemistry* 3, no. 6 (June 1, 1955): 466.

Food and Agriculture Organisation of the United Nations. "The Rise of Unregulated Livestock Production in East and Southeast Asia Prompts Health Concerns." February 6, 2017. http://www.fao.org/asiapacific/news/detail-events/en/c/469630/.

Food and Drug Administration. "Certification of Batches of Antibiotic and Antibiotic-Containing Drugs." 16 Fed. Reg. 3647 § (1951).

——. "National Antimicrobial Resistance Monitoring System (NARMS) Integrated Report 2012–2013." Washington, D.C.: Food and Drug Administration. Accessed April 13, 2016. http://www.fda.gov/downloads/AnimalVeterinary/SafetyHealth/Antimicrobial Resistance/NationalAntimicrobialResistanceMonitoringSystem/UCM453398.pdf.

——. "Report to the Commissioner of the Food and Drug Administration by the FDA Task Force on the Use of Antibiotics in Animal Feeds." Rockville, MD: Food and Drug Administration, 1972. http://hdl.handle.net/2027/coo.31924051104002.

Food and Drug Administration, Centers for Disease Control and Prevention, and U.S. Department of Agriculture. "On-Farm Antimicrobial Use and Resistance Data Collection Public Meeting, September 30, 2015 (FDA/CVM, CDC, and USDA)." Washington, D.C.: FDA/CVM, CDC, and USDA, November 27, 2015. https://www.regulations.gov/document?D=FDA-2015-N-2768-0011.

Food and Drug Administration, Office of Surveillance and Epidemiology. "Drug Use Review." Washington, D.C.: U.S. Department of Health and Human Services Food and Drug Administration, April 5, 2012. http://www.fda.gov/downloads/drugs/drugsafety/informationbydrugclass/ucm319435.pdf.

Food and Drug Administration, U.S. Department of Health and Human Services. "Enrofloxacin for Poultry: Opportunity for Hearing." 65 Fed. Reg. 64954 § (2000).

——. "National Antimicrobial Resistance Monitoring System 2010 Retail Meat Report." Washington, D.C.: Food and Drug Administration, March 2012. http://www.fda.gov/downloads/AnimalVeterinary/SafetyHealth/AntimicrobialResistance/NationalAnti microbialResistanceMonitoringSystem/UCM293581.pdf.

Food and Drug Administration, U.S. Department of Health, Education and Welfare. "Diamond Shamrock Chemical Co., et al.: Penicillin-Containing Premixes; Opportunity for a Hearing." 42 Fed. Reg. 43772 § (1977).

——. Pfizer, Inc., et al. "Tetracycline (Chlortetracycline and Oxytetracycline)-Containing Premixes; Opportunity for a Hearing." 42 Fed. Reg. 56264 § (1977).

——. "Tolerances and Exemptions From Tolerances for Pesticide Chemicals in or on Raw Agricultural Commodities; Tolerance for Residues of Chlortetracycline." 20 Fed. Reg. 8776 § (1955).

——. "Exemption From Certification of Antibiotic Drugs for Use in Animal Feed and of Animal Feed Containing Antibiotic Drugs." 18 Fed. Reg. 2335 § (1953). https://www.loc.gov/item/fr018077/.

Food Integrity Campaign. "Historic Filing: Farmer Sues Perdue for Violation of FSMA Whistleblower Protection Law." February 19, 2015. http://www.foodwhistleblower.org/historic-filing-farmer-sues-perdue-for-violation-of-fsma-whistleblower-protection-law/.

"Foster Farms—Road Trip." The Hall of Advertising. YouTube. March 7, 2015. Advertisement. https://www.youtube.com/watch?v=3n1x71G1DEQ.

Fox, Grace. "The Origins of UNRRA." Political Science Quarterly 65, no. 4 (December 1950): 561. doi:10.2307/2145664.

Frana, Timothy S., et al. "Isolation and Characterization of Methicillin-Resistant Staphylococcus aureus From Pork Farms and Visiting Veterinary Students." PLoS ONE 8, no. 1 (January 3, 2013): e53738.

Bibliography

Frappaolo, P. J., and G. B. Guest. "Regulatory Status of Tetracyclines, Penicillin and Other Antibacterial Drugs in Animal Feeds." *Journal of Animal Science* 62, suppl. 3 (1968): 86–92.

Freerksen, Enno. "Fundamentals of Mode of Action of Antibiotics in Animals." In *Proceedings of the First International Conference on the Use of Antibiotics in Agriculture, 19-21 October 1955*, 91–106. Washington, D.C.: National Academy of Sciences, 1956.

"Fresh Food Plan Found." *Odessa (TX) American*, January 6, 1956.

Friedlander, Blaine. "Robert C. Baker, Creator of Chicken Nuggets and Cornell Chicken Barbecue Sauce, Dies at 84." *Cornell Chronicle*, March 16, 2006. http://www.news.cornell.edu/stories/2006/03/food-and-poultry-scientist-robert-c-baker-dies-age-84.

"G7 Ise-Shima Leaders Declaration." Ministry of Foreign Affairs of Japan, May 27, 2016. http://www.mofa.go.jp/files/000160266.pdf.

"G20 Leaders' Communiqué." *People's Daily Online*, September 5, 2016. http://en.people.cn/n3/2016/0906/c90000-9111018.html.

Gannon, Arthur. "Georgia's Broiler Industry." *Georgia Review* 6, no. 3 (Fall 1952): 306–17.

Garrod, L. P. "Sources and Hazards to Man of Antibiotics in Foods." *Proceedings of the Royal Society of Medicine* 57 (November 1964): 1087–88.

Gastro-Enteritis (Tees-side), Pub. L. No. vol 762 cc1619–30 (April 11, 1968).

Gastro-Enteritis Outbreak (Tees-side), Pub. L. No. vol 760 cc146–9W (March 7, 1968).

Gates, Deborah. "Somerset Homeowners Clash With Poultry Farmer." Wilmington (DE). *Delaware News Journal*, July 26, 2014. http://www.delawareonline.com/story/news/local/2014/07/26/somerset-homeowners-clash-poultry-farmer/13226907/.

Gaunt, P. N., and L. J. Piddock. "Ciprofloxacin Resistant *Campylobacter spp.* in Humans: An Epidemiological and Laboratory Study." *Journal of Antimicrobial Chemotherapy* 37, no. 4 (April 1996): 747–57.

Gee, Kelsey. "Poultry's Tough New Problem: 'Woody Breast.'" *Wall Street Journal*, March 29, 2016, sec. B.

Geenen, P. L., et al. "Prevalence of Livestock-Associated MRSA on Dutch Broiler Farms and in People Living and/or Working on These Farms." *Epidemiology and Infection* 141, no. 5 (May 2013): 1099–1108.

Geijlswijk, Inge M. van, Dik J. Mevius, and Linda F. Puister-Jansen. "[Quantity of veterinary antibiotic use]." *Tijdschrift Voor Diergeneeskunde* 134, no. 2 (January 15, 2009): 69–73.

George, D. B., and A. R. Manges. "A Systematic Review of Outbreak and Non-Outbreak Studies of Extraintestinal Pathogenic *Escherichia coli* Causing Community-Acquired Infections." *Epidemiology and Infection* 138, no. 12 (December 2010): 1679–90.

Georgia Humanities Council, University of Georgia Press, University System of Georgia/GALILEO, and Office of the Governor. *New Georgia Encyclopedia*. http://www.georgiaencyclopedia.org/about-nge.

Gerhard, Gesine. "Food as a Weapon: Agricultural Sciences and the Building of a Greater German Empire." *Food, Culture and Society* 14, no. 3 (September 1, 2011): 335–51.

"Germ Survival in Face of Antibiotics." *Times (London)*. February 26, 1965.

Gibbs, Shawn G., et al. "Isolation of Antibiotic-Resistant Bacteria from the Air Plume Downwind of a Swine Confined or Concentrated Animal Feeding Operation." *Environmental Health Perspectives* 114, no. 7 (March 27, 2006): 1032–37.

Gisolfi, Monica Richmond. "From Crop Lien to Contract Farming: The Roots of Agribusiness in the American South, 1929-1939." *Agricultural History*, 2006, 167–89.

Giufre, M., et al. "*Escherichia coli* of Human and Avian Origin: Detection of Clonal Groups Associated with Fluoroquinolone and Multidrug Resistance in Italy." *Journal of Antimicrobial Chemotherapy* 67, no. 4 (April 1, 2012): 860–67.

"Global Animal Partnership Commits to Requiring 100 Percent Slower-Growing Chicken Breeds by 2024." *BusinessWire*, March 17, 2016. http://www.businesswire.com/news/home/20160317005528/en/Global-Animal-Partnership-Commits-Requiring-100-Percent.

Glynn, M. K., et al. "Emergence of Multidrug-Resistant *Salmonella enterica* Serotype Typhimurium DT104 Infections in the United States." *New England Journal of Medicine* 338, no. 19 (May 7, 1998): 1333–38.

Godley, Andrew C., and Bridget Williams. "The Chicken, the Factory Farm, and the Supermarket: The Emergence of the Modern Poultry Industry in Britain." In *Food Chains: From Farmyard to Shopping Cart*, edited by Warren Belasco and Roger Horowitz. Philadelphia: University of Pennsylvania Press, 2010.

Goldberg, Herbert S. "Evaluation of Some Potential Public Health Hazards from Non-Medical Uses of Antibiotics." In *Antibiotics in Agriculture: Proceedings of the University of Nottingham Ninth Easter School in Agricultural Science*, edited by M. Woodbine, 389–404. London: Butterworths, 1962.

Gordon, H. A. "The Germ-Free Animal: Its Use in the Study of 'Physiologic' Effects of the Normal Microbial Flora on the Animal Host." *American Journal of Digestive Diseases* 5 (October 1960): 841–67.

Gough, E. K., et al. "The Impact of Antibiotics on Growth in Children in Low and Middle Income Countries: Systematic Review and Meta-Analysis of Randomised Controlled Trials." *British Medical Journal* 348, no. 6 (April 15, 2014): g2267.

Grady, Denise. "Bacteria Concerns in Denmark Cause Antibiotics Concerns in U.S." *New York Times*, November 4, 1999.

Graham, Jay P., et al. "Growth Promoting Antibiotics in Food Animal Production: An Economic Analysis." *Public Health Reports* 122, no. 1 (2007): 79–87.

Graham, Jay P., et al. "Antibiotic Resistant Enterococci and Staphylococci Isolated from Flies Collected Near Confined Poultry Feeding Operations." *Science of the Total Environment* 407, no. 8 (April 2009): 2701–10.

Grave, K., et al. "Sales of Veterinary Antibacterial Agents in Nine European Countries During 2005–09: Trends and Patterns." *Journal of Antimicrobial Chemotherapy* 67, no. 12 (December 1, 2012): 3001–3008.

Grave, K., et al. "Comparison of the Sales of Veterinary Antibacterial Agents Between 10 European Countries." *Journal of Antimicrobial Chemotherapy* 65, no. 9 (September 1, 2010): 2037–40. doi:10.1093/jac/dkq247.

Greenwood, David. *Antimicrobial Drugs: Chronicle of a Twentieth Century Medical Triumph.* New York: Oxford University Press, 2008.

Gupta, Amita, et al., "Antimicrobial Resistance Among Campylobacter Strains, United States, 1997–2001." *Emerging Infectious Diseases* 10, no. 6 (2004). http://wwwnc.cdc.gov/eid/article/10/6/03-0635_article.htm.

Gupta, K., et al. "Managing Uncomplicated Urinary Tract Infection—Making Sense Out of Resistance Data." *Clinical Infectious Diseases* 53, no. 10 (November 15, 2011): 1041–42.

Gupta, K., et al. "Executive Summary: International Clinical Practice Guidelines for the Treatment of Acute Uncomplicated Cystitis and Pyelonephritis in Women: A 2010

Bibliography

Update by the Infectious Diseases Society of America and the European Society for Microbiology and Infectious Diseases." *Clinical Infectious Diseases* 52, no. 5 (March 1, 2011): 561–64.

Haley, Andrew P. *Turning the Tables: Restaurants and the Rise of the American Middle Class, 1880–1920.* Chapel Hill: University of North Carolina Press, 2011.

Hannah, Elizabeth Lyon, et al. "Molecular Analysis of Antimicrobial-Susceptible and -Resistant *Escherichia coli* From Retail Meats and Human Stool and Clinical Specimens in a Rural Community Setting." *Foodborne Pathogens and Disease* 6, no. 3 (April 2009): 285–95.

Hansen, Peter L., and Ronald Lester Mighell. *Economic Choices in Broiler Production.* Washington, D.C.: U.S. Department of Agriculture, 1956.

Harold, Laverne C., and Robert A. Baldwin. "Ecologic Effects of Antibiotics." *FDA Papers* 1 (February 1967): 20–24.

Harper, Abby L., et al. "An Overview of Livestock-Associated MRSA in Agriculture." *Journal of Agromedicine* 15, no. 2 (March 31, 2010): 101–104.

Harris, Marion S. "Home Demonstration." *Bennington (VT) Banner.* January 20, 1958.

Harrison, Ruth. *Animal Machines: The New Factory Farming Industry.* New York: Ballantine Books, 1966.

Hasman, Henrik, et al. "Detection of *mcr-1* Encoding Plasmid-Mediated Colistin-Resistant *Escherichia coli* Isolates From Human Bloodstream Infection and Imported Chicken Meat, Denmark 2015." *Eurosurveillance* 20, no. 49 (December 10, 2015).

Heederik, Dick. "Benchmarking Livestock Farms and Veterinarians." Slide presentation, SDa Autoriteit Dirgeneesmiddelen, Utrecht, August 9, 2013.

Heinzerling, Lisa. "Undue Process at the FDA: Antibiotics, Animal Feed, and Agency Intransigence." *Vermont Law Review* 37, no. 4 (2013): 1007–31.

Hennessy, T. W., et al. "A National Outbreak of *Salmonella enteritidis* Infections from Ice Cream." *New England Journal of Medicine* 334, no. 20 (May 16, 1996): 1281–86.

Herikstad, H., et al. "Emerging Quinolone-Resistant Salmonella in the United States." *Emerging Infectious Diseases* 3, no. 3 (September 1997): 371–72.

Herold, B. C., et al. "Community-Acquired Methicillin-Resistant *Staphylococcus aureus* in Children with No Identified Predisposing Risk." *Journal of the American Medical Association* 279, no. 8 (February 25, 1998): 593–98.

Hewitt, William L. "Penicillin-Historical Impact on Infection Control." *Annals of the New York Academy of Sciences* 145, no. 1 (1967): 212–15.

Hill, George, Thomson Prentice, Pearce Wright, and Thomas Stuttaford. "The Bitter Harvest." *Times (London),* March 4, 1987.

Hise, Kelley B. "History of PulseNet USA." Paper presented at the Association of Public Health Laboratories 14th Annual PulseNet Update Meeting, Chicago, August 31, 2010. https://www.aphl.org/conferences/proceedings/Documents/2010/2010_APHL_Pulse Net_Meeting/002-Hise.pdf.

Hobbs, B. C., et al. "Antibiotic Treatment of Poultry in Relation to *Salmonella typhimurium.*" *Monthly Bulletin of the Ministry of Health and the Public Health Laboratory Service* 19 (October 1960): 178–92.

Hoelzer, Karin. "Judicious Animal Antibiotic Use Requires Drug Label Refinements." Washington, D.C.: Pew Charitable Trusts, October 4, 2016. http://pew.org/2dqrjCo.

Hoffmann, Stanley. "The Effects of World War II on French Society and Politics." *French Historical Studies* 2, no. 1 (Spring 1961): 28–63.

Hogue, Allan, et al. "*Salmonella* Typhimurium DT104 Situation Assessment, December 1997." Washington, D.C.: U.S. Department of Agriculture Animal and Plant Health Inspection Service, December 1997. https://www.aphis.usda.gov/animal_health/emergingissues/downloads/dt104.pdf.

Holland, John. "After 75 Years, Foster Farms Remembers Its Path to Success." *Modesto (CA) Bee*, June 16, 2014. http://www.modbee.com/news/local/article3166439.html.

Holmberg, S. D., et al. "Drug-Resistant Salmonella from Animals Fed Antimicrobials." *New England Journal of Medicine* 311, no. 10 (September 6, 1984): 617–22.

Holmberg, S. D., et al. "Animal-to-Man Transmission of Antimicrobial-Resistant Salmonella: Investigations of U.S. Outbreaks, 1971-1983." *Science* 225, no. 4664 (August 24, 1984): 833–35.

Holmes, Alison H., et al. "Understanding the Mechanisms and Drivers of Antimicrobial Resistance." *Lancet* 387, no. 10014 (January 2016): 176–87.

Horowitz, Roger. "Making the Chicken of Tomorrow: Reworking Poultry as Commodities and as Creatures, 1945-1990." In *Industrializing Organisms: Introducing Evolutionary History*, edited by Philip Scranton and Susan Schrepfer. London: Routledge, 2004.

"How Safe Is That Chicken?." *Consumer Reports* (January 2010): 19. http://www.consumer reports.org/cro/magazine-archive/2010/january/food/chicken-safety/overview/chicken-safety-ov.htm.

Huijsdens, Xander W., et al. "Community-Acquired MRSA and Pig-Farming." *Annals of Clinical Microbiology and Antimicrobials* 5, no. 1 (2006): 26.

Huijsdens, X. W., et al. "Molecular Characterisation of PFGE Non-Typable Methicillin-Resistant *Staphylococcus aureus* in the Netherlands, 2007." *Eurosurveillance* 14, no. 38 (September 24, 2009).

"Human Food Safety and the Regulation of Animal Drugs: Twenty-Seventh Report." § House Committee on Government Operations, 1985.

Humane Society of the United States. "The Welfare of Animals in the Chicken Industry." New York: Humane Society of the United States, December 2013. http://www.humane society.org/assets/pdfs/farm/welfare_broiler.pdf.

———. "Welfare Issues With Selective Breeding for Rapid Growth in Broiler Chickens and Turkeys." New York: Humane Society of the United States, May 2014. http://www .humanesociety.org/assets/pdfs/farm/welfiss_breeding_chickens_turkeys.pdf.

Hylton, Wil S. "A Bug in the System." *New Yorker*, February 2, 2015.

Infectious Diseases Society of America. "Bad Bugs, No Drugs." Washington, D.C.: Infectious Diseases Society of America, July 2004. https://www.idsociety.org/uploadedFiles/IDSA/Policy_and_Advocacy/Current_Topics_and_Issues/Advancing_Product_Research_and_Development/Bad_Bugs_No_Drugs/Statements/As%20Antibiotic%20Discovery%20Stagnates%20A%20Public%20Health%20Crisis%20Brews.pdf.

"Infectious Drug Resistance." *New England Journal of Medicine* 275, no. 5 (August 4, 1966): 277.

Iovine, Nicole M., and Martin J. Blaser. "Antibiotics in Animal Feed and Spread of Resistant *Campylobacter* from Poultry to Humans." *Emerging Infectious Diseases* 10, no. 6 (June 2004): 1158–89.

Izdebski, R., A. et al. "Mobile *mcr-1*-Associated Resistance to Colistin in Poland." *Journal of Antimicrobial Chemotherapy* 71, no. 8 (August 2016): 2331–33.

Bibliography

Jakobsen, L., et al. "Is *Escherichia coli* Urinary Tract Infection a Zoonosis? Proof of Direct Link With Production Animals and Meat." *European Journal of Clinical Microbiology and Infectious Diseases* 31, no. 6 (June 2012): 1121–29.

Jakobsen, Lotte, et al. "*Escherichia coli* Isolates From Broiler Chicken Meat, Broiler Chickens, Pork, and Pigs Share Phylogroups and Antimicrobial Resistance With Community-Dwelling Humans and Patients With Urinary Tract Infection." *Foodborne Pathogens and Disease* 7, no. 5 (May 2010): 537–47.

Jalonick, Mary Clare. "Still No Recall of Chicken Tied to Outbreak of Antibiotic-Resistant *Salmonella*." Associated Press, May 28, 2014.

Janzen, Kristi Bahrenburg. "Loss of Small Slaughterhouses Hurts Farmers, Butchers and Consumers." *Farming Magazine* (2004).

Jess, Tine. "Microbiota, Antibiotics, and Obesity." *New England Journal of Medicine* 371, no. 26 (December 25, 2014): 2526–28.

Jevons, M. Patricia. " 'Celbenin'-Resistant Staphylococci." *British Medical Journal* no. 1 (January 14, 1961): 124–25.

Jiménez, A., et al. "Prevalence of Fluoroquinolone Resistance in Clinical Strains of Campylobacter Jejuni Isolated in Spain." *Journal of Antimicrobial Chemotherapy* 33, no. 1 (January 1994): 188–90.

Johnson, J. R., et al. "Epidemic Clonal Groups of *Escherichia coli* as a Cause of Antimicrobial-Resistant Urinary Tract Infections in Canada, 2002 to 2004." *Antimicrobial Agents and Chemotherapy* 53, no. 7 (July 1, 2009): 2733–39.

Johnson, J. R., et al. "Isolation and Molecular Characterization of Nalidixic Acid-Resistant Extraintestinal Pathogenic *Escherichia coli* From Retail Chicken Products." *Antimicrobial Agents and Chemotherapy* 47, no. 7 (July 1, 2003): 2161–68.

Johnson, James R., et al. "Contamination of Retail Foods, Particularly Turkey, From Community Markets (Minnesota, 1999–2000) With Antimicrobial-Resistant and Extraintestinal Pathogenic *Escherichia coli*." *Foodborne Pathogens and Disease* 2, no. 1 (2005): 38–49.

Johnson, James R., et al. "Similarity Between Human and Chicken *Escherichia coli* Isolates in Relation to Ciprofloxacin Resistance Status." *Journal of Infectious Diseases* 194, no. 1 (July 1, 2006): 71–78.

Johnson, James R., et al. "Antimicrobial-Resistant and Extraintestinal Pathogenic *Escherichia coli* in Retail Foods." *Journal of Infectious Diseases* 191, no. 7 (April 1, 2005): 1040–49.

Johnson, James R., and Thomas A. Russo. "Uropathogenic *Escherichia coli* as Agents of Diverse Non-Urinary Tract Extraintestinal Infections." *Journal of Infectious Diseases* 186, no. 6 (September 15, 2002): 859–64.

Johnson, James R., et al. "Antimicrobial Drug-Resistant *Escherichia coli* From Humans and Poultry Products, Minnesota and Wisconsin, 2002–2004." *Emerging Infectious Diseases* 13, no. 6 (June 2007): 838–46.

Johnson, James R., et al. "Molecular Analysis of *Escherichia coli* From Retail Meats (2002–2004) From the United States National Antimicrobial Resistance Monitoring System." *Clinical Infectious Diseases* 49, no. 2 (July 15, 2009): 195–201.

Johnson, Timothy J., et al. "Associations Between Multidrug Resistance, Plasmid Content, and Virulence Potential Among Extraintestinal Pathogenic and Commensal *Escherichia coli* From Humans and Poultry." *Foodborne Pathogens and Disease* 9, no. 1 (January 2012): 37–46.

Johnson, Timothy J., et al. "Examination of the Source and Extended Virulence Genotypes of *Escherichia coli* Contaminating Retail Poultry Meat." *Foodborne Pathogens and Disease* 6, no. 6 (August 2009): 657–67.

Jones, Harold W. "Report of a Series of Cases of Syphilis Treated by Ehrlich's Arsenobenzole at the Walter Reed General Hospital, District of Columbia." *Boston Medical and Surgical Journal* vol. CLXIV, no. 11 (March 16, 1911): 381-383.

Jørgensen, Peter S., et al. "Use Antimicrobials Wisely." *Nature News* 537, no. 7619 (September 8, 2016): 159.

Josephson, Paul. "The Ocean's Hot Dog: The Development of the Fish Stick." *Technology and Culture* 49, no. 1 (2008): 41–61.

Jou, Ruwen, et al. "Enrofloxacin in Poultry and Human Health." *American Journal of Tropical Medicine and Hygiene* 67 (2002): 533–38.

Jukes, T. H. "A Town in Harmony." *Chemical Week* (August 18, 1962).

——. "Adventures with Vitamins." *Journal of the American College of Nutrition* 7, no. 2 (April 1988): 93–99.

——. "Alar and Apples." *Science* 244, no. 4904 (May 5, 1989): 515.

——. "Antibacterial Agents in Animal Feeds." *Clinical Toxicology* 14, no. 3 (March 1979): 319–22.

——. "Antibiotics and Meat." *New York Times*, October 2, 1972.

——. "Antibiotics in Animal Feeds." *New England Journal of Medicine* 282, no. 1 (January 1, 1970): 49–50.

——. "Antibiotics in Feeds." *Science* 204, no. 4388 (April 6, 1979): 8. doi:10.1126/science.204.4388.8.

——. "Antibiotics in Nutrition." *Antibiotics in Nutrition* (1955).

——. "BST and Milk Production." *Science* 265, no. 5169 (July 8, 1994): 170.

——. "Drug-Resistant Salmonella From Animals Fed Antimicrobials." *New England Journal of Medicine* 311, no. 26 (December 27, 1984): 1699.

——. "Food Additives." *New England Journal of Medicine* 297, no. 8 (August 25, 1977): 427–30.

——. "How Safe Is Our Food Supply?" *Archives of Internal Medicine* 138, no. 5 (May 1978): 772–74.

——. "Medical Versus Animal Antibiotics in Resistance." *Nature* 313, no. 5999 (January 17, 1985): 186.

——. "Megavitamin Therapy." *Journal of the American Medical Association* 233, no. 6 (August 11, 1975): 550–51.

——. "Public Health Significance of Feeding Low Levels of Antibiotics to Animals." *Advances in Applied Microbiology* 16 (1973): 1–54.

——. "Searching for Magic Bullets: Early Approaches to Chemotherapy-Antifolates, Methotrexate—the Bruce F. Cain Memorial Award Lecture." *Cancer Research* 47, no. 21 (November 1, 1987): 5528–36.

——. "Some Historical Notes on Chlortetracycline." *Reviews of Infectious Diseases* 7, no. 5 (October 1985): 702–707.

——. "Today's Non-Orwellian Animal Farm." *Nature* 355, no. 6361 (February 13, 1992): 582.

Jukes, Thomas H. "Antibiotics in Meat Production." *Journal of the American Medical Association* 232, no. 3 (1975): 292–93.

——. "Antioxidants, Nutrition, and Evolution." *Preventive Medicine* 21, no. 2 (1992): 270–76.

Bibliography

——. "Carcinogens in Food and the Delaney Clause." *Journal of the American Medical Association* 241, no. 6 (1979): 617–19.

——. "Cyclamate Sweeteners." *Journal of the American Medical Association* 236, no. 17 (1976): 1987–89.

——. "DDT." *Journal of the American Medical Association* 229, no. 5 (1974): 571–73.

——. "Diethylstilbestrol in Beef Production: What Is the Risk to Consumers?" *Preventive Medicine* 5, no. 3 (1976): 438–53.

——. "Guest Opinions." *Professional Animal Scientist* 11, no. 4 (1995): 238–39.

——. "The Organic Food Myth." *Journal of the American Medical Association* 230, no. 2 (1974): 276–77.

——. "The Present Status and Background of Antibiotics in the Feeding of Domestic Animals." *Annals of the New York Academy of Sciences* 182, no. 1 (1971): 362–79.

——. "Vitamins, Metabolic Antagonists, and Molecular Evolution." *Protein Science* 6, no. 1 (1997): 254–56.

Kadariya, Jhalka, et al. "*Staphylococcus aureus* and Staphylococcal Food-Borne Disease: An Ongoing Challenge in Public Health." *BioMed Research International* 2014 (2014): 1–9.

Kaempffert, Waldemar. "Effectiveness of New Antibiotic, Aureomycin, Demonstrated Against Virus Diseases." *New York Times,* July 25, 1948.

Kaesbohrer, A., A. et al. "Emerging Antimicrobial Resistance in Commensal *Escherichia coli* With Public Health Relevance." *Zoonoses and Public Health* 59 (September 2012): 158–65.

Kampelmacher, E. H. "Some Aspects of the Non-Medical Uses of Antibiotics in Various Countries." In *Antibiotics in Agriculture: Proceedings of the University of Nottingham Ninth Easter School in Agricultural Science,* edited by M. Woodbine, 315–32. London: Butterworths, 1962.

Kaufman, Marc. "Ending Battle With FDA, Bayer Withdraws Poultry Antibiotic." *Washington Post,* September 9, 2005.

Kaufmann, A. F., et al. "Pontiac Fever: Isolation of the Etiologic Agent *(Legionella pneumophilia)* and Demonstration of Its Mode of Transmission." *American Journal of Epidemiology* 114, no. 3 (September 1981): 337–47.

Kempf, Isabelle, et al. "What Do We Know About Resistance to Colistin in Enterobacteriaceae in Avian and Pig Production in Europe?" *International Journal of Antimicrobial Agents* 42, no. 5 (November 2013): 379–83.

Kennedy, Donald S. " 'Antibiotics in Animal Feeds,' remarks to the National Advisory Food and Drug Committee, April 15, 1977, Rockville, Md." Donald Kennedy Personal Papers (SC0708), Department of Special Collections and University Archives, Stanford University Libraries, Stanford, CA.

——. "The Threat From Antibiotic Use on the Farm." *Washington Post,* August 22, 2013. http://www.washingtonpost.com/opinions/the-threat-from-antibiotic-...e-farm/2013 /08/22/c407ed72-0ab2-11e3-8974-f97ab3b3c677_story.html.

Kesternich, Iris, et al. "The Effects of World War II on Economic and Health Outcomes Across Europe." *Review of Economics and Statistics* 96, no. 1 (March 2014): 103–18. doi:10.1162/ REST_a_00353.

Khan, Lina. "Obama's Game of Chicken." *Washington Monthly* vol. 44, no. 11/12 (November-December 2012): 32-38. http://washingtonmonthly.com/magazine/novdec-2012/ obamas-game-of-chicken/.

Kieler, Ashlee. "Foster Farms Recalls Chicken After USDA Inspectors Finally Link It to Salmonella Case." *Consumerist,* July 7, 2014.

Kindy, Kimberly, and Brady Dennis. "Salmonella Outbreaks Expose Weaknesses in USDA Oversight of Chicken Parts." *Washington Post,* February 6, 2014.

Kirst, H. A., et al. "Historical Yearly Usage of Vancomycin." *Antimicrobial Agents and Chemotherapy* 42, no. 5 (May 1998): 1303–1304.

Kiser, J. S. "A Perspective on the Use of Antibiotics in Animal Feeds." *Journal of Animal Science* 42, no. 4 (1976): 1058–72.

Kline, E. F. "Maintenance of High Quality in Fish Fillets With Acronize." *Proceedings of the Annual Gulf and Caribbean Fisheries Institute* 10 (August 1958): 80–84.

Kline, Kelly E., et al. "Investigation of First Identified *mcr-1* Gene in an Isolate From a U.S. Patient—Pennsylvania, 2016." *Morbidity and Mortality Weekly Report* 65, no. 36 (September 16, 2016): 977–78.

Kluytmans, J. A. J. W., et al. "Extended-Spectrum-Lactamase-Producing *Escherichia coli* From Retail Chicken Meat and Humans: Comparison of Strains, Plasmids, Resistance Genes, and Virulence Factors." *Clinical Infectious Diseases* 56, no. 4 (February 15, 2013): 478–87.

Knowber, Charles R. "A Real Game of Chicken: Contracts, Tournaments, and the Production of Broilers." *Journal of Law, Economics, and Organization* 5, no. 2 (Autumn 1989): 271–92.

Kobell, Rona. "Poultry Mega-Houses Forcing Shore Residents to Flee Stench, Traffic." Seven Valleys (PA). *Bay Journal,* July 22, 2015. http://marylandreporter.com/2015/07/22/poultry-mega-houses-forcing-shore-residents-to-flee-stench-traffic/.

Kohler, A. R., et al. "Comprehensive Studies of the Use of a Food Grade of Chlortetracycline in Poultry Processing." *Antibiotics Annual,* 1956–57, 822–30.

Koike, S., et al. "Monitoring and Source Tracking of Tetracycline Resistance Genes in Lagoons and Groundwater Adjacent to Swine Production Facilities Over a 3-Year Period." *Applied and Environmental Microbiology* 73, no. 15 (August 1, 2007): 4813–23.

Krieger, Lisa M. "California Links Hollister Dairy to 1997 Outbreak of Salmonella." *San Jose (CA) Mercury News,* October 3, 1999.

Kristof, Nicholas. "Abusing Chickens We Eat." *New York Times,* December 3, 2014. https://www.nytimes.com/2014/12/04/opinion/nicholas-kristof-abusing-chickens-we-eat.html?_r=1.

Kumar, Kuldip, et al. "Antibiotic Use in Agriculture and Its Impact on the Terrestrial Environment." *Advances in Agronomy* 87 (2005): 1–54.

Larson, Clarence. "Pioneers in Science and Technology Series: Thomas Jukes." Center for Oak Ridge Oral History, March 29, 1988. Clarence E. Larson Science and Technology Oral History Collection, Collection C0079, Special Collections and Archives, George Mason University Libraries, Arlington, VA. http://cdm16107.contentdm.oclc.org/cdm/ref/collection/p15388coll1/id/522.

Laurence, William L. " 'Wonder Drug' Aureomycin Found to Spur Growth 50%." *New York Times,* April 10, 1950.

Laveck, G. D., and R. T. Ravenholt. "Staphylococcal Disease: An Obstetric, Pediatric, and Community Problem." *American Journal of Public Health and the Nation's Health* 46, no. 10 (October 1956): 1287–96.

Lawrence, Robert S., and Keeve E Nachman. "Letter From the Johns Hopkins Center for a Livable Future to James C. Stofko, Somerset County Health Department, Re Broiler Production." February 2015.

Bibliography

——. "Letter Fom the Johns Hopkins Center for a Livable Future to Lori A. Brewster, Health Officer, Wicomico County, Re Broiler Production," January 21, 2016.

Lax, Eric. *The Mold in Dr. Florey's Coat: The Story of the Penicillin Miracle.* 2nd ed. New York: Henry Holt and Co., 2004.

Laxminarayan, Ramanan, et al. "UN High-Level Meeting on Antimicrobials—What Do We Need?" *Lancet* 388, no. 10041 (July 2016): 218–20.

Laxminarayan, Ramanan, et al. "Access to Effective Antimicrobials: A Worldwide Challenge." *Lancet* 387, no. 10014 (January 2016): 168–75.

Laxminarayan, Ramanan, et al. "The Economic Costs of Withdrawing Antimicrobial Growth Promoters From the Livestock Sector." OECD Food, Agriculture and Fisheries Papers, February 23, 2015.

Leedom Larson, K. R., et al. "Methicillin-Resistant *Staphylococcus aureus* in Pork Production Shower Facilities." *Applied and Environmental Microbiology* 77, no. 2 (January 15, 2011): 696–98.

Leeson, Steven, and John D. Summers. *Broiler Breeder Production.* Guelph: University Books, 2000.

Lehmann, R. P. "Implementation of the Recommendations Contained in the Report to the Commissioner Concerning the Use of Antibiotics on Animal Feed." *Journal of Animal Science* 35, no. 6 (1972): 1340–41.

Leonard, Christopher. *The Meat Racket: The Secret Takeover of America's Food Business.* New York: Simon & Schuster, 2014.

Lepley, K. C., et al. "Dried Whole Aureomycin Mash and Meat and Bone Scraps for Growing-Fattening Swine." *Journal of Animal Science* 9, no. 4 (November 1950): 608–614.

Lesesne, Henry. "Antibiotic Now Keeps Poultry Fresh." *Terre Haute (IN) Tribune Star*, May 13, 1956.

——. "Pilgrims Wouldn't Know '56 Bird." *Salem (OH) News*, November 20, 1956.

——. "Poultrymen Hear About 'Acronize PD' From Food Technologist at Session." *Florence (SC) Morning News*, February 8, 1957.

Les fermiers de Loué: des hommes et des volailles, petites et grandes histoires. Le Mans, France: Syvol, 1999.

Levere, Jane L. "A Cook-Off Among Chefs to Join Delta's Kitchen." *New York Times,* July 21, 2013. http://www.nytimes.com/2013/07/22/business/media/a-cook-off-among-chefs-to-join-deltas-kitchen.html.

Leverstein-van Hall, M. A., et al. "Dutch Patients, Retail Chicken Meat and Poultry Share the Same ESBL Genes, Plasmids and Strains." *Clinical Microbiology and Infection* 17, no. 6 (June 2011): 873–80.

Levitt, Tom. " 'I Don't See a Problem': Tyson Foods CEO on Factory Farming and Antibiotic Resistance." *Guardian (Manchester)*, April 5, 2016. https://www.theguardian.com/sustainable-business/2016/apr/05/tyson-foods-factory-farming-antibiotic-resistance-donnie-smith.

Levy, S. B., et al. "Changes in Intestinal Flora of Farm Personnel After Introduction of a Tetracycline-Supplemented Feed on a Farm." *New England Journal of Medicine* 295, no. 11 (September 9, 1976): 583–88.

——. "Spread of Antibiotic-Resistant Plasmids from Chicken to Chicken and From Chicken to Man." *Nature* 260, no. 5546 (March 4, 1976): 40–42.

Levy, S. B., and L. McMurry. "Detection of an Inducible Membrane Protein Associated With R-Factor-Mediated Tetracycline Resistance." *Biochemical and Biophysical Research Communications* 56, no. 4 (February 27, 1974): 1060–68.

Levy, Sharon. "Reduced Antibiotic Use in Livestock: How Denmark Tackled Resistance." *Environmental Health Perspectives* 122, no. 6 (June 1, 2014): A160–65. doi:10.1289/ehp.122-A160.

Levy, Stuart B. *The Antibiotic Paradox*. Cambridge, MA: Da Capo Press, 2002.

——. "Playing Antibiotic Pool: Time to Tally the Score." *New England Journal of Medicine* 311, no. 10 (September 6, 1984): 663–64.

——. Testimony Before the Subcommittee on Health of the U.S. House of Representatives Committee on Energy and Commerce (2010). 111th Cong., 2nd sess.

Levy, Stuart B., and Bonnie Marshall. "Antibacterial Resistance Worldwide: Causes, Challenges and Responses." *Nature Medicine* 10, no. 12s (December 2004): S122–29.

Liakopoulos, Apostolos, et al. "The Colistin Resistance *mcr-1* Gene Is Going Wild." *Journal of Antimicrobial Chemotherapy* 71, no. 8 (August 2016): 2335–36.

Linder, Marc. "I Gave My Employer a Chicken That Had No Bone: Joint Firm-State Responsibility for Line-Speed-Related Occupational Injuries." *Case Western Reserve Law Review* 46 (1995): 33.

Linton, A. H. "Antibiotic Resistance: The Present Situation Reviewed." *Veterinary Record* 100, no. 17 (April 23, 1977): 354–60.

——. "Has Swann Failed?" *Veterinary Record* 108, no. 15 (April 11, 1981): 328–31.

Linton, K. B., et al. "Antibiotic Resistance and Transmissible R-Factors in the Intestinal Coliform Flora of Healthy Adults and Children in an Urban and a Rural Community." *Journal of Hygiene* 70, no. 1 (March 1972): 99–104.

Literak, Ivan, et al. "Broilers as a Source of Quinolone-Resistant and Extraintestinal Pathogenic *Escherichia coli* in the Czech Republic." *Microbial Drug Resistance* 19, no. 1 (February 2013): 57–63.

Liu, Yi-Yun, et al. "Emergence of Plasmid-Mediated Colistin Resistance Mechanism *mcr-1* in Animals and Human Beings in China: A Microbiological and Molecular Biological Study." *Lancet Infectious Diseases* 16, no. 2 (February 2016): 161–68.

Loudon, I. "Deaths in Childbed from the Eighteenth Century to 1935." *Medical History* 30, no. 1 (January 1986): 1–41.

Love, D. C., et al. "Feather Meal: A Previously Unrecognized Route for Reentry Into the Food Supply of Multiple Pharmaceuticals and Personal Care Products (PPCPs)." *Environmental Science and Technology* 46, no. 7 (April 3, 2012): 3795–3802.

Love, David C., et al. "Dose Imprecision and Resistance: Free-Choice Medicated Feeds in Industrial Food Animal Production in the United States." *Environmental Health Perspectives* 119, no. 3 (October 28, 2010): 279–83.

Love, John F. *McDonald's: Behind the Arches*. Rev. ed. New York: Bantam, 1995.

Lyhs, Ulrike, et al. "Extraintestinal Pathogenic *Escherichia coli* in Poultry Meat Products on the Finnish Retail Market." *Acta Veterinaria Scandinavica* 54 (November 16, 2012): 64.

Lyons, Richard D. "Backers of Laetrile Charge a Plot Is Preventing the Cure of Cancer." *New York Times*, July 13, 1977.

——. "F.D.A. Chief Heading for Less Trying Job." *New York Times*, June 17, 1979.

Bibliography

MacDonald, James. "Technology, Organization, and Financial Performance in U.S. Broiler Production." *Economic Information Bulletin* (June 2014). https://www.ers.usda.gov/ publications/pub-details/?pubid=43872.

———. "The Economic Organization of U.S. Broiler Production." *Economic Information Bulletin*. (June 2008). https://www.ers.usda.gov/publications/pub-details/?pubid= 44256.

MacDonald, James M., and William D. McBride. "The Transformation of U.S. Livestock Agriculture: Scale, Efficiency, and Risks." *Economic Information Bulletin* (January 2009). https://www.ers.usda.gov/publications/pub-details/?pubid=44294.

Machlin, L. J., et al. "Effect of Dietary Antibiotic Upon Feed Efficiency and Protein Requirement of Growing Chickens." *Poultry Science* 31, no. 1 (January 1, 1952): 106–109.

Maddox, John. "Obituary: Thomas Hughes Jukes (1906-99)." *Nature* 402, no. 6761 (1999): 478.

Madsen, Lillie L. "Acronizing Process Almost Doubles Poultry Shelf Life." *Statesman*, September 13, 1956.

Maeder, Thomas. *Adverse Reactions*. New York: Morrow, 1994.

Maitland, A. I. "Why Has Swann Failed?" Letter to the editor. *British Medical Journal* 280, no. 6230 (June 21, 1980): 1537.

Majowicz, Shannon E., et al. "The Global Burden of Nontyphoidal *Salmonella* Gastroenteritis." *Clinical Infectious Diseases* 50, no. 6 (March 15, 2010): 882–89.

Malhotra-Kumar, Surbhi, et al. "Colistin Resistance Gene *mcr-1* Harboured on a Multidrug Resistant Plasmid." *Lancet Infectious Diseases* 16, no. 3 (March 2016): 283–84.

Malik, Rohit. "Catch Me if You Can: Big Food Using Big Tobacco's Playbook? Applying the Lessons Learned From Big Tobacco to Attack the Obesity Epidemic." Food and Drug Law Seminar Paper, Harvard Law School, 2010. http://nrs.harvard.edu/urn-3:HUL. InstRepos:8965631.

Manges, A. R., and J. R. Johnson. "Food-Borne Origins of *Escherichia coli* Causing Extraintestinal Infections." *Clinical Infectious Diseases* 55, no. 5 (September 1, 2012): 712–19.

Manges, A. R., et al. "Widespread Distribution of Urinary Tract Infections Caused by a Multidrug-Resistant *Escherichia coli* Clonal Group." *New England Journal of Medicine* 345, no. 14 (October 4, 2001): 1007–13.

Manges, A. R., et al. "The Changing Prevalence of Drug-Resistant *Escherichia coli* Clonal Groups in a Community: Evidence for Community Outbreaks of Urinary Tract Infections." *Epidemiology and Infection* 134, no. 2 (August 19, 2005): 425.

Manges, Amee R., et al. "Retail Meat Consumption and the Acquisition of Antimicrobial Resistant *Escherichia coli* Causing Urinary Tract Infections: A Case-Control Study." *Foodborne Pathogens and Disease* 4, no. 4 (2007): 419–31.

Manges, Amee R., et al. "Endemic and Epidemic Lineages of *Escherichia coli* That Cause Urinary Tract Infections." *Emerging Infectious Diseases* 14, no. 10 (October 2008): 1575–83.

Margach, James. "Antibiotics Curbs Will Be Tough." *Sunday Times (London)*, November 16, 1969.

Marler, Bill. "A Forgotten Foster Farms Salmonella Heidelberg Outbreak." Marler Blog, March 5, 2014. http://www.marlerblog.com/legal-cases/a-forgotten-foster-farms-salmonella-heidelberg-outbreak/.

——. "Final Demand Letter to Ron Foster, President, Foster Farms Inc., in re: 2013 Foster Farms Chicken Salmonella Outbreak, Client: Rick Schiller," April 15, 2014.

——. "Publisher's Platform: WWFFD? (What Would Foster Farms Do?)." *Food Safety News*, April 5, 2014. http://www.foodsafetynews.com/2014/04/wwffd-what-would-foster -farms-do/.

Marshall, B. M., and S. B. Levy. "Food Animals and Antimicrobials: Impacts on Human Health." *Clinical Microbiology Reviews* 24, no. 4 (October 1, 2011): 718–33.

Marshall, Joseph, and Robert C. Baker. "New Marketable Poultry and Egg Products: 12. Chicken Sticks." Agricultural Economics Research Publications. Ithaca, NY: Departments of Agricultural Economics and Poultry Husbandry, Cornell University, 1963.

Marston, et al. "Antimicrobial Resistance." *Journal of the American Medical Association* 316, no. 11 (September 20, 2016): 1193.

Martin, Douglas. "Robert C. Baker, Who Reshaped Chicken Dinner, Dies at 84." *New York Times,* March 16, 2006. http://www.nytimes.com/2006/03/16/nyregion/robert-c -baker-who-reshaped-chicken-dinner-dies-at-84.html.

Martinez, Steve. "A Comparison of Vertical Coordination in the U.S. Poultry, Egg, and Pork Industries." *Agriculture Information Bulletin* (May 2002).

McEwen, Scott A., and Paula J. Fedorka-Cray. "Antimicrobial Use and Resistance in Animals." *Clinical Infectious Diseases* 34, Supp. 3 (June 1, 2002): S93–106.

McGann, Patrick, et al. "*Escherichia coli* Harboring *mcr-1* and *bla* CTX-M on a Novel IncF Plasmid: First Report of *mcr-1* in the United States." *Antimicrobial Agents and Chemotherapy* 60, no. 7 (July 2016): 4420–21.

McGeown, D., et al. "Effect of Carprofen on Lameness in Broiler Chickens." *Veterinary Record,* June 12, 1999, 668–71.

McGowan, John P. and A.R.G. Emslie. "Rickets in Chickens, With Special Reference to Its Nature and Pathogenesis." *Biochemical Journal* 28, no. 4 (1934): 1503–12.

McKenna, Carol. "Ruth Harrison: Campaigner Revealed the Grim Realities of Factory Farming—and Inspired Britain's First Farm Animal Welfare Legislation." *Guardian (Manchester),* July 5, 2000. https://www.theguardian.com/news/2000/jul/06/ guardianobituaries.

Mediavilla, José R., et al. "Colistin- and Carbapenem-Resistant *Escherichia coli* Harboring *mcr-1* and *bla*NDM-5, Causing a Complicated Urinary Tract Infection in a Patient From the United States." *mBio* 7, no. 4 (August 30, 2016).

Mellon, Margaret, et al. "Hogging It: Estimates of Antimicrobial Use in Livestock." Cambridge, MA: Union of Concerned Scientists, January 2001. http://www.ucsusa.org/ food_and_agriculture/our-failing-food-system/industrial-agriculture/hogging-it -estimates-of.html#.WKN-AvONtF8.

"The Men Who Fought It." *Evening Gazette (Middlesbrough),* March 7, 1968.

Mendelson, Marc, et al. "Maximising Access to Achieve Appropriate Human Antimicrobial Use in Low-Income and Middle-Income Countries." *Lancet* 387, no. 10014 (January 2016): 188–98.

Metsälä, Johanna, et al. "Mother's and Offspring's Use of Antibiotics and Infant Allergy to Cow's Milk:" *Epidemiology* 24, no. 2 (March 2013): 303–9.

Michigan Farm Bureau. "Comment on the Judicious Use of Medically Important Antimicrobial Drugs in Food-Producing Animals—Draft Guidance." Regulations.gov, September 3, 2010. https://www.regulations.gov/document?D=FDA-2010-D-0094-0405.

Bibliography

Miles, Tricia D., et al. "Antimicrobial Resistance of *Escherichia coli* Isolates From Broiler Chickens and Humans." *BMC Veterinary Research* 2 (February 6, 2006): 7.

Ministry of Economic Affairs. "Reduced and Responsible: Policy on the Use of Antibiotics in Food-Producing Animals in the Netherlands." Utrecht: Ministry of Economic Affairs, Netherlands, February 2014. http://www.government.nl/files/documents -and-publications/leaflets/2014/02/28/reduced-and-responsible-use-of-antibiotics-in -food-producing-animals-in-the-netherlands/use-of-antibiotics-in-food-producing -animals-in-the-netherlands.pdf.

Mintz, E. "A Riddle Wrapped in a Mystery Inside an Enigma: Brainerd Diarrhoea Turns 20." *Lancet* 362, no. 9401 (December 20, 2003): 2037–38.

"Miracle Drugs Get Down to Earth," *Business Week*, no. 1417 (October 27, 1956): 139–40.

Mølbak, K., et al. "An Outbreak of Multidrug-Resistant, Quinolone-Resistant *Salmonella* Enterica Serotype Typhimurium DT104." *New England Journal of Medicine* 341, no. 19 (November 4, 1999): 1420–25.

Moore, P. R., and A. Evenson. "Use of Sulfasuxidine, Streptothricin, and Streptomycin in Nutritional Studies with the Chick." *Journal of Biological Chemistry* 165, no. 2 (October 1946): 437–41.

Moorin, Rachael E., et al. "Long-Term Health Risks for Children and Young Adults After Infective Gastroenteritis." *Emerging Infectious Diseases* 16, no. 9 (September 2010): 1440–47.

Mrak, Emil M. "Food Preservation." In *Proceedings of the First International Conference on the Use of Antibiotics in Agriculture, 19-21 October 1955*, 223–30. Washington, D.C.: National Academy of Sciences, 1956.

Mueller, N. T., et al. "Prenatal Exposure to Antibiotics, Cesarean Section and Risk of Child- hood Obesity." *International Journal of Obesity* 39, no. 4 (April 2015): 665–70.

Murphy, O. M., et al. "Ciprofloxacin-Resistant Enterobacteriaceae." *Lancet* 349, no. 9057 (April 5, 1997): 1028–29.

Nadimpalli, Maya, et al. "Persistence of Livestock-Associated Antibiotic-Resistant *Staphylo- coccus aureus* Among Industrial Hog Operation Workers in North Carolina Over 14 Days." *Occupational and Environmental Medicine* 72, no. 2 (February 2015): 90–99.

Nandi, S., et al. "Gram-Positive Bacteria Are a Major Reservoir of Class 1 Antibiotic Resis- tance Integrons in Poultry Litter." *Proceedings of the National Academy of Sciences* 101, no. 18 (May 4, 2004): 7118–22.

National Chicken Council. "Broiler Chicken Industry Key Facts 2016." National Chicken Council. Accessed April 18, 2016. http://www.nationalchickencouncil.org/ about-the-industry/statistics/broiler-chicken-industry-key-facts/.

———. "Per Capita Consumption of Poultry and Livestock, 1965 to Estimated 2016, in Pounds." National Chicken Council. Accessed April 18, 2016. http://www.national chickencouncil.org/about-the-industry/statistics/per-capita-consumption-of-poultry -and-livestock-1965-to-estimated-2012-in-pounds/

———. "U.S. Broiler Performance." National Chicken Council. Accessed April 18, 2016. http://www.nationalchickencouncil.org/about-the-industry/statistics/u-s -broiler -performance/

National Institute for Public Health and the Environment, and Stichting Werkgroep Anti- bioticabeleid. "Nethmap/MARAN 2013: Consumption of Antimicrobial Agents and Antimicrobial Resistance Among Medically Important Bacteria in the Netherlands;

Monitoring of Antimicrobial Resistance and Antibiotic Usage in Animals in the Netherlands in 2012." Nijmegen, March 9, 2013. http://www.swab.nl/swab/cms3.nsf/uploads/ADFB2606CCFDF6E4C1257BDB0022F93F/$FILE/Nethmap_2013%20def_web.pdf.

National Pork Producers Council. "Dear Subway Management Team and Franchisee Owners." Wall Street Journal, October 28, 2015. Advertisement. http://www.pork.org/wp-content/uploads/2015/10/102815_lettertosubway_final_print_wsj.pdf.

National Research Council. "Effects on Human Health of Subtherapeutic Use of Antimicrobials in Animal Feeds." Washington, D.C.: National Academy of Sciences, 1980. http://public.eblib.com/choice/publicfullrecord.aspx?p=3376953.

——. Proceedings of the First International Conference on the Use of Antibiotics in Agriculture, 19–21 October 1955. Washington, D.C.: National Academy of Sciences, 1956.

——. "The Use of Drugs in Food Animals: Benefits and Risks." Washington, D.C.: National Academy of Sciences, 1999. http://www.nap.edu/catalog/5137/the-use-of-drugs-in-food-animals-benefits-and-risks.

——. "The Use of Drugs in Animal Feeds: Proceedings of a Symposium." Washington, D.C.: National Academy of Sciences, 1969. http://catalog.hathitrust.org/Record/001516883.

Natural Resources Defense Council. "Going Mainstream: Meat and Poultry Raised Without Routine Antibiotics Use." December 2015. https://www.nrdc.org/sites/default/files/antibiotic-free-meats-CS.pdf.

Nature-Times News Service. "The Resistant Tees-Side Bacterium." Times (London). December 28, 1967.

Neeling, A. J. de, et al. "High Prevalence of Methicillin Resistant Staphylococcus aureus in Pigs." Veterinary Microbiology 122, no. 3–4 (June 21, 2007): 366–72.

Nelson, J. M., et al. "Fluoroquinolone-Resistant Campylobacter Species and the Withdrawal of Fluoroquinolones From Use in Poultry: A Public Health Success Story." Clinical Infectious Diseases 44, no. 7 (April 1, 2007): 977–80.

Nelson, Jennifer M., et al. "Prolonged Diarrhea Due to Ciprofloxacin-Resistant Campylobacter Infection." Journal of Infectious Diseases 190, no. 6 (September 15, 2004): 1150–57.

Nelson, Mark L., and Stuart B. Levy. "The History of the Tetracyclines." Annals of the New York Academy of Sciences 1241, no. 1 (December 2011): 17–32.

Neushul, P. "Science, Government, and the Mass Production of Penicillin." Journal of the History of Medicine and Allied Sciences 48, no. 4 (October 1993): 371–95.

"New Broiler Process Plan Keeps Meat Fresher Longer." Florence (SC) Morning News, February 21, 1956.

"New Philosophy in Administration of Food and Drug Laws Involved in Miller Pesticide Bill." Journal of Agricultural and Food Chemistry 1, no. 9 (July 2, 1953): 601.

"New Poultry Process Will Be Used at Chehalis Plant." (Centralia, WA) Daily Chronicle, July 10, 1956.

"New Process Helps Preserve Freshness of Poultry, Fish." (San Rafael, CA) Daily Independent Journal, December 7, 1955.

Ng, H., et al. "Antibiotics in Poultry Meat Preservation: Development of Resistance among Spoilage Organisms." Applied Microbiology 5, no. 5 (September 1957): 331–33.

Nicholson, Arnold. "More White Meat for You." Saturday Evening Post, August 9, 1947.

Bibliography

Nilsson, O., et al. "Vertical Transmission of *Escherichia coli* Carrying Plasmid-Mediated AmpC (pAmpC) Through the Broiler Production Pyramid." *Journal of Antimicrobial Chemotherapy* 69, no. 6 (June 1, 2014): 1497–1500. doi:10.1093/jac/dku030.

Njoku-Obi, A. N., et al. "A Study of the Fungal Flora of Spoiled Chlortetracycline Treated Chicken Meat." *Applied Microbiology* 5, no. 5 (September 1957): 319–21.

Norman, Lloyd. "G.O.P. to Open Inquiry into Meat Famine." *Chicago Tribune*, September 29, 1946, sec. 1.

"Obituaries: E. S. Anderson: Bacteriologist Who Predicted the Problems Associated With Human Resistance to Antibiotics." *Times (London)*, March 27, 2006. http://www.the times.co.uk/tto/opinion/obituaries/article2086666.ece.

"Obituaries: E. S. Anderson: Ingenious Microbiologist Who Investigated How Bacteria Become Resistant to Antibiotics." (London, UK) *Independent*, March 23, 2006. http://www.independent.co.uk/news/obituaries/e-s-anderson-6105831.html.

O'Brien, Thomas F. "Emergence, Spread, and Environmental Effect of Antimicrobial Resistance: How Use of an Antimicrobial Anywhere Can Increase Resistance to Any Antimicrobial Anywhere Else." *Clinical Infectious Diseases* 34, Suppl. 3 (June 1, 2002): S78–84.

O'Connor, Clare. "Chick-Fil-A CEO Cathy: Gay Marriage Still Wrong, but I'll Shut Up About It and Sell Chicken," March 19, 2014. https://www.forbes.com/sites/clare oconnor/2014/03/19/chick-fil-a-ceo-cathy-gay-marriage-still-wrong-but-ill-shut-up -about-it-and-sell-chicken/#496eed632fcb.

Office of Technology Assessment, U.S. Congress. "Drugs in Livestock Feed." Office of Technology Assessment, U.S. Congress, 1979. http://hdl.handle.net/2027/umn.31951003054358w.

Ollinger, Michael, et al. "Structural Change in the Meat, Poultry, Dairy, and Grain Processing Industries." Economic Research Report. U.S. Department of Agriculture Economic Research Service, March 2005. https://www.ers.usda.gov/publications/pub-details/ ?pubid=45671.

O'Neill, Molly. "Rare Breed." *Saveur*, October 14, 2009.

Osterholm, M. T., et al. "An Outbreak of a Newly Recognized Chronic Diarrhea Syndrome Associated With Raw Milk Consumption." *Journal of the American Medical Association* 256, no. 4 (July 25, 1986): 484–90.

O'Sullivan, Kevin. "Seven-Year-Old Ian Reddin's Food Poisoning Put Family Life on Hold." *Irish Times (Dublin)*, June 7, 1999.

Oregon Public Health Division. "Summary of *Salmonella* Heidelberg Outbreaks Involving PFGE Patterns SHEX-005 and 005a. Oregon, 2004–2012." Portland: Oregon Health Authority, June 20, 2014. https://public.health.oregon.gov/DiseasesConditions/ CommunicableDisease/Outbreaks/Documents/Outbreak%20Report_2012-2394_and relatedinvestigations_heidelberg.pdf.

Ottke, Robert Crittenden, and Charles Franklin Niven, Jr. Preservation of meat. U.S. Patent 3057735 A, filed January 25, 1957, and issued October 9, 1962. http://www.google.com/ patents/US3057735.

Overdevest, Ilse. "Extended-Spectrum B-Lactamase Genes of *Escherichia coli* in Chicken Meat and Humans, the Netherlands." *Emerging Infectious Diseases* 17, no. 7 (July 2011): 1216–22.

Parsons, Heidi. "Foster Farms Official Shares Data Management Tips, *Salmonella* Below 5%." *Food Quality News*, November 12, 2014. http://www.foodqualitynews.com/content/ view/print/989474.

"Pass the 'Acronized' Chicken, Please!" (Algona, IA). *Kossuth County Advance,* May 28, 1957.

Paterson, David L, and Patrick N. A. Harris. "Colistin Resistance: A Major Breach in Our Last Line of Defence." *Lancet Infectious Diseases* 16, no. 2 (February 2016): 132–33.

Paxton, H., et al. "The Gait Dynamics of the Modern Broiler Chicken: A Cautionary Tale of Selective Breeding." *Journal of Experimental Biology* 216, no. 17 (September 1, 2013): 3237–48.

"Penicillin's Finder Assays Its Future." *New York Times,* June 26, 1945.

Penn State Extension. "Primary Breeder Companies—Poultry." Accessed February 13, 2014. http://extension.psu.edu/animals/poultry/links/breeder-companies.

Perreten, Vincent, et al. "Colistin Resistance Gene *mcr-1* in Avian-Pathogenic *Escherichia coli* in South Africa." *Antimicrobial Agents and Chemotherapy* 60, no. 7 (July 2016): 4414–15.

Pew Campaign on Human Health and Industrial Farming. "Record-High Antibiotic Sales for Meat and Poultry Production." July 17, 2013. http://pew.org/1YkUC8K.

Pew Charitable Trusts. "The Business of Broilers: Hidden Costs of Putting a Chicken on Every Grill." Washington, D.C.: Pew Charitable Trusts, December 20, 2013. http://www.pewtrusts.org/~/media/legacy/uploadedfiles/peg/publications/report/business ofbroilersreportthepewcharitabletrustspdf.pdf.

——. "Comment on the Judicious Use of Medically Important Antimicrobial Drugs in Food-Producing Animals—Draft Guidance." Regulations.gov, August 27, 2010. https://www.regulations.gov/document?D=FDA-2010-D-0094-0398.

——. "Weaknesses in FSIS's Salmonella Regulation: How Two Recent Outbreaks Illustrate a Failure to Protect Public Health." Washington, D.C.: Pew Charitable Trusts, December 2013. http://www.pewtrusts.org/~/media/legacy/uploadedfiles/phg/content_level_pages/reports/fsischickenoutbreakreportv6pdf.pdf.

Pew Commission on Industrial Farm Animal Production. "Putting Meat on the Table: Industrial Farm Animal Production in America." Baltimore, MD: Johns Hopkins Bloomberg School of Public Health, April 2008. http://www.jhsph.edu/research/centers-and-institutes/johns-hopkins-center-for-a-livable-future/_pdf/news_events/PCIFAPFin.pdf.

Pew Environment Group. "Big Chicken: Pollution and Industrial Poultry Production in America." Washington, D.C.: Pew Charitable Trusts, July 27, 2011. http://www.pewtrusts.org/~/media/legacy/uploadedfiles/peg/publications/report/pegbigchicken july2011pdf.pdf.

Phillips, I., et al. "Epidemic Multiresistant *Escherichia coli* Infection in West Lambeth Health District." *Lancet* 1, no. 8593 (May 7, 1988): 1038–41.

Piddock, L. J. "Quinolone Resistance and *Campylobacter* spp." *Journal of Antimicrobial Chemotherapy* 36, no. 6 (December 1995): 891–98.

Plantz, Bruce. "Consumer Misconceptions Dangerous for American Agriculture." WATTAgNet.com, January 27, 2016. http://www.wattagnet.com/articles/25742-consumer-misconceptions-dangerous-for-american-agriculture.

Podolsky, Scott. *The Antibiotic Era: Reform, Resistance, and the Pursuit of a Rational Therapeutics.* Baltimore, MD: Johns Hopkins University Press, 2014.

Poirel, Laurent, and Patrice Nordmann. "Emerging Plasmid-Encoded Colistin Resistance: The Animal World as the Culprit?" *Journal of Antimicrobial Chemotherapy* 71, no. 8 (August 2016): 2326–27.

President Barack Obama. (Executive Order 13676: Combating Antibiotic-Resistant Bacteria." 79 *Fed. Reg.*, 56931 § (2014). https://www.gpo.gov/fdsys/pkg/FR-2014-09-23/pdf/2014-22805.pdf.

President of the General Assembly. "Draft Political Declaration of the High-Level Meeting of the General Assembly on Antimicrobial Resistance." New York: United Nations, September 21, 2016. http://www.un.org/pga/71/wp-content/uploads/sites/40/2016/09/DGACM_GAEAD_ESCAB-AMR-Draft-Political-Declaration-1616108E.pdf.

———. "Programme of the High Level Meeting on Antibiotic Resistance." New York: United Nations, September 19, 2016. http://www.un.org/pga/71/wp-content/uploads/sites/40/2015/08/HLM-on-Antimicrobial-Resistance-19-September-2016.pdf.

President's Council of Advisors on Science and Technology, Executive Office of the President. "Report to the President on Combating Antibiotic Resistance." Washington, D.C.: President's Council of Advisors on Science and Technology, September 18, 2014. https://obamawhitehouse.archives.gov/sites/default/files/microsites/ostp/PCAST/pcast_amr_sept_2014_final.pdf.

PR Newswire. "After Eliminating Human Antibiotics In Chicken Production in 2014, Perdue Continues Its Leadership." July 8, 2015. http://www.prnewswire.com/news-releases/after-eliminating-human-antibiotics-in-chicken-production-in-2014-perdue-continues-its-leadership-role-to-reduce-all-antibiotic-use--human-and-animal-300110015.html.

Price, Lance B., et al. "Elevated Risk of Carrying Gentamicin-Resistant *Escherichia coli* among U.S. Poultry Workers." *Environmental Health Perspectives* 115, no. 12 (December 2007): 1738–42.

Price, Lance B., et al. "The Persistence of Fluoroquinolone-Resistant *Campylobacter* in Poultry Production." *Environmental Health Perspectives* 115, no. 7 (March 19, 2007): 1035–39.

Price, Lance B., et al. "*Staphylococcus aureus* CC398: Host Adaptation and Emergence of Methicillin Resistance in Livestock." *mBio* 3, no. 1 (2012).

Pringle, Peter. *Experiment Eleven: Dark Secrets behind the Discovery of a Wonder Drug.* New York: Walker Books, 2012.

"Problems in the Poultry Industry. Part I," § Subcommittee No. 6 of the Select Committee on Small Business, House of Representatives, 85th Cong., 1st sess., pursuant to H. Res. 56. (1957).

"Problems in the Poultry Industry. Part II," § Subcommittee No. 6 of the Select Committee on Small Business, House of Representatives, 85th Cong., 1st sess., pursuant to H. Res. 56. (1957).

"Problems in the Poultry Industry. Part III," § Subcommittee No. 6 of the Select Committee on Small Business, House of Representatives, 85th Cong., 1st sess. pursuant to H. Res. 56. (1957).

Public Broadcasting System, *Frontline.* "Who's Responsible for That Manure? Poisoned Waters." April 21, 2009. http://www.pbs.org/wgbh/pages/frontline/poisonedwaters/themes/chicken.html.

"Public Husbandry." *Sunday Times (London),* July 14, 1968.

"The QSR 50: The Top 50 Brands in Quick Service and Fast Casual." *QSR Magazine,* August 3, 2015. https://www.qsrmagazine.com/reports/qsr50-2015-top-50-chart.

"Quality Market." *(Helena, MT) Independent Record,* April 18, 1957. Advertisement.

Radhouani, Hajer, et al. "Potential Impact of Antimicrobial Resistance in Wildlife, Environment and Human Health." *Frontiers in Microbiology* 5 (2014).

Ramchandani, Meena, et al. "Possible Animal Origin of Human-Associated, Multidrug-Resistant, Uropathogenic *Escherichia coli.*" *Clinical Infectious Diseases* 40, no. 2 (January 15, 2005): 251–57.

Rapoport, Melina, et al. "First Description of *mcr-1*-Mediated Colistin Resistance in Human Infections Caused by *Escherichia coli* in Latin America: Table 1." *Antimicrobial Agents and Chemotherapy* 60, no. 7 (July 2016): 4412–13.

Ravenholt, R. T., et al. "Staphylococcal Infection in Meat Animals and Meat Workers." *Public Health Reports* 76 (October 1961): 879–88.

Recommendations to the Commissioner for the Control of Foodborne Human Salmonellosis: The Report of the FDA Salmonella Task Force. Washington, D.C.: 1973.

Reed, Lois. "Our Readers Speak: Likes Letter From Lauretta Walkup." *(Butte, MT) Standard-Post,* December 5, 1959.

Reese, Frank. "On Animal Husbandry for Poultry Production." December 2014. http://goodshepherdpoultryranch.com/wp-content/uploads/2015/06/frankreesetreatise.pdf.

The Regulation of Animal Drugs by the Food and Drug Administration: Hearings Before a Subcommittee of the Committee on Government Operations, House of Representatives, 99th Cong., 1st sess., July 24 , 25, 1985.

"Report of the Special Meeting on Urgent Food Problems, Washington, D.C., May 20–27, 1946." Washington, D.C.: Food and Agriculture Organization of the United Nations, June 6, 1946.

Review on Antimicrobial Resistance. "Antimicrobial Resistance: Tackling a Crisis for the Health and Wealth of Nations." London: Review on Antimicrobial Resistance, December 2014. https://amr-review.org/sites/default/files/AMR%20Review%20Paper%20-%20Tackling%20a%20crisis%20for%20the%20health%20and%20wealth%20of%20nations_1.pdf.

Reynolds, L. A., and E. M. Tansey. "Foot and Mouth Disease: The 1967 Outbreak and Its Aftermath." London: Wellcome Trust Centre for the History of Medicine at UCL, 2003.

Rickes, E. L., et al. "Comparative Data on Vitamin B_{12} From Liver and From a New Source, *Streptomyces griseus.*" *Science* 108, no. 2814 (December 3, 1948): 634–35.

Riley, Lee W., and Amee R. Manges. "Epidemiologic Versus Genetic Relatedness to Define an Outbreak-Associated Uropathogenic Escherichia coli Group." *Clinical Infectious Diseases* 41, no. 4 (August 15, 2005): 567–70

Riley, Lee W., et al. "Obesity in the United States–Dysbiosis From Exposure to Low-Dose Antibiotics?" *Frontiers in Public Health* 1 (2013).

Rinsky, Jessica L., et al. "Livestock-Associated Methicillin and Multidrug Resistant *Staphylococcus aureus* Is Present Among Industrial, Not Antibiotic-Free Livestock Operation Workers in North Carolina." *PLoS ONE* 8, no. 7 (July 2, 2013): e67641.

Ritz, Casey W., and William C. Merka. "Maximizing Poultry Manure Use Through Nutrient Management Planning." University of Georgia Extension, July 30, 2004. http://extension.uga.edu/publications/detail.cfm?number=B1245.

Robinson, Timothy P., et al. "Animal Production and Antimicrobial Resistance in the Clinic." *Lancet* 387, no. 10014 (January 2016): e1–3.

Rogers, Richard T. "Broilers: Differentiating a Commodity." In *Industry Studies,* edited by Larry L. Duetsch, 3rd ed., 59–95. Armonk, NY: M. E. Sharpe, 2002.

Bibliography

Roth, Anna. "What You Need to Know About the Corporate Shift to Cage-Free Eggs." *Civil Eats*, January 28, 2016. http://civileats.com/2016/01/28/what-you-need-to-know -about-the-corporate-shift-to-cage-free-eggs/.

Rountree, P. M., and B. M. Freeman. "Infections Caused by a Particular Phage Type of *Staphylococcus aureus*." *Medical Journal of Australia* 42, no. 5 (July 30, 1955): 157–61.

Ruhe, J. J., and A. Menon. "Tetracyclines as an Oral Treatment Option for Patients With Community Onset Skin and Soft Tissue Infections Caused by Methicillin-Resistant *Staphylococcus aureus*." *Antimicrobial Agents and Chemotherapy* 51, no. 9 (September 1, 2007): 3298–3303.

Rule, Ana M., et al. "Food Animal Transport: A Potential Source of Community Exposures to Health Hazards From Industrial Farming (CAFOs)." *Journal of Infection and Public Health* 1, no. 1 (2008): 33–39.

Russell, Cristine. "Research Links Human Illness, Livestock Drugs." *Washington Post*, September 6, 1984, sec. A.

Russell, J. B., and A. J. Houlihan. "The Ionophore Resistance of Ruminal Bacteria and Its Relationship to Other Forms of Antibiotic Resistance." Paper presented at the Cornell Nutrition Conference for Feed Manufacturers, East Syracuse, NY, October 21, 2003. https://naldc.nal.usda.gov/catalog/20731.

Russo, T. A., and J. R. Johnson. "Proposal for a New Inclusive Designation for Extraintestinal Pathogenic Isolates of *Escherichia coli*: ExPEC." *Journal of Infectious Diseases* 181, no. 5 (May 2000): 1753–54.

Russo, Thomas A., and James R. Johnson. "Medical and Economic Impact of Extraintestinal Infections due to *Escherichia coli*: Focus on an Increasingly Important Endemic Problem." *Microbes and Infection* 5, no. 5 (April 2003): 449–56.

Ruzauskas, Modestas, and Lina Vaskeviciute. "Detection of the *mcr-1* Gene in *Escherichia coli* Prevalent in the Migratory Bird Species *Larus argentatus*." *Journal of Antimicrobial Chemotherapy* 71, no. 8 (August 2016): 2333–34.

Saberan, Abdi, and Olivier Deck. *Landes en toute liberté*. Lavaur, France: Edition AVFL, 2005.

"Safeway." *Bend (OR) Bulletin*, December 10, 1959. Advertisement.

Salyers, Abigail A., and Dixie D. Whitt. *Revenge of the Microbes: How Bacterial Resistance Is Undermining the Antibiotic Miracle*. Washington, D.C.: ASM Press, 2005.

Sanchez, G. V., et al. "Trimethoprim-Sulfamethoxazole May No Longer Be Acceptable for the Treatment of Acute Uncomplicated Cystitis in the United States." *Clinical Infectious Diseases* 53, no. 3 (August 1, 2011): 316–17.

Sanders, Robert. "Outspoken UC Berkeley Biochemist and Nutritionist Thomas H. Jukes Has Died at Age 93." University of California, Berkeley, November 10, 1999.

Sanderson Farms. *The Truth About Chicken—Supermarket*. 2016. https://www.youtube.com/ watch?time_continue=9&v=3BdgVvJOWiQ.

Sannes, Mark R., et al. "Predictors of Antimicrobial-Resistant *Escherichia coli* in the Feces of Vegetarians and Newly Hospitalized Adults in Minnesota and Wisconsin." *Journal of Infectious Diseases* 197, no. 3 (February 1, 2008): 430–34.

Sawyer, Gordon. *Northeast Georgia: A History*. Mount Pleasant, SC: Arcadia Publishing, 2001. https://www.arcadiapublishing.com/Products/9780738523705.

———. *The Agribusiness Poultry Industry: A History of Its Development*. New York: Exposition Press, 1971.

Saxon, Wolfgang. "Anne Miller, 90, First Patient Who Was Saved by Penicillin." *New York Times*, June 9, 1999.

Sayer, Karen. "Animal Machines: The Public Response to Intensification in Great Britain, C. 1960-C. 1973." *Agricultural History* 87, no. 4 (September 1, 2013): 473–501.

Scallan, Elaine, et al. "Foodborne Illness Acquired in the United States—Major Pathogens." *Emerging Infectious Diseases* 17, no. 1 (January 2011): 7–15.

Schell, Orville. *Modern Meat: Antibiotics, Hormones and the Pharmaceutical Farm*. New York: Random House 1984.

Schmall, Emily. "The Cult of Chick-Fil-A." *Forbes*, July 6, 2007. http://www.forbes.com/forbes/2007/0723/080.html.

Schmidt, C. J., et al. "Comparison of a Modern Broiler Line and a Heritage Line Unselected Since the 1950s." *Poultry Science* 88, no. 12 (December 1, 2009): 2610–19.

Schuessler, Ryan. "Maryland Residents Fight Poultry Industry Expansion." *Al Jazeera America*. http://america.aljazeera.com/articles/2015/11/23/maryland-residents-fight-poultry-industry-expansion.html.

Schulfer, Anjelique, and Martin J. Blaser. "Risks of Antibiotic Exposures Early in Life on the Developing Microbiome." *PLOS Pathogens* 11, no. 7 (July 2, 2015): e1004903.

Scully, Matthew. *Dominion: The Power of Man, the Suffering of Animals, and the Call to Mercy*. New York: St. Martin's Press, 2002.

Seeger, Karl C., et al. "The Results of the Chicken-of-Tomorrow 1948 National Contest." Newark: University of Delaware Agricultural Experiment Station, USDA, July 1948. http://hdl.handle.net/2027/uc1.b2825459.

Shanker, Deena. "Just Months After Big Pork Said It Couldn't Be Done, Tyson Is Raising up to a Million Pigs Without Antibiotics." Quartz.com, February 24, 2016. https://qz.com/624270/just-months-after-big-pork-said-it-couldnt-be-done-tyson-is-raising-up-to-a-million-pigs-without-antibiotics/.

Sharpless, Rebecca. Reimert Thorolf Ravenholt. Population and Reproductive Health Oral History Project, Sophia Smith Collection, Smith College, Northampton, MA, July 18, 2002. https://www.smith.edu/library/libs/ssc/prh/transcripts/ravenholt-trans.pdf.

Sheikh, Ali Ahmad, et al. "Antimicrobial Resistance and Resistance Genes in *Escherichia coli* Isolated From Retail Meat Purchased in Alberta, Canada." *Foodborne Pathogens and Disease* 9, no. 7 (July 2012): 625–31.

Shrader, H. L. "The Chicken-of-Tomorrow Program: Its Influence on 'Meat-Type' Poultry Production." *Poultry Science* 31, no. 1 (January 1952): 3–10.

Shrimpton, D. H. "The Use of Chlortetracycline (Aureomycin) to Retard the Spoilage of Poultry Carcasses." *Journal of the Science of Food and Agriculture* 8 (August 1957): 485–89.

Sieburth, J. M., et al. "Effect of Antibiotics on Intestinal Microflora and on Growth of Turkeys and Pigs." *Proceedings of the Society for Experimental Biology and Medicine, Society for Experimental Biology and Medicine* 76, no. 1 (January 1951): 15–18.

Simões, Roméo Rocha, et al. "Seagulls and Beaches as Reservoirs for Multidrug-Resistant *Escherichia coli*." *Emerging Infectious Diseases* 16, no. 1 (January 2009): 110–12.

Singer, Randall S. "Urinary Tract Infections Attributed to Diverse ExPEC Strains in Food Animals: Evidence and Data Gaps." *Frontiers in Microbiology* 6 (2015): 28. doi:10.3389/fmicb.2015.00028.

Bibliography

Singer, Randall S., et al. "Modeling the Relationship Between Food Animal Health and Human Foodborne Illness." *Preventive Veterinary Medicine* 79, no. 2–4 (May 2007): 186–203.

Singer, Randall S., and Charles L. Hofacre. "Potential Impacts of Antibiotic Use in Poultry Production." *Avian Diseases* 50, no. 2 (June 2006): 161–72.

Singh, Pallavi, et al. "Characterization of Enteropathogenic and Shiga Toxin-Producing *Escherichia coli* in Cattle and Deer in a Shared Agroecosystem." *Frontiers in Cellular and Infection Microbiology* 5 (April 1, 2015).

Skov, Robert L., and Dominique L. Monnet. "Plasmid-Mediated Colistin Resistance (*mcr-1* Gene): Three Months Later, the Story Unfolds." *Eurosurveillance* 21, no. 9 (March 3, 2016).

Sloane, Julie. "I Turned My Father's Tiny Egg Farm Into a Poultry Powerhouse and Became the Face of an Industry." *Fortune Small Business*, September 1, 2003. http://money.cnn.com/magazines/fsb/fsb_archive/2003/09/01/350797/.

Smaldone, Giorgio, et al. "Occurrence of Antibiotic Resistance in Bacteria Isolated from Seawater Organisms Caught in Campania Region: Preliminary Study." *BMC Veterinary Research* 10 (July 15, 2014): 161.

Smith, E. L. "The Discovery and Identification of Vitamin B_{12}." *British Journal of Nutrition* 6, no. 1 (1952): 295–299.

Smith, H. W. "Why Has Swann Failed?" Letter to the editor. *British Medical Journal* 280, no. 6230 (June 21, 1980): 1537.

Smith, H. W., and M. A. Lovell. "*Escherichia coli* Resistant to Tetracyclines and to Other Antibiotics in the Faeces of U.K. Chickens and Pigs in 1980." *Journal of Hygiene* 87, no. 3 (December 1981): 477–83.

Smith, J. L., et al. "Impact of Antimicrobial Usage on Antimicrobial Resistance in Commensal *Escherichia coli* Strains Colonizing Broiler Chickens." *Applied and Environmental Microbiology* 73, no. 5 (March 1, 2007): 1404–14.

Smith, K. E., et al. "Quinolone-Resistant *Campylobacter jejuni* Infections in Minnesota, 1992-1998. Investigation Team." *New England Journal of Medicine* 340, no. 20 (May 20, 1999): 1525–32.

Smith, Page, and Charles Daniel. *The Chicken Book*. Athens: University of Georgia Press, 2000.

Smith, R., and J. Coast. "The True Cost of Antimicrobial Resistance." *British Medical Journal* 346, no. 11 (March 11, 2013): f1493.

Smith, S. P., et al. "Temporal Changes in the Prevalence of Community-Acquired Antimicrobial-Resistant Urinary Tract Infection Affected by *Escherichia coli* Clonal Group Composition." *Clinical Infectious Diseases* 46, no. 5 (March 1, 2008): 689–95.

Smith, Tara C., et al. "Methicillin-Resistant *Staphylococcus aureus* in Pigs and Farm Workers on Conventional and Antibiotic-Free Swine Farms in the USA." *PLoS ONE* 8, no. 5 (May 7, 2013): e63704.

Smith, Tara C., and Shylo E. Wardyn. "Human Infections with *Staphylococcus aureus* CC398." *Current Environmental Health Reports* 2, no. 1 (March 2015): 41–51.

Sneeringer, Stacey, et al. "Economics of Antibiotic Use in U.S. Livestock Production." Economic Research Report. Washington, D.C.: Economic Research Service, U.S. Department of Agriculture, November 2015. https://www.ers.usda.gov/webdocs/publications/err200/55528_err200_summary.pdf.

Sobel, Jeremy, et al. "Investigation of Multistate Foodborne Disease Outbreaks." *Public Health Reports* 117, no. 1 (February 2002): 8–19.

Solomons, I. A. "Antibiotics in Animal Feeds—Human and Animal Safety Issues." *Journal of Animal Science* 46, no. 5 (1978): 1360–1368.

Song, Qin, et al. "Optimization of Fermentation Conditions for Antibiotic Production by Actinomycetes YJ1 Strain against *Sclerotinia sclerotiorum*." *Journal of Agricultural Science* 4, no. 7 (May 21, 2012).

Soule, George. "Chicken Explosion." *Harper's Magazine* 222 (April 1961): 77–79.

Souverein, Dennis, et al. "Costs and Benefits Associated With the MRSA Search and Destroy Policy in a Hospital in the Region Kennemerland, the Netherlands." *PLOS ONE* 11, no. 2 (February 5, 2016): e0148175.

Spake, Amanda. "Losing the Battle of the Bugs (Cover)." *U.S. News & World Report*, May 10, 1999.

———. "O Is for Outbreak (Cover)." *U.S. News & World Report*, November 24, 1997.

Stamm, W. E. "An Epidemic of Urinary Tract Infections?" *New England Journal of Medicine* 345, no. 14 (October 4, 2001): 1055–57.

Stevenson, G. W., and Holly Born. "The 'Red Label' Poultry System in France: Lessons for Renewing an Agriculture-of-the-Middle in the United States." In *Remaking the North American Food System: Strategies for Sustainability*, edited by C. Clare Hinrichs and Thomas A. Lyson, 144–62. Lincoln: University of Nebraska Press, 2007.

Stokstad, E. L. R. "Antibiotics in Animal Nutrition." *Physiological Reviews* 34, no. 1 (1954): 25–51.

Stokstad, E. L. R., and T. H. Jukes. "Effect of Various Levels of Vitamin B_{12} Upon Growth Response Produced by Aureomycin in Chicks." *Proceedings of the Society for Experimental Biology and Medicine. Society for Experimental Biology and Medicine* 76, no. 1 (January 1951): 73–76.

———. "Further Observations on the 'Animal Protein Factor.'" *Proceedings of the Society for Experimental Biology and Medicine* 73, no. 3 (March 1, 1950): 523–28.

———. "The Multiple Nature of the Animal Protein Factor." *Journal of Biological Chemistry* 180, no. 2 (September 1949): 647–54.

Stokstad, E. L. R., et al. "The Multiple Nature of the Animal Protein Factor." *Journal of Biological Chemistry* 180, no. 2 (September 1, 1949): 647–54.

Stone, I. F. "Fumbling with Famine." *Nation*, March 23, 1946.

Striffler, Steve. *Chicken: The Dangerous Transformation of America's Favorite Food*. New Haven, CT: Yale University Press, 2007.

Strom, Stephanie. "Into the Family Business at Perdue." *New York Times*, July 31, 2015.

———. "Perdue Sharply Cuts Antibiotic Use in Chickens and Jabs at Its Rivals." *New York Times*, July 31, 2015.

Subcommittee on Dairy and Poultry of the Committee on Agriculture, House of Representatives. *Impact of Chemical and Related Drug Products and Federal Regulatory Processes. Hearings Before the Subcommittee on Dairy and Poultry of the Committee on Agriculture, House of Representatives*, 95th Cong., 1st sess. (1977).

Subcommittee on Health of the Committee on Labor and Public Welfare and the Subcommittee on Administrative Practice and Procedure of the Committee on the Judiciary, U.S. Senate. *Preclinical and Clinical Testing by the Pharmaceutical Industry, 1976. Joint Hearings*, 94th Cong., 2nd sess. (1975).

Bibliography

Subcommittee on Legislation Affecting the Food and Drug Administration of the Committee on Labor and Public Welfare, U.S. Senate. *Mandatory Poultry Inspection. Hearings, on S. 3176, a Bill to Amend the Federal Food, Drug, and Cosmetic Act, so as to Prohibit the Movement in Interstate or Foreign Commerce of Unsound, Unhealthful, Diseased, Unwholesome, or Adulterated Poultry or Poultry Products,* 84th Cong., 2nd sess. (May 9, 10, 1956).

Subcommittee on Oversight and Investigations, House Committee on Interstate and Foreign Commerce, on Food and Drug Administration. *Antibiotics in Animal Feeds, Hearings,* 95th Cong., 1st sess. (September 19, 23, 1977). http://hdl.handle.net/2027/umn.31951d00283261m.

Sun, M. "Antibiotics and Animal Feed: A Smoking Gun." *Science* 225, no. 4668 (September 21, 1984): 1375.

———. "In Search of Salmonella's Smoking Gun." *Science* 226, no. 4670 (October 5, 1984): 30–32.

———. "New Study Adds to Antibiotic Debate." *Science* 226, no. 4676 (November 16, 1984): 818.

———. "Use of Antibiotics in Animal Feed Challenged." *Science* 226, no. 4671 (October 12, 1984): 144–46.

Sunde, Milton. "Seventy-Five Years of Rising American Poultry Consumption: Was It Due to the Chicken of Tomorrow Contest?" *Nutrition Today* 38, no. 2 (2003): 60–62.

Surgeon-General's Office, United States. "Report of the Surgeon-General of the Army to the Secretary of War for the Fiscal Year Ending June 30, 1921." Washington D.C.: Government Printing Office, June 30, 1921.

Swann, Michael Meredith, and Joint Committee on the Use of Antibiotics in Animal Husbandry and Veterinary Medicine. *Report Presented to Parliament by the Secretary of State for Social Services, the Secretary of State for Scotland, the Minister of Agriculture, Fisheries and Food and the Secretary of State for Wales by Command of Her Majesty.* London: HMSO, November 1969.

Swartz, Morton N. "Human Diseases Caused by Foodborne Pathogens of Animal Origin." *Clinical Infectious Diseases* 34, Suppl. 3 (June 1, 2002): S111–22.

Tarr, Adam. "California Firm Recalls Chicken Products Due to Possible *Salmonella* Heidelberg Contamination." U.S. Department of Agriculture, July 12, 2014. https://www.fsis.usda.gov/wps/portal/fsis/topics/recalls-and-public-health-alerts/recall-case-archive/archive/2014/recall-044-2014-release.

Tarr, H. L. A., et al. "Antibiotics in Food Processing, Experimental Preservation of Fish and Beef With Antibiotics." *Journal of Agricultural and Food Chemistry* 2, no. 7 (1954): 372–75.

Tartof, S. Y., et al. "Analysis of a Uropathogenic *Escherichia coli* Clonal Group by Multilocus Sequence Typing." *Journal of Clinical Microbiology* 43, no. 12 (December 1, 2005): 5860–64.

Teillant, Aude, et al. "Potential Burden of Antibiotic Resistance on Surgery and Cancer Chemotherapy Antibiotic Prophylaxis in the USA: A Literature Review and Modelling Study." *Lancet Infectious Diseases* 15, no. 12 (December 2015): 1429–37.

Ternhag, Anders, et al. "Short- and Long-Term Effects of Bacterial Gastrointestinal Infections." *Emerging Infectious Diseases* 14, no. 1 (January 2008): 143–48.

Thatcher, F. S., and A. Loit. "Comparative Microflora of Chlor-Tetracycline-Treated and Nontreated Poultry with Special Reference to Public Health Aspects." *Applied Microbiology* 9 (January 1961): 39–45.

Thorpe, Cheleste M. "Shiga Toxin-Producing *Escherichia coli* Infection." *Clinical Infectious Diseases* 38, no. 9 (May 1, 2004): 1298–1303.

"305,000 K-12 Students in Chicago Offered Chicken Raised Without Antibiotics." *Sustainable Food News*, November 1, 2011. https://www.sustainablefoodnews.com/printstory. php?news_id=14362.

Threlfall, E. J., et al. "High-Level Resistance to Ciprofloxacin in *Escherichia coli*." *Lancet* 349, no. 9049 (February 8, 1997): 403.

Threlfall, E. J., et al. "Increasing Spectrum of Resistance in Multiresistant Salmonella Typhimurium." *Lancet* 347, no. 9007 (April 13, 1996): 1053–54.

Threlfall, E. J., et al. "Plasmid-Encoded Trimethoprim Resistance in Multiresistant Epidemic *Salmonella* Typhimurium Phage Types 204 and 193 in Britain." *British Medical Journal* 280, no. 6225 (May 17, 1980): 1210–11.

Threlfall, E. J., et al. "Increasing Incidence of Resistance to Trimethoprim and Ciprofloxacin in Epidemic *Salmonella* Typhimurium DT104 in England and Wales." *Eurosurveillance* 2, no. 11 (November 1997): 81–84.

———. "Multiresistant *Salmonella* Typhimurium DT 104 and *Salmonella bacteraemia*." *Lancet* 352, no. 9124 (July 25, 1998): 287–88.

———. "Spread of Multiresistant Strains of *Salmonella* Typhimurium Phage Types 204 and 193 in Britain." *British Medical Journal* 2, no. 6143 (October 7, 1978): 997.

Titus, Andrea. "The Burden of Antibiotic Resistance in Indian Neonates." Center for Disease Dynamics, Economics and Policy, August 18, 2012. http://www.cddep.org/blog/posts/visualization_series_burden_antibiotic_resistance_indian_neonates.

Tobey, Mark B. "Public Workshops: Agriculture and Antitrust Enforcement Issues in Our 21st Century Economy." Washington, D.C.: U.S. Department of Justice, March 2012. https://www.justice.gov/atr/events/public-workshops-agriculture-and-antitrust-enforcement-issues-our-21st-century-economy-10.

Toossi, Mitra. "A Century of Change: The U.S. Labor Force, 1950-2050." *Monthly Labor Review* 125 (2002): 15.

"Town and Country Market." *(Ukiah, CA) Daily Journal*, August 22, 1956. Advertisement.

Trasande, L., et al. "Infant Antibiotic Exposures and Early-Life Body Mass." *International Journal of Obesity* 37, no. 1 (January 2013): 16–23.

Travers, Karin, and Barza Michael. "Morbidity of Infections Caused by Antimicrobial-Resistant Bacteria." *Clinical Infectious Diseases* 34, Suppl. 3 (June 1, 2002): S131–34.

"Trouble on the Farm." *Economist*, November 15, 1969.

Trust for America's Health. "Comment on the Judicious Use of Medically Important Antimicrobial Drugs in Food-Producing Animals—Draft Guidance." Regulations.gov, August 24, 2010. https://www.regulations.gov/document?D=FDA-2010-D-0094-0365.

Tucker, Anthony. "Anti-Anti-Antibiotics." *Guardian (Manchester)*, January 30, 1968.

———. "Obituary: ES Anderson: Brilliant Bacteriologist Who Foresaw the Public Health Dangers of Genetic Resistance to Antibiotics." *Guardian (Manchester)*, March 22, 2006.

Tucker, Robert A. "History of the U.S. Food and Drug Administration." Interview with Donald Kennedy, Ph.D., June 17, 1996. http://www.fda.gov/downloads/AboutFDA/WhatWeDo/History/OralHistories/SelectedOralHistoryTranscripts/UCM265233.pdf.

Bibliography

"Two Hands for Donald Kennedy." *New York Times*, July 2, 1979.

"Two Years Pass, Nothing Done." *Sunday Times (London)*, April 28, 1968.

Tyson Foods. "Tyson Foods' New Leaders Position Company for Future Growth." News release, February 21, 2017. http://www.tysonfoods.com/media/news-releases/2017/02/tyson-foods-new-leaders-position-company-for-future-growth.

"Tyson Fresh Meats Launches Open Prairie Natural Pork." TysonFoods.com. February 22, 2016. http://www.tysonfoods.com/media/news-releases/2016/02/natural-pork-launch.aspx.

United Nations Secretary-General. "Secretary-General's Remarks to High-Level Meeting on Antimicrobial Resistance [as Delivered]." United Nations, September 21, 2016. https://www.un.org/sg/en/content/sg/statement/2016-09-21/secretary-generals-remarks-high-level-meeting-antimicrobial.

United Press International. "Now Using Antibiotics to Keep Meat Fresh." *Pantagraph*, August 12, 1956.

———. "Science Finding Ways to Stall Food Spoilage." *Ottawa Journal*, November 10, 1956.

"USDA Takes New Steps to Fight *E. coli*, Protect the Food Supply." U.S. Department of Agriculture, September 13, 2011. https://www.usda.gov/wps/portal/usda/usdahome?contentidonly=true&contentid=2011/09/0400.xml.

U.S. Department of Agriculture. *A Graphic Summary of Farm Animals and Animal Products (Based Largely on the Census of 1940)*. Washington, D.C.: 1943. http://hdl.handle.net/2027/uva.x030450594.

U.S. Department of Agriculture, Bureau of Agricultural Economics, and Bureau of the Census, U.S. Department of Commerce. "United States Census of Agriculture 1950: A Graphic Summary." 1952. http://agcensus.mannlib.cornell.edu/AgCensus/getVolumeTwoPart.do?volnum=5&year=1950&part_id=1081&number=6&title=Agriculture%201950%20-%20A%20Graphic%20Summary.

U.S. Department of Agriculture, Food Safety and Inspection Service. "Nationwide Broiler Chicken Microbiological Baseline Data Collection Program, July 1994–June 1995." April 1996. http://agris.fao.org/agris-search/search.do?recordID=US201300313983.

U.S. Department of Agriculture, National Agricultural Statistics Service. "Agricultural Resource Management Survey Broiler Highlights 2011." Accessed December 9, 2016. https://www.nass.usda.gov/Surveys/Guide_to_NASS_Surveys/Ag_Resource_Management/ARMS_Broiler_Factsheet/.

———. "Poultry Slaughter 2014 Annual Summary." February 2015. http://usda.mannlib.cornell.edu/MannUsda/viewDocumentInfo.do?documentID=1497.

———. "Poultry Slaughter 2015 Annual Summary." February 2016. http://usda.mannlib.cornell.edu/MannUsda/viewDocumentInfo.do?documentID=1497.

U.S. Government Accountability Office. "Animal Agriculture: Waste Management Practices." Washington, D.C., July 1, 1999. http://www.gao.gov/products/RCED-99-205.

U.S. Poultry & Egg Association. "Industry Economic Data." Accessed May 31, 2015. https://www.uspoultry.org/economic_data/.

Van Boeckel, Thomas P., et al. "Global Trends in Antimicrobial Use in Food Animals." *Proceedings of the National Academy of Sciences of the United States of America* 112, no. 18 (May 5, 2015): 5649–54.

Van Cleef, B. A. G. L., et al. "High Prevalence of Nasal MRSA Carriage in Slaughterhouse Workers in Contact with Live Pigs in the Netherlands." *Epidemiology and Infection* 138, no. 5 (May 2010): 756–63.

Van Loo, Inge, et al. "Emergence of Methicillin-Resistant *Staphylococcus aureus* of Animal Origin in Humans." *Emerging Infectious Diseases* 13, no. 12 (December 2007): 1834–39.

Van Rijen, M. M. L., et al. "Methicillin-Resistant *Staphylococcus aureus* Epidemiology and Transmission in a Dutch Hospital." *Journal of Hospital Infection* 72, no. 4 (August 2009): 299–306. doi:10.1016/j.jhin.2009.05.006.

Van Rijen, M. M. L., et al. "Increase in a Dutch Hospital of Methicillin-Resistant *Staphylococcus aureus* Related to Animal Farming." *Clinical Infectious Diseases* 46, no. 2 (January 15, 2008): 261–63.

Vasquez, Amber M., et al. "Investigation of *Escherichia coli* Harboring the *mcr-1* Resistance Gene—Connecticut, 2016." *Morbidity and Mortality Weekly Report* 65, no. 36 (September 16, 2016): 979–80.

Vaughn, Reese H., and George F. Stewart. "Antibiotics as Food Preservatives." *Journal of the American Medical Association* 174, no. 10 (1960): 1308–1310.

Velázquez, J. B., et al. "Incidence and Transmission of Antibiotic Resistance in Campylobacter Jejuni and Campylobacter Coli." *Journal of Antimicrobial Chemotherapy* 35, no. 1 (January 1995): 173–78.

Vermont Department of Health. "Disease Control Bulletin September 1998: Salmonella Typhimurium DT104." Accessed February 18, 2016. http://www.healthvermont.gov/pubs/disease_control/1998/1998-09.aspx.

Veterinary Correspondent. "Hazards in Feed Additives." *Times (London)*, March 3, 1969.

Vickers, H. R., L. Bagratuni, and S. Alexander. "Dermatitis Caused by Penicillin in Milk." *Lancet* 1, no. 7016 (February 15, 1958): 351–52.

Vidaver, Anne K. "Uses of Antimicrobials in Plant Agriculture." *Clinical Infectious Diseases* 34, Suppl. 3 (June 1, 2002): S107–10.

Vieira, Antonio R., et al. "Association Between Antimicrobial Resistance in *Escherichia coli* Isolates From Food Animals and Blood Stream Isolates From Humans in Europe: An Ecological Study." *Foodborne Pathogens and Disease* 8, no. 12 (December 2011): 1295–1301.

Vincent, Caroline, et al. "Food Reservoir for *Escherichia coli* Causing Urinary Tract Infections." *Emerging Infectious Diseases* 16, no. 1 (January 2010): 88–95.

"Viruses Begin to Yield." *New York Times*, July 23, 1948.

Vitenskapkomiteen for mattrygghet (Norwegian Scientific Committee for Food Safety). "The Risk of Development of Antimicrobial Resistance With the Use of Coccidiostats in Poultry Diets," December 14, 2015. http://www.english.vkm.no/eway/default.aspx?pid=278&trg=Content_6390&Content_6390=6393:2093761::0:6745:1:::0:0.

Voetsch, Andrew C., et al. "FoodNet Estimate of the Burden of Illness Caused by Nontyphoidal Salmonella Infections in the United States." *Clinical Infectious Diseases* 38 Suppl. 3 (April 15, 2004): S127-134.

Vos, Margreet C., and Henri A. Verbrugh. "MRSA: We Can Overcome, but Who Will Lead the Battle?" *Infection Control and Hospital Epidemiology* 26, no. 2 (2005): 117–120.

Voss, Andreas, et al. "Methicillin-Resistant *Staphylococcus aureus* in Pig Farming." *Emerging Infectious Diseases* 11, no. 12 (December 2005): 1965-1966.

Vukina, T., and W. E. Foster. "Efficiency Gains in Broiler Production Through Contract Parameter Fine Tuning." *Poultry Science* 75, no. 11 (November 1996): 1351–58.

Walker, Homer W., and John C. Ayres. "Antibiotic Residuals and Microbial Resistance in Poultry Treated with Tetracyclines." *Journal of Food Science* 23, no. 5 (1958): 525–31.

Bibliography

Walker, J. C. "Pioneer Leaders in Plant Pathology: Benjamin Minge Duggar." *Annual Review of Phytopathology* 20, no. 1 (1982): 33–39.

Wang, Zuoyue. *In Sputnik's Shadow: The President's Science Advisory Committee and Cold War America.* New Brunswick, NJ: Rutgers University Press, 2008.

Warren, Don C. "A Half-Century of Advances in the Genetics and Breeding Improvement of Poultry." *Poultry Science* 37, no. 1 (January 1958): 3–20.

Watanabe, T. "Infective Heredity of Multiple Drug Resistance in Bacteria." *Bacteriological Reviews* 27 (March 1963): 87–115.

Watanabe, T., and T. Fukasawa. "Episome-Mediated Transfer of Drug Resistance in Enterobacteriaceae. I. Transfer of Resistance Factors by Conjugation." *Journal of Bacteriology* 81 (May 1961): 669–78.

Watts, J. Craig. "Easing the Plight of Poultry Growers." *(Raleigh, NC) News and Observer,* August 24, 2011.

WBOC-16. "Somerset County Approves New Poultry House Regulations." August 10, 2016. http://www.wboc.com/story/32732399/somerset-county-approves-new-poultry-house -regulations.

Wegener, H. C., et al. "Use of Antimicrobial Growth Promoters in Food Animals and *Enterococcus faecium* Resistance to Therapeutic Antimicrobial Drugs in Europe." *Emerging Infectious Diseases* 5, no. 3 (June 1999): 329–35.

Weiser, H. H., et al. "The Use of Antibiotics in Meat Processing." *Applied Microbiology* 2, no. 2 (March 1954): 88–94.

Welch, H. "Antibiotics in Food Preservation; Public Health and Regulatory Aspects." *Science* 126, no. 3284 (December 6, 1957): 1159–61.

———. "Problems of Antibiotics in Food as the Food and Drug Administration Sees Them." *American Journal of Public Health and the Nation's Health* 47, no. 6 (June 1957): 701–5.

Werner, F. J. M., and SDa Executive Board. "Usage of Antibiotics in Livestock in the Netherlands in 2012." Utrecht: SDa Autoriteit Dirgeneesmiddelen, July 2013. http://www .autoriteitdiergeneesmiddelen.nl/en/home.

Wertheim, H. F. L., et al. "Low Prevalence of Methicillin-Resistant *Staphylococcus aureus* (MRSA) at Hospital Admission in the Netherlands: The Value of Search and Destroy and Restrictive Antibiotic Use." *Journal of Hospital Infection* 56, no. 4 (April 2004): 321–25.

Westgren, Randall E. "Delivering Food Safety, Food Quality, and Sustainable Production Practices: The Label Rouge Poultry System in France." *American Journal of Agricultural Economics* 81, no. 5 (December 1, 1999): 1107–11.

"Whale Steak for Dinner." *Science News-Letter* 70, no. 20 (1956): 315–15.

White, David G., et al, eds. *Frontiers in Antimicrobial Resistance: A Tribute to Stuart B. Levy.* Washington, D.C.: American Society for Microbiology, 2005.

White-Stevens, Robert, et al. "The Use of Chlortetracycline-Aureomycin in Poultry Production." *Cereal Science Today,* September 1956.

"Why Did These 30 Babies Die? Asks MP." *Sunday Times (London),* April 13, 1969.

"Why Has Swann Failed?" Editorial. *British Medical Journal* 280, no. 6225 (May 17, 1980): 1195–96.

Williams Smith, H. "The Effects of the Use of Antibiotics on the Emergence of Antibiotic-Resistant Disease-Producing Organisms in Animals." In *Antibiotics in Agriculture:*

Proceedings of the University of Nottingham Ninth Easter School in Agricultural Science, edited by M. Woodbine, 374–88. London: Butterworths, 1962.

"With Its New Farm and Home Division, Cyanamid Is Placing Increasing Stress on Consumer Agricultural Chemicals." *Journal of Agricultural and Food Chemistry* 5, no. 9 (September 1957): 712–13.

Witte, W. "Impact of Antibiotic Use in Animal Feeding on Resistance of Bacterial Pathogens in Humans." *Ciba Foundation Symposium* 207 (1997): 61–71.

———. "Selective Pressure by Antibiotic Use in Livestock." *International Journal of Antimicrobial Agents* 16 Suppl. 1 (November 2000): S19-24.

Woodbine, Malcolm, ed. *Antibiotics in Agriculture: Proceedings of the 9th Easter School in Agricultural Science, 1962, University of Nottingham.* London: Butterworths, 1962.

"Woody Breast Condition." Poultry Site, September 16, 2014. http://www.thepoultrysite.com/articles/3274/woody-breast-condition/.

World Health Organization. "The Evolving Threat of Antimicrobial Resistance: Options for Action." Geneva: World Health Organization, 2012. http://apps.who.int/iris/bitstream/10665/44812/1/9789241503181_eng.pdf.

———. "The Public Health Aspects of the Use of Antibiotics in Food and Feedstuffs: Report of an Expert Committee [Meeting Held in Geneva from 11 to 17 December 1962]." WHO Technical Report. Geneva, 1963. http://www.who.int/iris/handle/10665/40563.

———. "Use of Quinolones in Food Animals and Potential Impact on Human Health." Geneva, 1998. http://www.who.int/foodsafety/publications/quinolones/en/.

Wrenshall, Charlton Lewis, et al. Method of treating fresh meat. U.S. Patent 2942982 A, filed November 2, 1956, and issued June 28, 1960. http://www.google.com/patents/US2942982.

Wright, E. D., and R. M. Perinpanayagam. "Multiresistant Invasive *Escherichia coli* Infection in South London." *Lancet* 1, no. 8532 (March 7, 1987): 556–57.

Wulf, M., and A. Voss. "MRSA in Livestock Animals: An Epidemic Waiting to Happen?" *Clinical Microbiology and Infection* 14, no. 6 (June 2008): 519–21.

Wulf, M. W. H., et al. "Infection and Colonization With Methicillin Resistant *Staphylococcus aureus* ST398 versus Other MRSA in an Area With a High Density of Pig Farms." *European Journal of Clinical Microbiology and Infectious Diseases* 31, no. 1 (January 2012): 61–65.

Wulf, Mireille, et al. "Methicillin-Resistant *Staphylococcus aureus* in Veterinary Doctors and Students, the Netherlands." *Emerging Infectious Diseases* 12 no. 12 (December 2006): 1939–41.

Wulf, M. W. H., et al. "MRSA Carriage in Healthcare Personnel in Contact With Farm Animals." *Journal of Hospital Infection* 70, no. 2 (October 2008): 186–90.

Xavier, Basil Britto, et al. "Identification of a Novel Plasmid-Mediated Colistin-Resistance Gene, *mcr-2* , in *Escherichia coli*, Belgium, June 2016." *Eurosurveillance* 21, no. 27 (July 7, 2016).

Yao, Xu, et al. "Carbapenem-Resistant and Colistin-Resistant *Escherichia coli* Co-Producing NDM-9 and MCR-1." *Lancet Infectious Diseases* 16, no. 3 (March 2016): 288–89.

Yong, Dongeun, et al. "Characterization of a New Metallo-β-Lactamase Gene, *bla*NDM-1, and a Novel Erythromycin Esterase Gene Carried on a Unique Genetic Structure in

Bibliography

Klebsiella pneumoniae Sequence Type 14 from India." *Antimicrobial Agents and Chemotherapy* 53, no. 12 (December 2009): 5046–54.

You, Y., et al. "Detection of a Common and Persistent *tet*(L)-Carrying Plasmid in Chicken-Waste-Impacted Farm Soil." *Applied and Environmental Microbiology* 78, no. 9 (May 1, 2012): 3203–13.

You, Yaqi, and Ellen K. Silbergeld. "Learning from Agriculture: Understanding Low-Dose Antimicrobials as Drivers of Resistome Expansion." *Frontiers in Microbiology* (2014).

Zuidhof, M. J., et al. "Growth, Efficiency, and Yield of Commercial Broilers From 1957, 1978, and 2005." *Poultry Science* 93, no. 12 (December 1, 2014): 2970–82.

Zuraw, Lydia. "FSIS Emails Reveal 'Snapshot' of Foster Farms Investigation." *Food Safety News*, June 16, 2015. http://www.foodsafetynews.com/2015/06/fsis-emails-reveal -snapshot-of-foster-farms-investigation/.

INDEX